Simple Arcadia for Beginners

Using Capella

SIMPLE ▶

Arcadia for Beginners

Using Capella

Pascal Roques
David Hetherington

Asatte Press

Austin, Texas

Simple Arcadia for Beginners
Using Capella

Pascal Roques, David Hetherington

First Edition

Version 1.0.0

ISBN 978-1-937468-14-9

http://AsattePress.com

Table of Contents

Foreword

MBSE is a way to implement co-engineering between disciplines and to develop complex systems with greater agility and efficiency. But how can we ensure that systems engineers embrace the cultural change of MBSE? And how can we smooth the learning curve which is usually too steep for systems engineers who do not necessarily have a software engineering background? One of the main answers is to use a tool implementing both a method and a modeling language with appropriate semantic guidance for systems architects.

Arcadia and *Capella* fit with this description and are now widely deployed on multiple operational projects in small and large organizations. As a systems engineer in the Automotive industry for more than 20 years, I had the opportunity to deploy *Arcadia* and *Capella* in a variety of projects and departments. My main takeaways were:

- The *Capella* intuitive interface and its usability allow an easy and fast adoption by the systems engineers.

- *Arcadia* and *Capella* transform engineering practices by breaking internal silos and providing better ways to communicate and understand each other.

- Getting support from an experienced MBSE trainer and coach is critical to avoid usual pitfalls found in any MBSE transformation process.

Pascal Roques is a senior consultant and trainer with over 30 years of experience. Having Pascal on board to support my teams in their MBSE journey was an obvious choice. Pascal has led more than 150 MBSE training courses in many countries for more than 2,000 engineers. Our first contact was in 2016 and since then, Pascal has conducted multiple training courses for my company always with very positive feedback.

Pascal is also the author of several books on modeling with UML and SysML, as well as the author of one of the first books on *Arcadia* and *Capella* – *Systems Architecture Modeling with the Arcadia Method, A Practical Guide to Capella*. This new book *Simple Arcadia for Beginners: Using Capella* is designed to be a companion to the previous one. Instead of showing finished representations, the book provides step-by-step guidance on model construction as well as the selection and completion of the appropriate diagrams.

Jerome Montigny
August 2022

Chapter 1 – Introduction

This book is for beginners. This book assumes that you have just downloaded a copy of *Capella* and are anxious to get started, but otherwise don't know too much about model-based systems engineering and don't have much experience using *Capella* or any other similar tool. The purpose of this book is to help you get through the initial learning curve and get you on your way to becoming proficient at system modeling.

How this Book is Organized

This book builds on the two key previous works about Arcadia:

- **Pascal Roques** – *Systems Architecture Modeling with the Arcadia Method: A Practical Guide to Capella (Implementation of Model Based System Engineering), 1st Edition.* ISTE Press - Elsevier, 2017

- **Jean-Luc Voirin** – *Model-based System and Architecture Engineering with the Arcadia Method, 1st Edition.* ISTE Press - Elsevier, 2017

While the previous books were a bit more focused on the **Arcadia** method, this book will focus more on the use of the *Capella* tool. [1] Also, the previous books were published in 2017 and based on version 1.1 of *Capella*. This book is based on version 6.1.0 of *Capella* and as such more closely reflects the current tool and thinking about how to best apply the *Arcadia* method.

This book covers all of the major *Arcadia* diagram types, showing detailed steps to construct simple models that demonstrate key concepts for each diagram type. If you work your way through all of the examples in this book, you should be ready to tackle more complex modeling problems as they come your way.

This book is organized as follows:

- **Chapter 1** – Introduction. This chapter.

- **Chapter 2** – This chapter will get you up and running with the tool. We will learn how to create models. We will create a few basic diagrams. After completing this chapter, you will have a feel for the basic structure of a model as well as a feel for how *Capella* works.

- **Chapter 3** – In this chapter we will discuss the high-level goals and purpose of model-based systems engineering before we delve deeper into the *Arcadia* method and into the *Capella* tool.

- **Chapter 4** to **Chapter 12** – In these chapters we dig deeper into each of the *Arcadia* diagram types. We will cover several of the most common usages of each diagram type. At the end of each chapter there will be a section listing references for further study about the topics presented in the chapter.

- **Chapter 13** – This final chapter discusses some of the challenges in helping an organization introduce model-based systems engineering techniques.

---

[1] For more information on these books, see the *Arcadia and Capella Books* section on page 373.

The back matter section of the book contains a number of useful appendices as well as some standard reference sections such as a glossary and an index.

About the Examples

As you work your way through this book, you will find that the example systems presented are simple to the point of silliness. We start with a doorbell. After that, most of the book concerns the development of a clock radio. Nothing that you encounter will involve more than high school mathematics. No advanced engineering expertise will be needed to understand the examples.

This focus on extremely simple examples is actually something that we have both learned from years of consulting. Technically complex examples are fun for the author but simply distracting for beginning readers.

Note that the coverage of the tool will evolve slightly as the book progresses. Starting in chapter 2, we will show every single mouse click needed to make the tool run. Nothing will be skipped or left as "an exercise for the reader". As the book progresses, however, we will gradually skip explanations of mundane operations explained in earlier chapters. As such, we recommend that you work your way through the book in the order that topics are presented.

About the Diagrams

The "Show every mouse-click" approach mentioned in the previous section requires a lot of annotated screen capture diagrams. Here we have had to make some tradeoffs to show as much detail as possible while keeping the print version of this book to a manageable size.

For initial study, the best experience will come from reading either the print edition or reading the electronic edition on a larger, high-quality device such as the *Kindle Fire HD 10* tablet.

One of the advantages of the Kindle, iOS, and Android devices is that the images can be zoomed. Simply press firmly on the image for a few seconds and a menu will appear allowing the image to be zoomed and scrolled as needed. This function is powerful enough to zoom way in to see the fine detail of an icon in a screen capture.

Both the *Kindle for PC* application and the *Kindle Cloud Reader* are excellent choices if you have a dual monitor arrangement. [2] Read the book in full screen mode on one monitor while working with the tool on the other. The images will be quite easy to read on a normal 24-inch monitor.

Alternatively, a *Kindle Fire HD 10* tablet or similar iOS or Android device combined with a single monitor running the tool should work well.

For more information, please consult the Asatte Press application note:

http://url4ap.net/App-Note-eReader.

---

[2] The *Kindle Cloud Reader* is available here: https://read.amazon.com/

Downloading the Examples

All of the examples in this book can be downloaded from:

https://url4ap.net/Capella-1ED-Examples.

The download file is a standard zip file. Once you unpack the file, you will find that it is organized in a hierarchy that matches the book.

Figure 1-1 – Example file in directory for section of book

Figure 1-1 shows the *Capella* file for the clock radio example model from Chapter 4 in the section *Missions* on page 58.

Note: The individual example zip files like the clock radio example model shown above do *not* need to be unzipped. The tool will import then directly as described in *Importing from an Archive File* on page 364.

Tool Version

The screen captures in this book were produced with *Capella* versions up to and including 6.1.0.

Permitted Use of the Examples

1) The book and examples are provided "As Is", without warranty of any kind, express or implied.

2) The book and examples can be used freely for all types of individual study and classroom instruction with the following restrictions:

    1) Copyright laws are respected. Students must purchase or rent individual copies of the book.

    2) The book and examples cannot be bundled or redistributed in any fashion without the express written permission of Asatte Press, Inc.

    3) Instructors must direct students to the Asatte Press, Inc website to download the examples.

    4) The examples cannot be stored on a local server and indirectly distributed.

    5) It is acceptable for the examples to be stored as part of the configuration management process for an instructional course as long as this storage is not used as a method of indirect redistribution or re-bundling.

Consultants and instructors at private and public educational institutions are welcome to use this book and the example files for the book in classes including instruction delivered for a fee provided the conditions of use above are respected. In other words, if you are a consultant you are welcome to use the books and examples as part of your paid service as long as you make it clear to your clients where the material is coming from and ask your clients to actively purchase books and download the examples themselves.

For questions about special arrangements, long-term licensing, bulk book purchases, teaching materials, and other uses that don't seem to fit into the conditions above, please contact:

Arcadia.for.Beginners.Capella.2023@asattepress.com

Style Conventions

To the greatest extent possible, this book follows United States spelling conventions and *The Chicago Manual of Style* [CMS17th] for style and formatting.

Periods inside quotation marks at the end of a sentence are one exception. This book follows the British convention (periods outside the quotation marks) because the American convention adds periods to text such as variable names that should not include a period.

Customer Support

Please address any concerns about the physical condition of your copy of the book, refunds, or other sales-related matters to the retailer who sold you the book.

Reports of typographical errors and technical questions can be directed to:

Arcadia.for.Beginners.Capella.2023@asattepress.com

Complimentary Printed Book for Typo Submission
Asatte Press has put a lot of time and effort into proofreading and checking this book. Nevertheless, there are probably a few remaining typographical errors in the text. Our plan is to periodically make minor corrective updates to the image used to print the book as well as to any electronic versions. If you find a typographical error, please send an e-mail to the address above pointing out the text that seems to be in error and suggest a correction. If your suggestion is accepted, after the next printing update, we will be happy to mail you a complimentary copy of the corrected print edition. Don't forget to include your mailing address in the e-mail.

Chapter 2 – Quick Start

Ready to start? We are going to dive right in, do a quick pass through the tool, and create a simple model. Don't worry too much if the *Arcadia* concepts presented in this chapter seem strange; we will be back to explain them in more detail in subsequent chapters.

Introducing the Tool

Before we start constructing our first **model**, there are a few routine housekeeping matters to take care of.

Starting Capella

Since *Capella* is a Java program that does not come with a Windows installation program, you will need to create a desktop shortcut for it.

If you have not created such a desktop shortcut yet, see Figure A-7 on page 359.

As soon as you start the tool, it will prompt you to enter a workspace for your project files. There is some complexity involved in selecting a workspace location, especially if you are going to use a source code control tool like Git. If you have not already done so, before you start the tool, review *Deciding Where to Put Your Project Files* on page 356

Double-click on the desktop icon to start the tool. A "Capella Launcher" window will appear which will prompt you to select a directory for the *Eclipse* workspace. *Capella* manages only a single workspace at once, but it is possible to switch between several workspaces.

Figure 2-1 – Capella icon

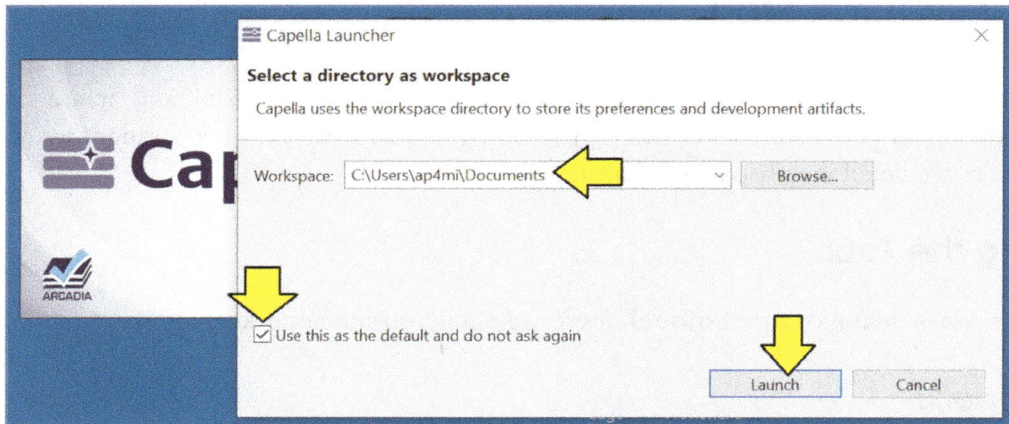

Figure 2-2 – Select workspace

Enter the directory where you would like new projects stored as mentioned above.

If you will be using the same directory every time, you will want to select the checkbox to make that directory the default and not ask again.

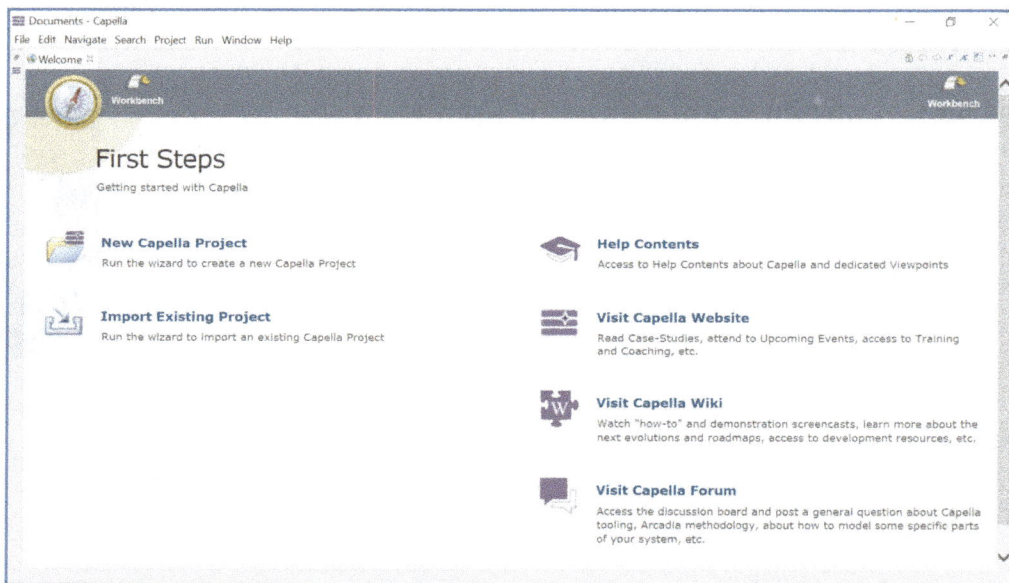

Figure 2-3 – Welcome screen

The welcome screen presents a handy list of information for beginners. On the left, there are commands to either create a new project or to import an existing project. On the right, there are links to help files and other useful online content.

Creating a Model

Note: if you would prefer to load the example file rather than creating a new model, you will need to import the example model from its "*.zip" archive file. *Capella* does not provide a simple "file open" operation. For more information see: *Archiving and Importing Capella Projects* on page 363.

Now we are ready to create our first **model**.

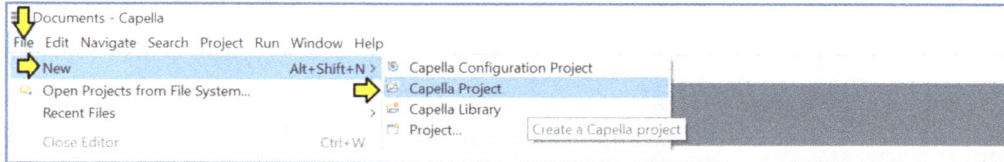

Figure 2-4 – Create a project

From the main menu, select "File", then "New", and then "Capella Project".

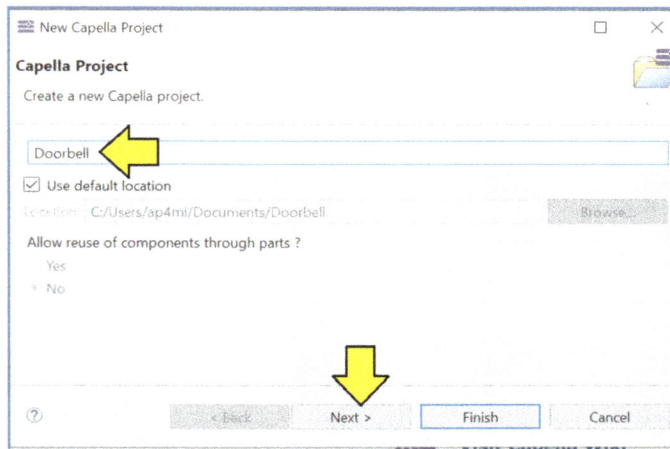

Figure 2-5 – Name the project

Our first project will be a doorbell system. Name the project: "Doorbell" and click "Next".

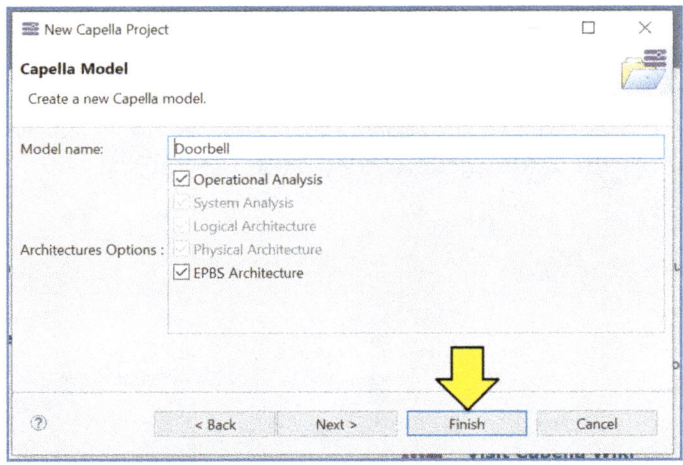

Figure 2-6 – Select options

The next panel allows you to select which levels of the *Arcadia* method you want to include in your model. For the moment, accept the defaults and click "Finish".

This step will create the subdirectory for the model (only) within the workspace folder which contains the model but can also contain other information.

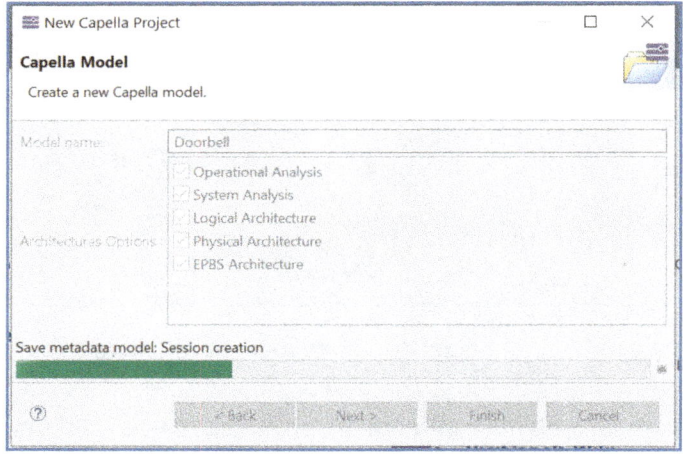

Figure 2-7 – Model progress indicator

A model progress indicator will appear briefly.

Figure 2-8 – You may need to dismiss the welcome screen

Where is my model?! You may need to dismiss the welcome screen. Click the small "x" at the right of the "Welcome" tab.

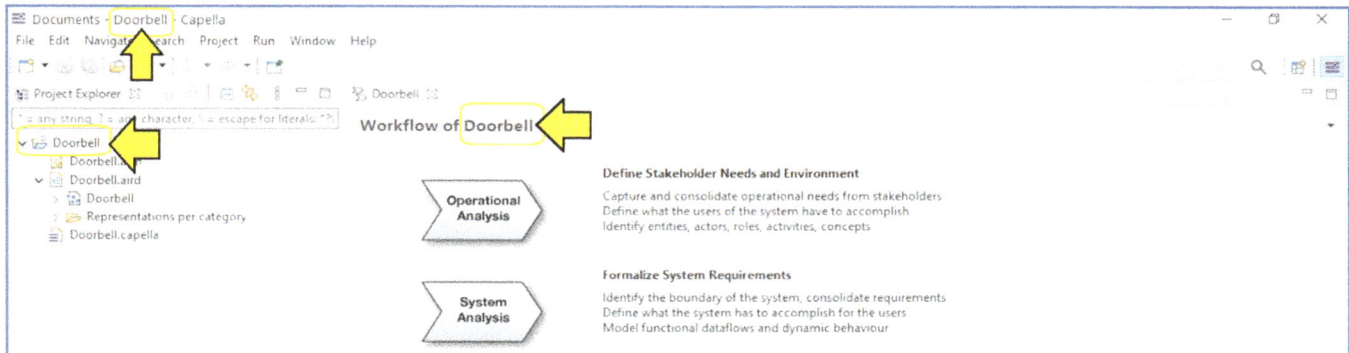

Figure 2-9 – The doorbell model appears

The doorbell model appears. Notice that the model has been named to match the model file name.

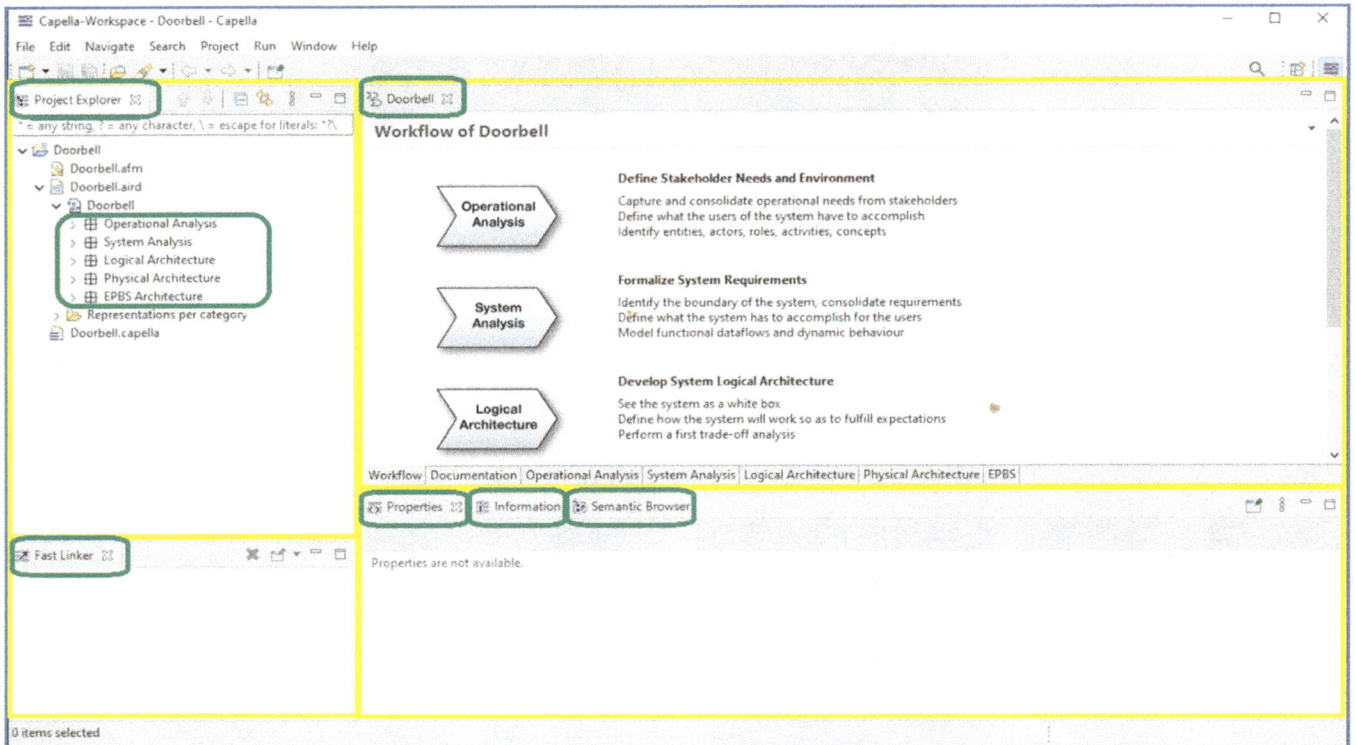

Figure 2-10 – Default tool layout

Capella is based on the *Eclipse* platform. As such, windows and views can generally be rearranged to suit your personal taste. However, let's take a closer look at the default view of the tool.

1) In the top-left corner of the screen, we can see a pane called the **Project Explorer**. This pane presents a hierarchical view of the model than can be expanded and explored similar to the file explorer user interfaces in Windows and other operating systems. In the project explorer, we see that the tool has already created 5 packages mapping to the Arcadia levels.

2) In the center of the screen, we can see a big **Editor** pane containing the **Activity Explorer** (in our case, named: "Workflow of Doorbell") and the future diagrams. The activity explorer contains icons for the different *Arcadia* levels and will be the entry point for the creation of diagrams. The editor pane is a tabbed view layout that will show diagrams as tabs as we create them.

3) There are three predefined tabs in the bottom pane: **Properties**, **Information**, **Semantic Browser**.

 • The properties view displays the features of the model element currently selected in the project explorer or in a diagram. You can also use the properties view to display the features of a diagram by selecting the diagram in the project explorer or by clicking in the background of the diagram itself.

 • The information view displays model validation results and miscellaneous additional messages, validation traces, and similar console output.

- The semantic browser displays all relationships entering and leaving the model element currently selected in the project explorer or in a diagram.

4) In the bottom-left corner of the screen, we can see a pane called the **Fast Linker**. In this book, we will not be making much use of this pane. Generally, we close it to make more room for the project explorer.

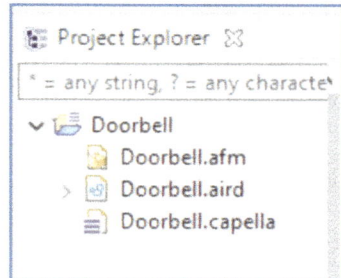

Figure 2-11 – Closer Look at the Project Explorer

Taking a closer look at the project explorer, we can see that it presents information as the contents of three files:

- **"*.afm"** – The *.afm file is an eclipse generated file and cannot be browsed from the tool.

- **"*.capella"** – The *.capella file contains the actual semantic model elements. However, this file cannot be browsed from the tool.

- **"*.aird"** – The *.aird file contains the representations (views of the model, through diagrams or tables) and is the only file we can browse and explore.

What if the Project Explorer isn't there?

What if the project explorer isn't visible on your screen? This kind of thing happens all the time with this sort of highly configurable tool.

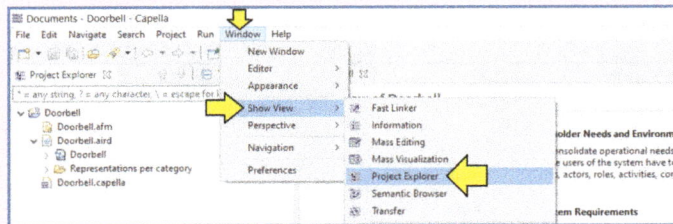

This problem can be solved by selecting "Window", then selecting "Show View" and then selecting "Project Explorer".

As mentioned previously, *Capella* is based on the *Eclipse* platform. At this point we will introduce some more precise terminology related to the underlying *Eclipse* platform:

1) The default layout shown in Figure 2-10 on page 10 is actually an *Eclipse* **perspective**.

2) *Capella* consists of one single *Eclipse* perspective, which is loaded by default when *Capella* is started.

3) The perspective defines the initial set of actions and panes that appears in the **workbench** window.

4) A perspective is a group of **views** and **editors** in the workbench window.

5) A view is a visual component within the workbench. It is typically used to navigate a list or hierarchy of information (such as the resources in the workbench) or display properties for the active editor. It can support editors and provides information and alternative presentation. Modifications made in a view are saved immediately if the view is not attached to an editor. For example, the project explorer and other navigation views display projects and other resources the user is working with, while the information view displays information that provides feedback to several *Capella* "complex" processes.

6) An editor is also a visual component within the workbench. It is typically used to edit or browse a resource. Editors are launched by (double-)clicking on a resource in a view. Modifications made in an editor follow an open-save-close model.

How do I Restore the Default Layout?

In the normal course of using *Capella*, views will be opened, moved, resized, and closed. The default layout can be restored with the "Window, Perspective, Reset Perspective" menu operation.

Understanding the Arcadia Template

In fact, the new model shown in Figure 2-9 is not really empty. Because we accepted the default options in Figure 2-6 on page 8, the model is already laid out with a structure matching the five *Arcadia* levels. Subfolders have been created automatically, and some model elements have been predefined.

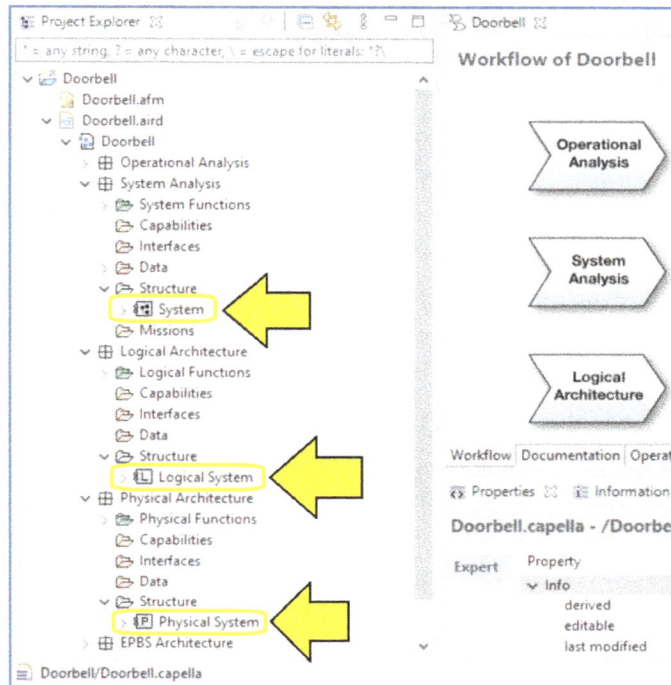

Figure 2-12 – Predefined elements

For example:

- A **system** has been created at the System Analysis level.
- A **logical system** has been created at the Logical Architecture level.
- A **physical system** has been created at the Physical Architecture level.

Depending on the options selected, the tool will create other types of elements as well.

- **Operational Analysis** – "What the users of the system need to accomplish". Operational analysis is quite similar to business process modeling. At this level, the system is not (yet) recognized as a model element.

- **System Analysis** – "What the system has to accomplish for the users". The system is seen as a black box containing no other structural elements, only allocated functions and interactions with external users or systems.

- **Logical Architecture** – "How the system will work to fulfill expectations". The logical architecture identifies logical components inside the system, the relationships between the logical components, and the content of the logical components. The logical architecture is implementation and technology independent.

- **Physical Architecture** – "How the system will be developed and built". The physical architecture defines the final architecture of the system, and how the system must be implemented, taking into account technological choices.

- **End Product Breakdown Structure (EPBS)** – "What is expected from the provider of each component". The physical components are grouped into **configuration items** for the purposes of formal configuration management and supplier contracting.

The activity explorer serves as the focal point for organizing *Arcadia* modeling activities. You will want to keep it open at all times.

What if the Activity Explorer isn't there?
What if the activity explorer isn't visible on your screen?

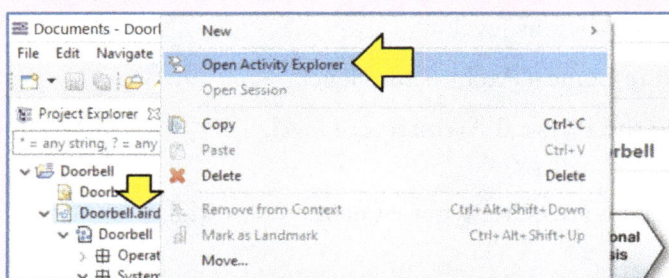

This problem can be solved by right-clicking on the *.aird file in the project explorer and selecting "Open Activity Explorer".

Arcadia contains five levels. Where should we start? *Arcadia* itself does not impose any particular sequence of modeling; it can be used top-down, bottom-up, or middle-out. For this "Quick Start" to get readers as familiar with the tool and method as rapidly as possible, we will start in the middle at the logical architecture level.

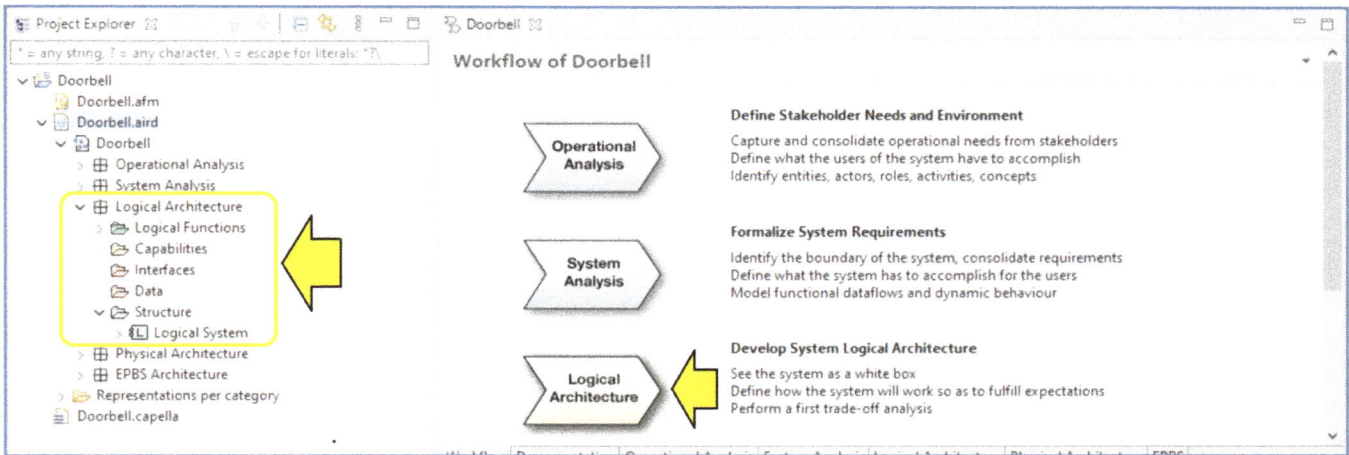

Figure 2-13 – Expand Logical Architecture

In the project explorer collapse all levels except the logical architecture level. Move the mouse over to the activity explorer and hover over the "Logical Architecture" icon. The icon will turn pale yellow/green.

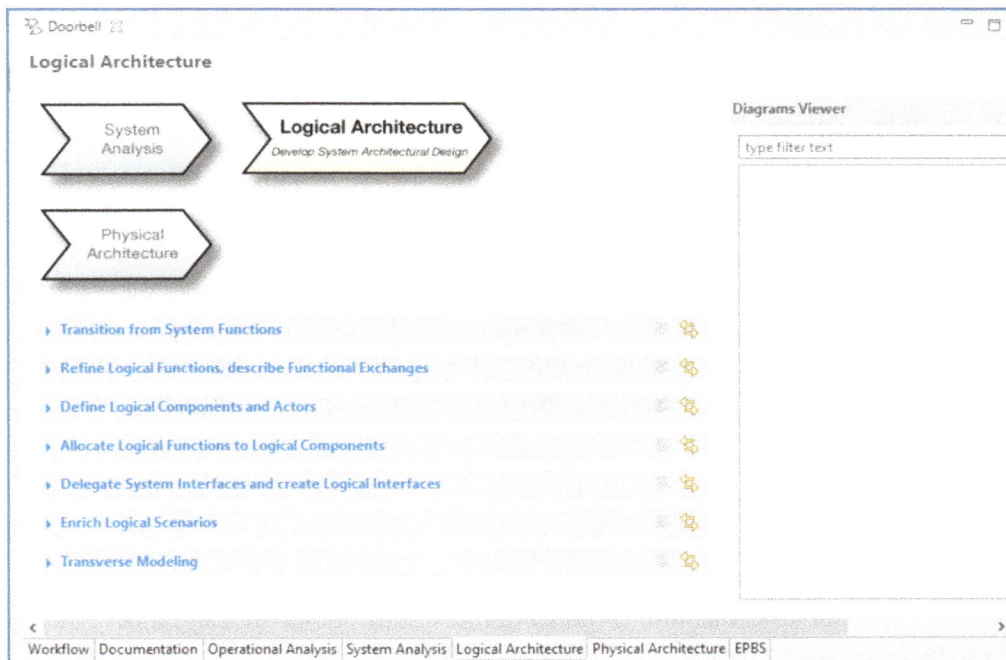

Figure 2-14 – Logical architecture activities

If you click on the "Logical Architecture" icon the activity explorer will expand to display a list of methodology activities related to logical architecture.

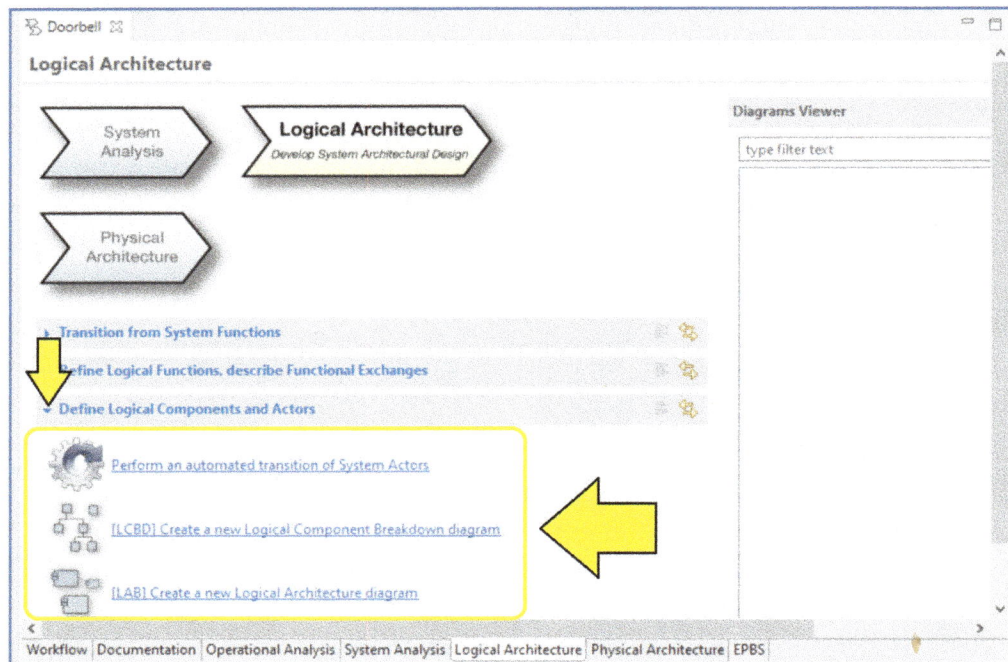

Figure 2-15 – Specific steps for an activity

If you select one of the activities, the activity explorer will expand to display a list of possible steps and diagrams to complete that activity. Note that the steps can be completed in any order.

Adding an Architecture Blank Diagram

Comparing Arcadia and SysML

If you are interested in comparing *Arcadia* and *Capella* with SysML, the doorbell system example we are working through in this chapter is very similar to the doorbell system example presented in the three *SysML for Beginners* books also published by Asatte Press. The quick start tutorials for those three books containing the SysML doorbell system examples are available from the Asatte Press website:

https://asattepress.com/Downloads/Tutorials.html

Now we are ready to add our first diagram. *Arcadia* defines many types of diagrams. The first diagram we will add is an **architecture blank diagram** – a very representative and important diagram for defining the elements of a system, as well as their functions and functional exchanges.

The main purpose of an architecture blank diagram is to show the allocation of **functions** to **components**. Architecture blank diagrams are used at all levels. We will be creating a "Logical Architecture Blank [LAB]" diagram.

As mentioned above, the purpose of logical architecture is to show "How the system will work to fulfill expectations". At the logical architecture level, functions and components are independent of technology and implementation. As such, the diagram shows the allocation of **logical functions** to **logical components**. At other architectural levels, the allocation might be to operational entities, the system, or other elements such as external actors.

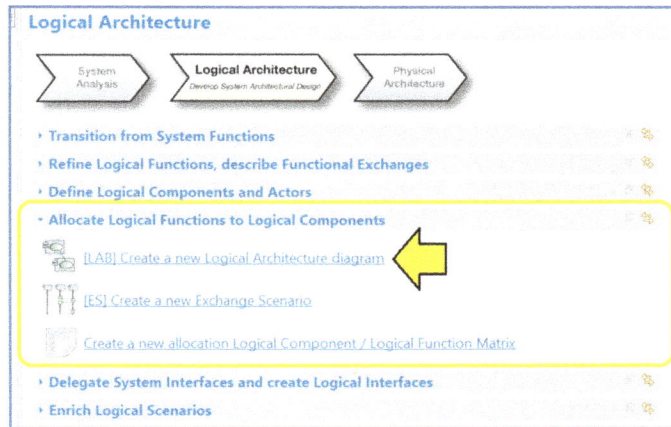

Figure 2-16 – Create a new Logical Architecture Diagram

In the "Logical Architecture" section of the expand the "Allocate Logical Functions to Logical Components" activity and select the "[LAB] Create a new Logical Architecture diagram" step.

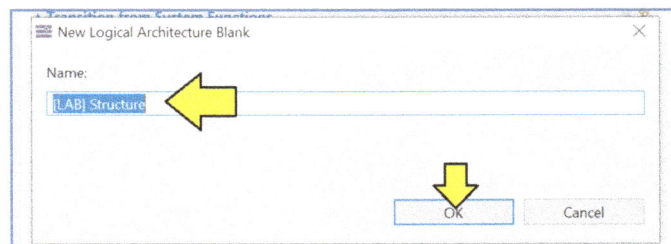

Figure 2-17 – Accept default name

A panel will appear which will allow you to name the diagram. For the moment, just accept the default name and click "OK".

For a quick start exercise like this one, the default diagram name will be completely adequate. However, we have two suggestions regarding *Arcadia* diagram names:

1) Make the diagram names unique. That is, once you have more than one "[LAB]" diagram, take the time to give each one a unique, meaningful name.

2) Keep the prefix. *Arcadia* and *Capella* will allow you to name the diagrams anything you please. However, experienced users are used to the reading the prefix "[LAB]" and other similar *Arcadia* prefixes to quickly recognize diagram types.

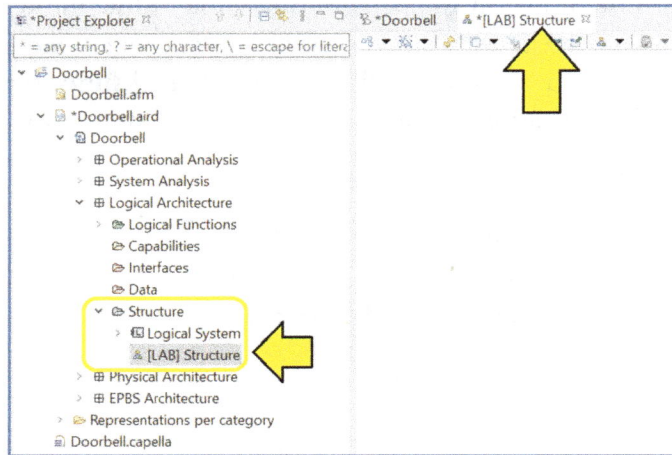

Figure 2-18 – Diagram appears in the Structure package

Capella will automatically place the new diagram in the "Structure" package.

The design of *Arcadia* and *Capella* is rather different than the design of SysML tools. Package names are predefined. Diagrams are automatically placed in predefined locations in the project explorer. Individual model elements are also mostly placed in predefined packages as well, with just a few exceptions that we will be introducing as we encounter them later in the book.

Initially, this approach using predefined packages may seem a little inflexible. The modeler may feel like his individual creativity is being suppressed. However, experience with many models created by many modelers in large organizations shows that modelers allowed to freely create package structures will tend to create a chaotic and bewildering variety of package structures. For large organizations, it is much more efficient to have all the models structured the same way so that everyone knows where to look for a certain type of information regardless of who created the model.

The newly created diagram is empty. To the right of the diagram, you will find the **Palette** which contains tools for creating content in the diagram.

Each diagram type has a specific palette. However, the layout is consistent across diagram types. There are several sections which can be expanded or collapsed as needed.

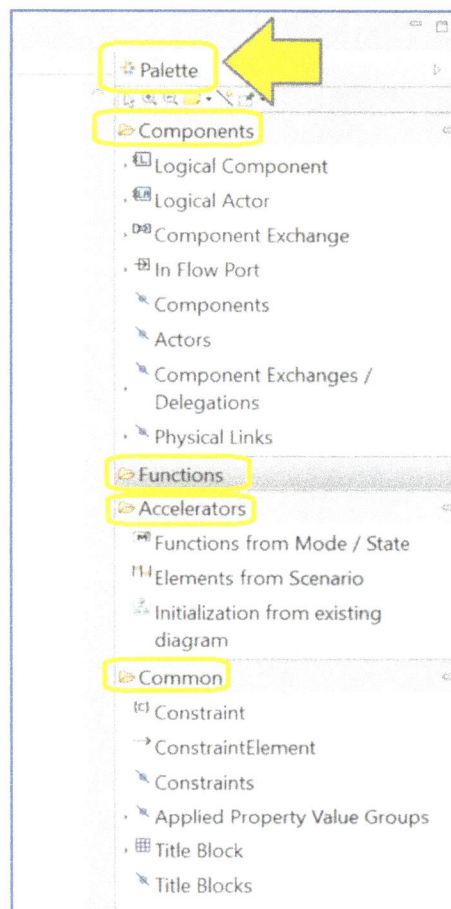

Figure 2-19 – The Palette

Full-Screen Mode

If you need more screen space to view or work on your diagram, you can simply double-click on the diagram tab to hide most of the windows. Double-clicking on the tab again restores the normal view.

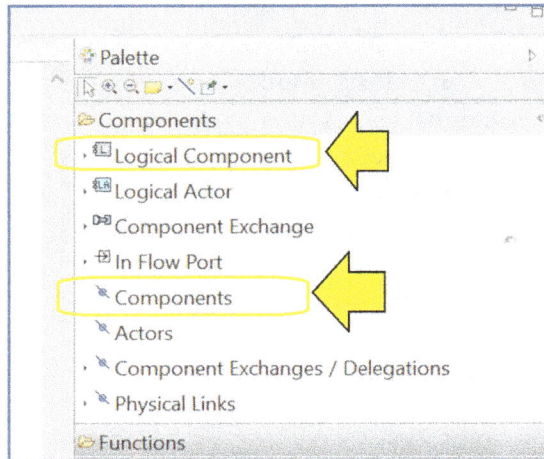

Figure 2-20 – Two kinds of component tools

Looking more closely at the palette we will find a section for components. In this section, there are two different types of tools for a component:

- **"Logical Component"** – This tool creates a new component in the model and also displays it in the diagram.

- **"Components"** – This tool does *not* create a new component in the model. This tool merely adds a graphical representation to the diagram for a component that already exists in the model.

Let's try out the "Components" tool first.

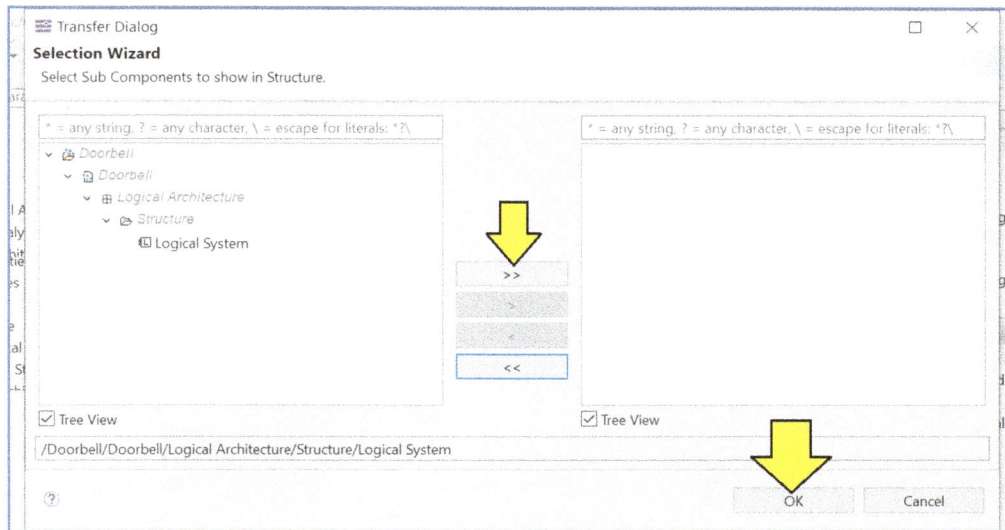

Figure 2-21 – Selection Wizard

Select the "Components" tool in the palette and click in the diagram. The "Selection Wizard" will appear. *Capella* has predefined a default "Logical System" component which is visible in the left pane. Click the double arrow in the middle of the wizard to move it to the right pane (and hence into the diagram). Click: "OK".

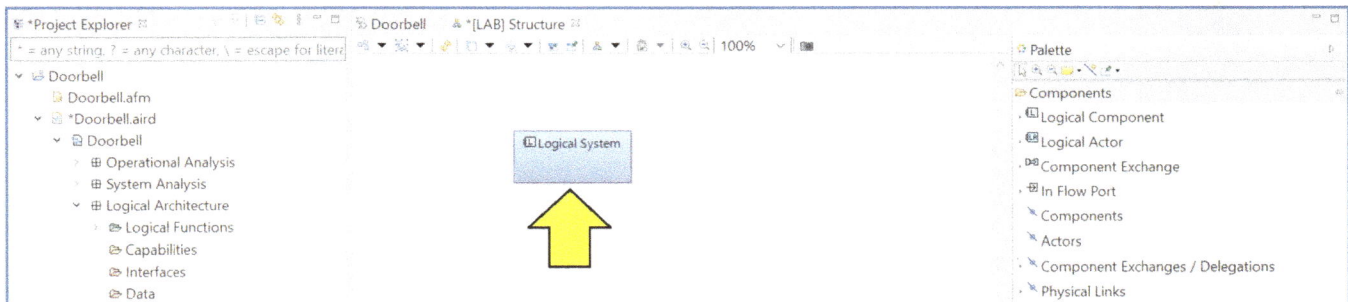

Figure 2-22 – The component appears in the diagram

The "Logical System" will appear in the diagram in the default color and size. The default color is fine, but this thing is our entire system. We are going to want to make it larger so we can put some other things inside of it.

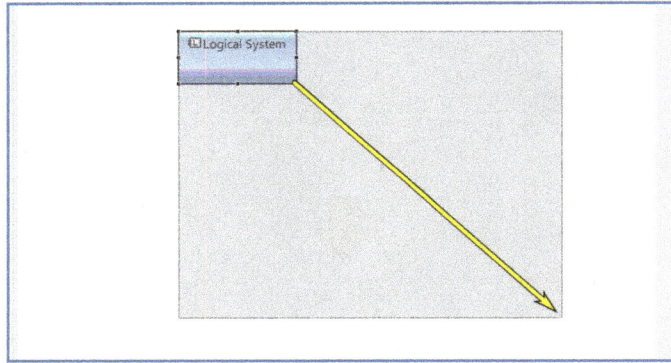

Figure 2-23 – Drag the corner to expand the component

Click to select the component. Drag one corner of the component to make it larger.

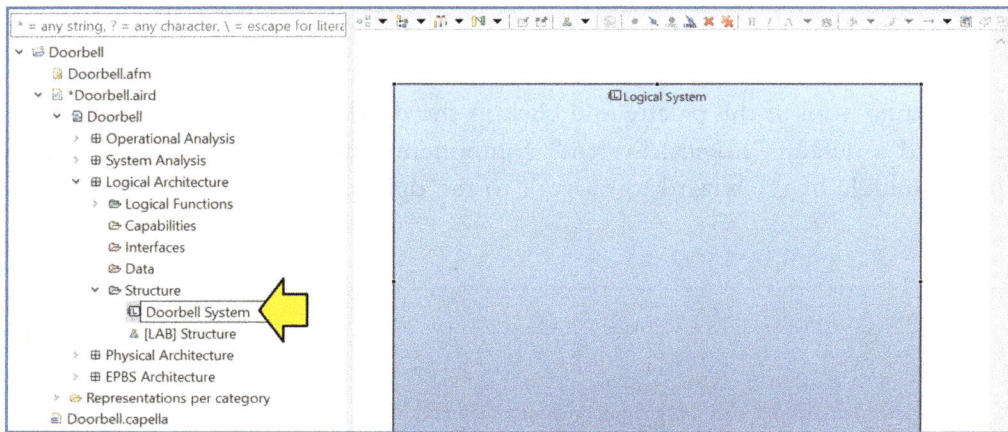

Figure 2-24 – Rename Logical System

Select the logical system in the project explorer, press "F2" and rename it: "Doorbell System".

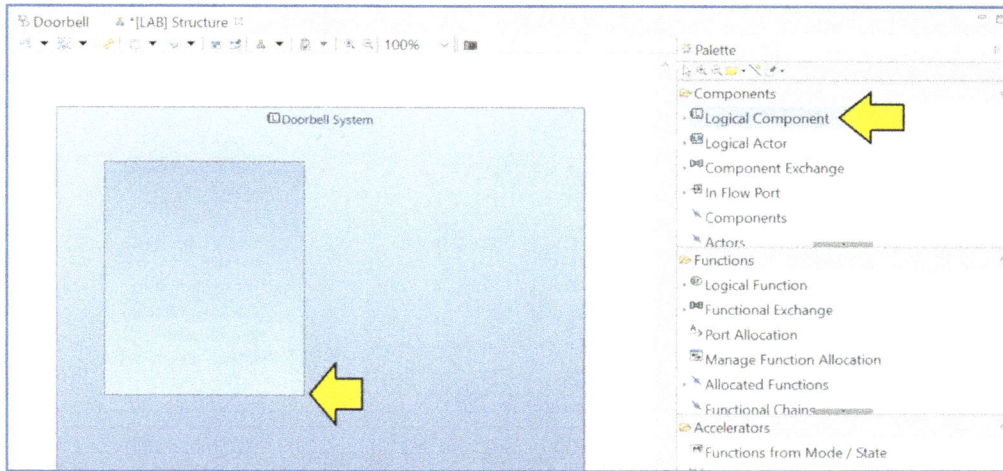

Figure 2-25 – Create nested logical component

Select the "Logical Component" tool in the palette, click in the diagram, and drag to make a larger outline.

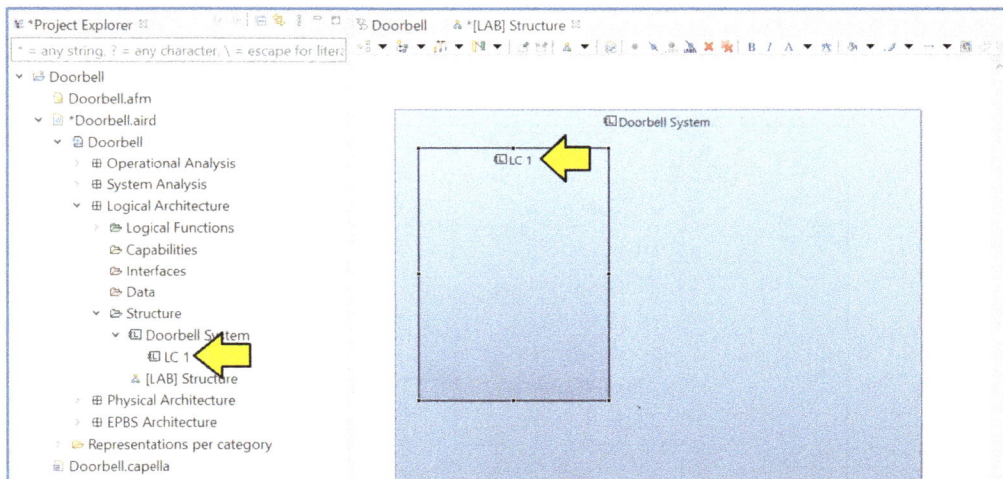

Figure 2-26 – LC1 has been created

Capella has created an element called: "LC1" in both the diagram and the project explorer. In other words, a new model element has been created.

A logical component is a structural element within the system, which can be connected with other logical components and external actors. A logical component can have one or more **Logical Functions** allocated to it. A logical component can also be sub-divided into sub-components.

Use the F2 key to rename "LC1" to "Doorbell Button Unit".

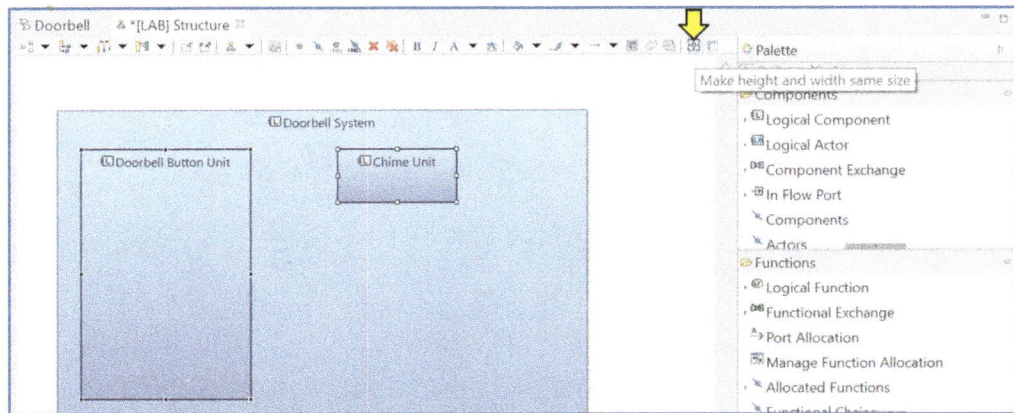

Figure 2-27 – Make Chime Unit same size

Create another logical component called: "Chime Unit". After you create the new component, leave it selected and the doorbell button unit component as well. Near the right of the toolbar along the top of the diagram, you will find an icon whose purpose is to make things the same size. Click this icon.

Notice that the order in which you select things is important. The last item selected will be the reference for the other selected item when you use an alignment or sizing function.

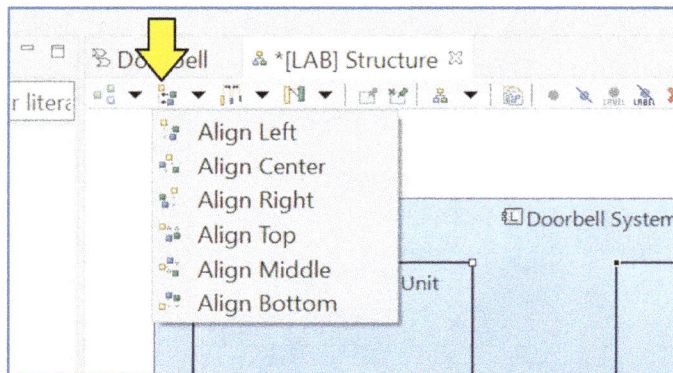

Figure 2-28 – Alignment tool

There is also a tool for aligning elements as well as another tool next to it for spacing elements.

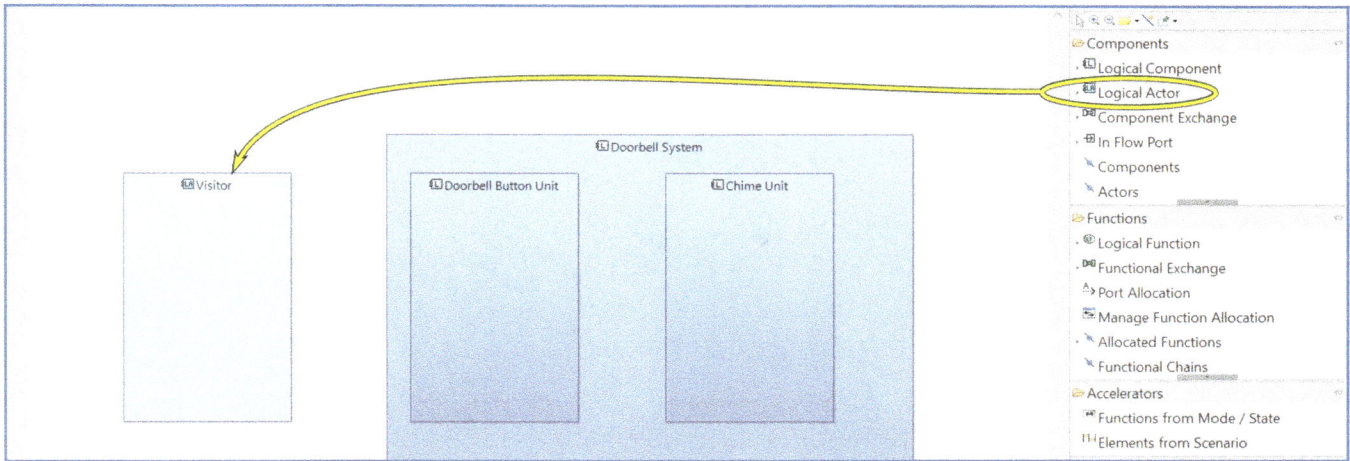

Figure 2-29 – Add visitor

Next, we can add a **Logical Actor** to the diagram. Select the "Logical Actor" tool in the palette and click in the diagram. Name the new element: "Visitor". Make it the same size as the doorbell button and chime units. External actors are always light blue, by default to distinguish them from internal components.

We can now add logical functions inside the logical components and the logical actor.

A logical function represents a behavior or service provided by a logical component or by a logical actor. A logical function owns **Function Ports** that allow it to communicate with the other logical functions. A logical function can be sub-divided into logical sub-functions.

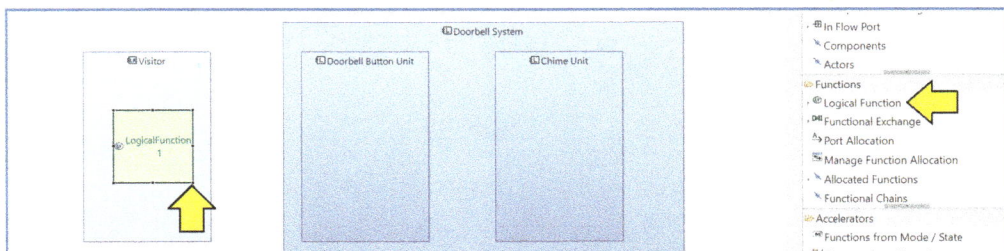

Figure 2-30 – Add Logical Function

Select the "Logical Function" tool in the "Functions" section of the palette. Click inside the visitor element and size the green box as shown.

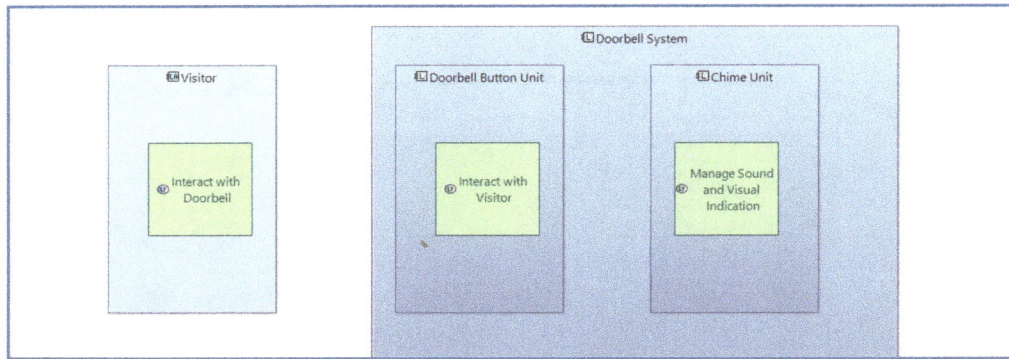

Figure 2-31 – Add two more logical functions

Rename the first logical function: "Interact with Doorbell". Add two more logical functions named: "Interact with Visitor" and "Manage Sound and Visual Indication" inside the doorbell button unit and the chime unit respectively. Make all three logical functions the same size.

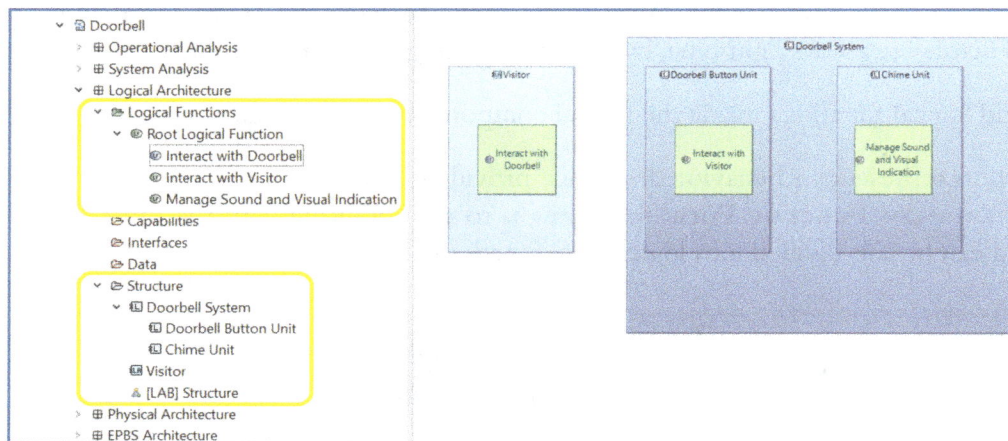

Figure 2-32 – Logical functions in the model

Notice that the logical functions are nested inside the logical components in the diagram, but not in the project explorer. The diagram represents **allocation** rather than containment. That is, the logical functions are allocated to their respective logical components, not contained within them.

If you look very carefully at the logical functions, you will notice that each one contains a small oval icon which in turn contains the letters: "LF". If you look even more closely, you will notice that the LF icon for the actor function is blue, distinguishing it from the other LF function icons which are green.

Next, we will introduce the concept of a **Functional Exchange**. A functional exchange represents a unidirectional exchange of information or matter between two functions.

If we were really doing formal systems engineering on a large system, we would carefully derive the sequence of steps in a functional exchange from the system operational concept. In later chapters, we will introduce some more rigorous techniques for making this sort of air-tight derivation. Right now, however, we are just interested in getting familiar with the tool, so we will use the following simple outline of the visitor/doorbell interaction to model the functional exchange:

1) As the visitor approaches the door, the doorbell button will be glowing dimly.

2) After the visitor pushes the button, the chime will sound inside, and the doorbell button will begin flashing brightly on and off.

3) When the chime is finished, the doorbell button will return to the glowing dimly state.

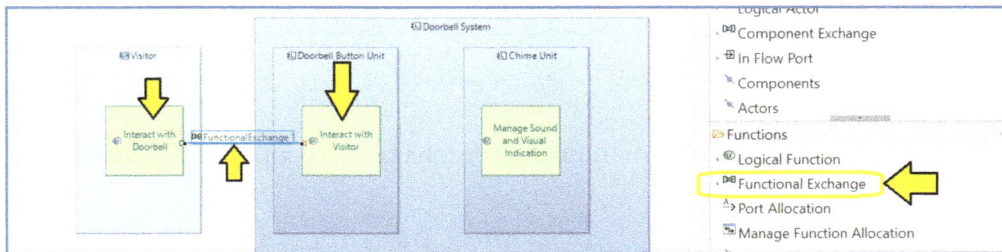

Figure 2-33 – Draw a functional exchange

Select "Functional Exchange" in the palette. Next, click once inside the logical function "Interact with Doorbell" and once inside the logical function "Interact with Visitor" to draw a functional exchange between the two.

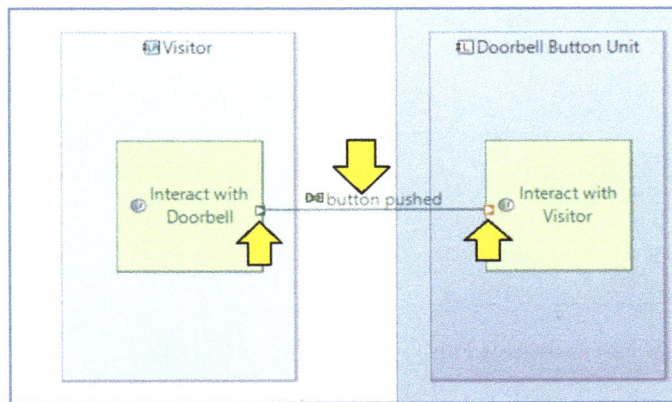

Figure 2-34 – Rename functional flow

Rename the functional flow: "button pushed". Notice:

• *Capella* has created ports on each logical function for the functional flow.

• Each port contains a small arrow to indicate the direction of the flow.

When creating functional flows, always click the "from" logical function first and the "to" logical function second.

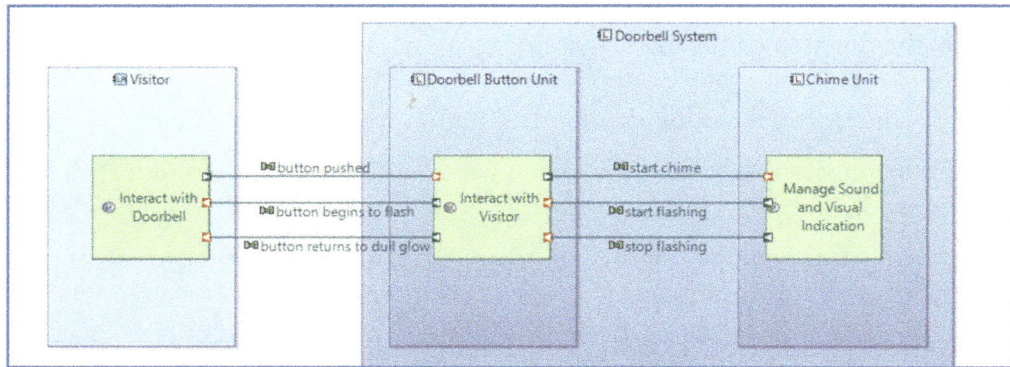

Figure 2-35 – Completed functional exchanges

Using the simple outline, complete a set of functional exchanges to support the desired behavior of the doorbell system.

If we now look at the project explorer again, we will not (initially) find the functional exchanges or the ports. In order to prevent the project explorer view from becoming overwhelmingly complicated, *Capella* has filter functions with default settings that suppress some detail – such as ports and functional exchanges. These default filter settings can be modified to show or hide these details in the project explorer.

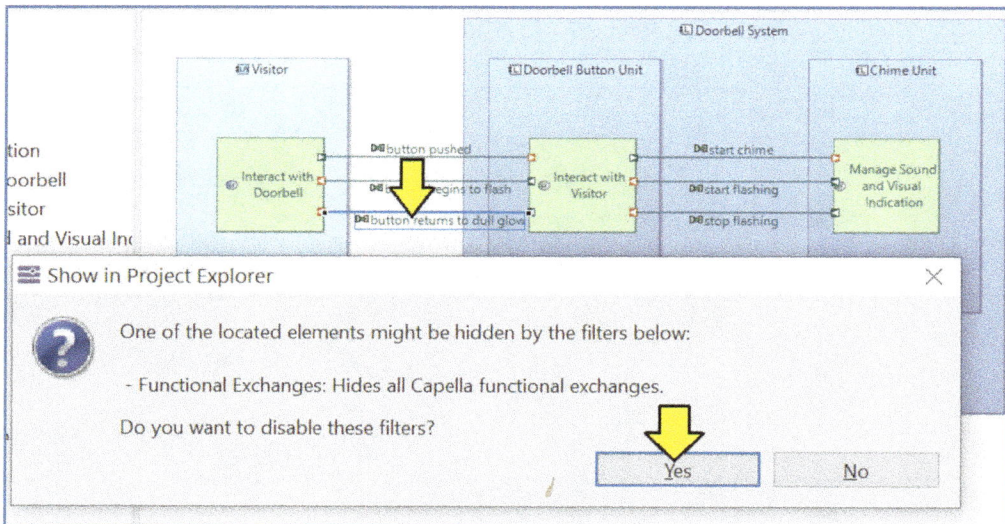

Figure 2-36 – Display functional exchanges

Select one of the functional exchanges in the diagram. Press F8. A panel will appear that allows you to disable the filter that suppresses the display of functional exchanges in the project explorer. Click: "Yes".

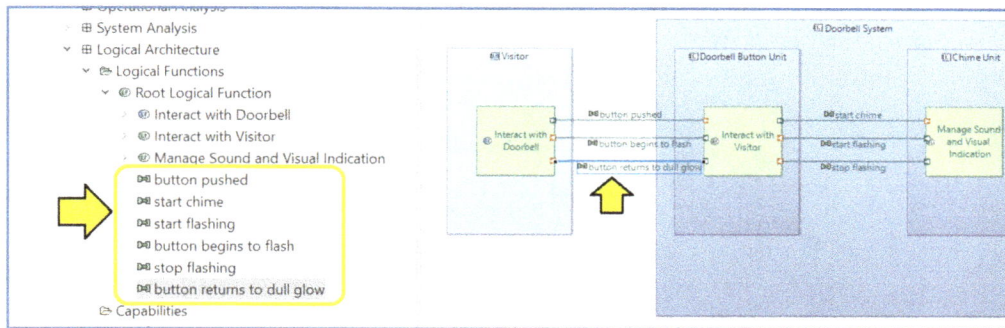

Figure 2-37 – Functional exchanges are now visible

The functional exchanges are now visible in the project explorer.

Now, let's take a closer look at the semantic browser. The semantic browser becomes more and more useful as the model gets larger. However, we can already see that it provides much more than the project explorer. For example, the allocation relationship is not visible in the project explorer because the project explorer is designed to only display containment and the allocation relationship is not a containment relationship. Likewise, inputs and outputs of functions are not shown in the project explorer either. The semantic browser is useful for exploring this sort of relationship detail. Selecting any model element – either in a diagram or in the project explorer – will cause the semantic browser to present a visualization of all the relevant model links.

Let's use the semantic browser to explore the details of a functional exchange. If the semantic browser is not open, use the "Show View" menu to display it as shown in *What if the Project Explorer isn't there?* on page 12.

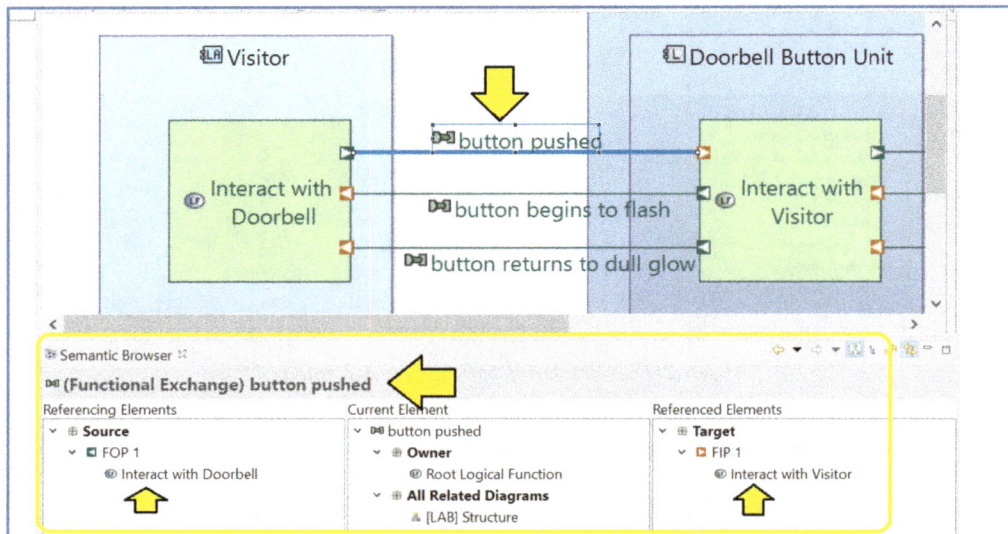

Figure 2-38 – Semantic browser for a functional exchange

Select the "button pushed" functional exchange. We can see that the name of the selected model element is displayed at the top of the semantic browser. Underneath the selected element, the semantic browser shows three columns:

1) **Referencing Elements** – In the first column, we have a list of model elements that reference the current model element. In this case the "Interact with Doorbell" logical function has an outgoing port that connects to the current element.

2) **Current Element** – The second column shows information about the current element.

3) **Referenced Elements** – In the third column, we have a list of model elements that are referenced by the current model element. In this case the "Interact with Visitor" logical function has an incoming port that the current element connects to.

Next, let's take a look at the semantic browser display for a logical function.

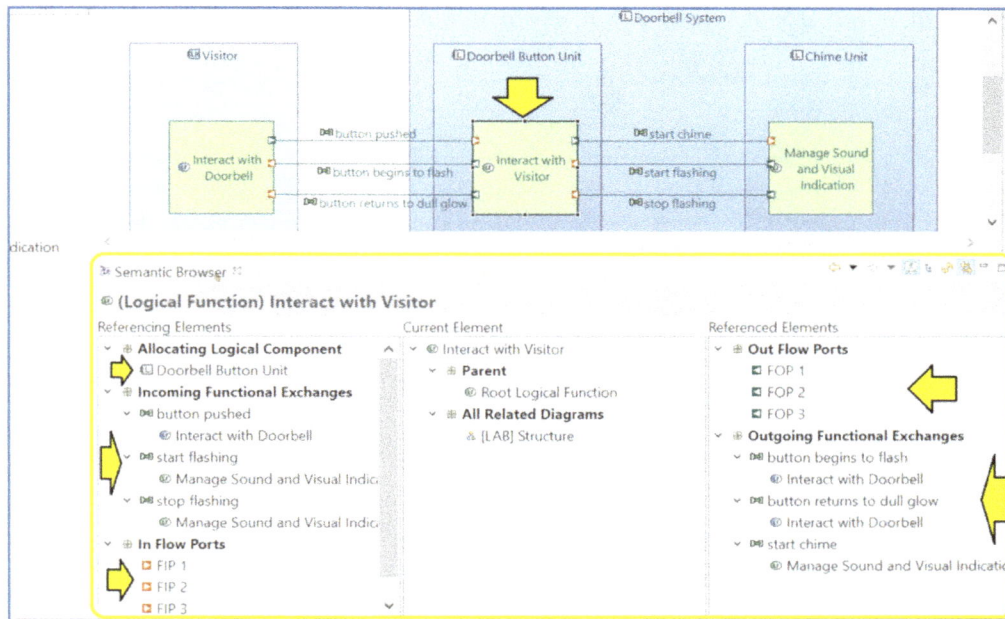

Figure 2-39 – Semantic browser for a logical function

Select the "Interact with Visitor" logical function. Again, we see that the name of the selected model element is displayed at the top of the semantic browser with three columns underneath:

1) **Referencing Elements –**

   • The allocation relationship for the logical function is shown here.

   • Incoming functional exchanges are shown here.

   • Input ports that belong to the logical function are shown here.

2) **Current Element –** The second column shows information about the current element.

3) **Referenced Elements –**

   • Outgoing functional exchanges are shown here.

   • Output ports that belong to the logical function are shown here.

Notice that the left/center/right columns of the semantic browser are in *semantic* order. Just because the outgoing ports are in the rightmost column of the semantic browser does *not* mean that those ports are on the right side of the model element in the diagram!

Adding a Scenario Diagram

Now we are ready to add a second diagram in our model: a **scenario diagram**. Once again, *Arcadia* defines many types of scenario diagrams. We will be creating an **Exchange Scenario [ES]** diagram. This diagram will represent a simple sequence of messages between the visitor and the two logical components.

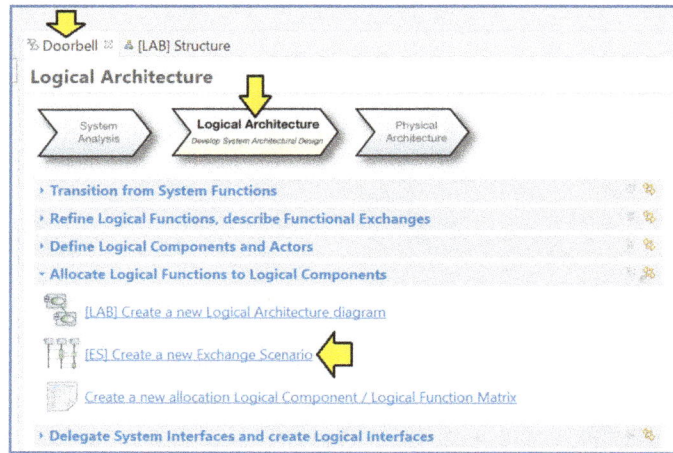

Figure 2-40 – Create exchange scenario

1) Click the "Doorbell" tab to return to the activity explorer. [3]

2) Make sure that you are still in the "Logical Architecture" section.

3) Expand the "Allocate Logical Functions to Logical Components" activity list.

4) Click on: "[ES] Create a new Exchange Scenario".

A window will appear to allow you to name the diagram. What should we name it? There are several factors we should consider in choosing a name for an exchange scenario:

1) *Capella* scenarios are tightly tied to the **capability** concept. Scenarios must belong to capabilities and will be stored in the related capability section of the project explorer.

2) In most system designs, we will create several different scenarios for each system capability, such as "Main Success Scenario", "Alternate Scenario", and "Error Scenario".

3) When you create a scenario, if you are not adding the scenario to an existing capability, *Capella* automatically creates a capability to own the scenario and gives the capability and its exchange scenario a default name of: "CapabilityRealization 1" or just "Capability 1" depending on which level of analysis you are working at.

4) Having the capability and its scenarios all identified with the same name is probably not what we will want in our final model. We will be back to clean that up in a moment. For the moment, we will want to give the scenario a name that fits an activity, typically a verb phrase.

5) Experienced *Capella* users recognize the "[ES]" prefix. In fact, we should enhance the prefix by adding a "L" for "Logical" making it: "[LES]". [4]

---

[3]If you have accidentally closed the activity explorer, see *What if the Activity Explorer isn't there?* on page 14.

[4] Here we are actually working around a small inconsistency in *Capella*. In other areas of the tool, the letter for the architectural level would be added to the prefix automatically.

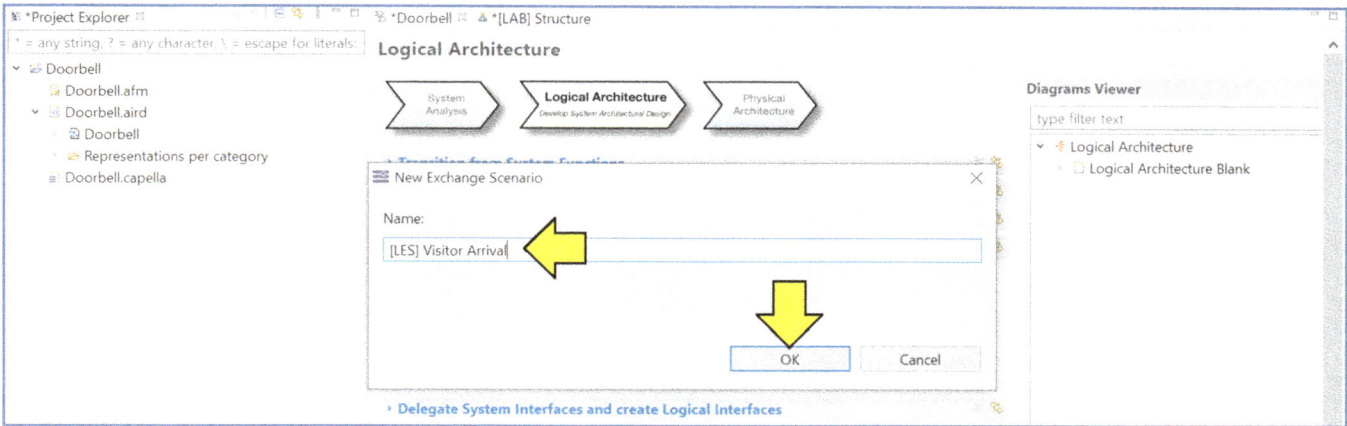

Figure 2-41 – Name the scenario

Name the scenario: "[LES] Visitor Arrival". Click: "OK".

We will discuss capabilities in more detail in later chapters of this book. For the moment, let's clean up the less-than-meaningful default name that was assigned by *Capella*.

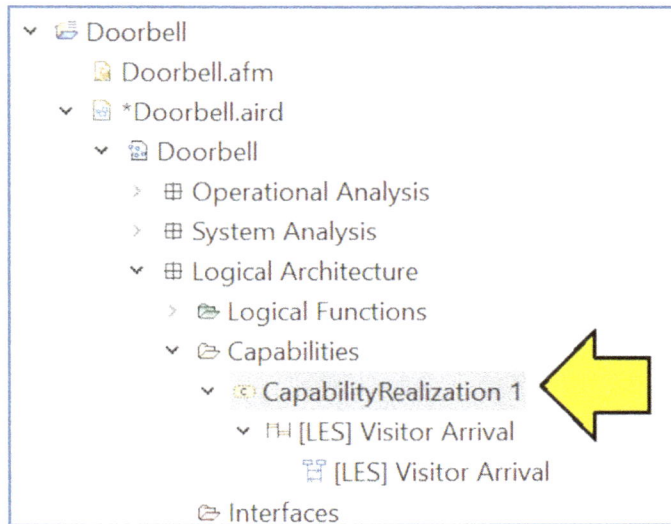

Figure 2-42 – Rename the capability

Find the capability in the project explorer and rename it: "Visitor Handling". The choice of capability name is a personal preference, but many architects use noun phrases for capabilities to distinguish them from the verb phrases used to name functions.

Note: We will cover capabilities in more detail in later chapters of the book.

What if the Palette isn't there?

What if the palette isn't visible on your screen? This kind of thing happens all the time with this sort of highly configurable tool.

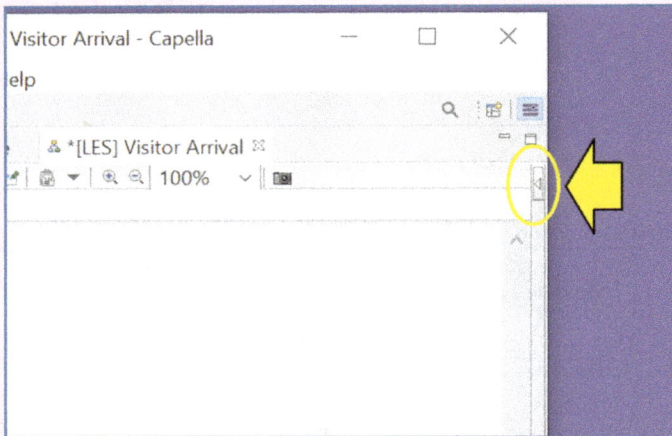

If the palette is hidden, you will find a small arrow icon in the upper right-hand corner of the diagram pane. Clicking on this arrow icon will expand the palette again.

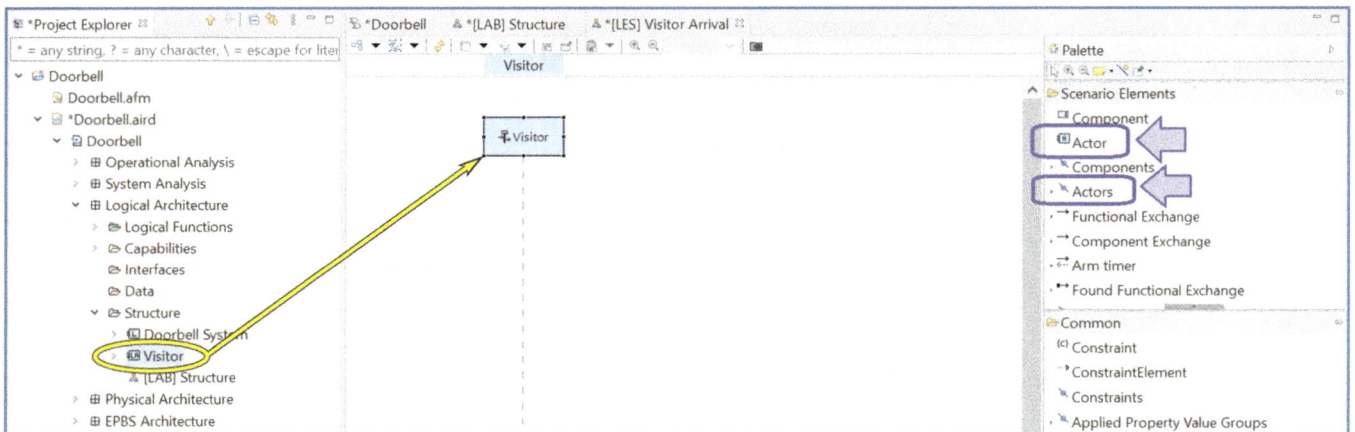

Figure 2-43 – Drag visitor to diagram

As before, icons are available in the palette to either create new model elements or add references to existing model elements in the diagram. In this case, however, we are going to use another diagramming technique: simply click on the visitor actor in the palette and drag the element into the diagram.

Capella has created something called an **instance role** for the visitor element. The *Arcadia* "instance role" concept is similar to the "lifeline" concept of UML and SysML.

• At the top is the element itself.

- Below that is a dashed line. The dashed line indicates the passage of time in the "life" of the element, starting at the top and progressing downward in time.

Note: in this book we will be mostly referring to the dashed line portion of this diagramming element. We will be using the somewhat more intuitive UML/SysML term *lifeline* rather than the formally correct, but less understandable *Arcadia* term: "instance role".

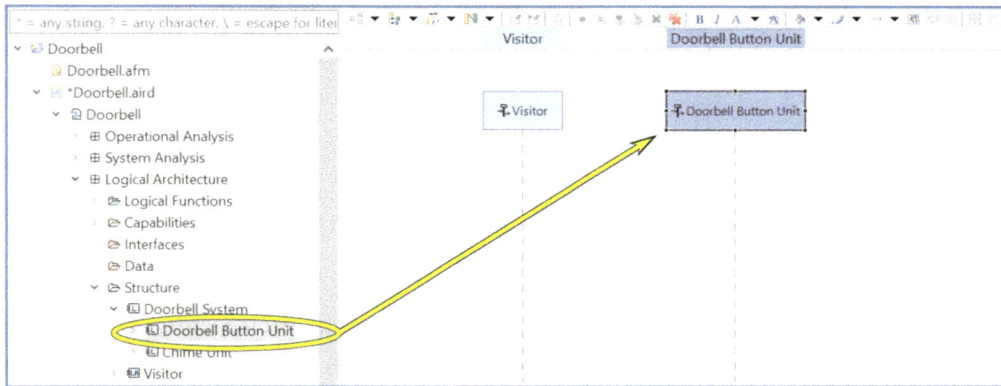

Figure 2-44 – Drag the doorbell button unit to the diagram

Drag the doorbell button unit to the diagram. The instance role (the top of the lifeline) for a component looks a little different than the instance for the actor: dark blue instead of light blue. The colors and their meanings match the colors shown in Figure 2-29 on page 25. You can resize the header box of the lifeline by selecting it and dragging the corners of the box around as needed. You can also change the horizontal position of the lifeline by dragging it back-and-forth as needed.

Drag the chime unit block to the diagram as well. We now have three instance roles (lifelines) for three elements.

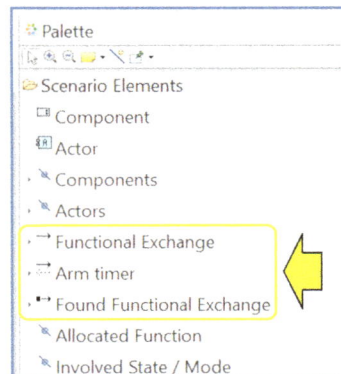

Figure 2-45 – Message types

Now we are ready to start creating the actual interaction between these three elements. In the palette you will find icons for several different types of messages. We will cover the differences between these types of messages in more detail in the chapter on scenario diagrams. For now, we will use the first type: "Functional Exchange". Each message in the scenario diagram will refer to a functional exchange between logical functions allocated to the relevant component or actor.

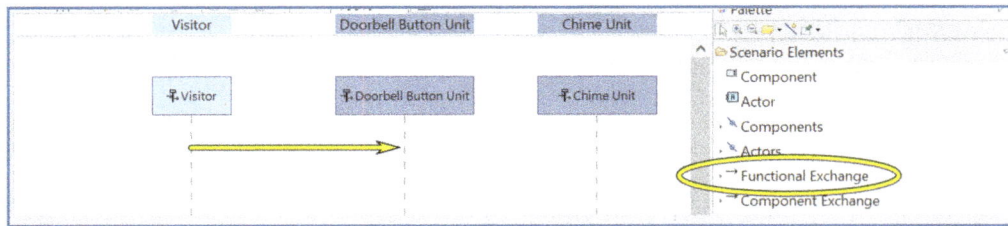

Figure 2-46 – Draw a functional exchange message

Select "Functional Exchange" in the palette, click on the lifeline for the visitor, and drag rightward to draw a message arrow from the visitor to the doorbell button unit.

Capella automatically opens a window prompting you to select one of the existing functional exchanges. In our case, only one functional exchange is relevant: the one that goes from "Interact with DoorBell" (which is allocated to "Visitor") to "Interact with Visitor" (which is allocated to "DoorBell Button Unit").

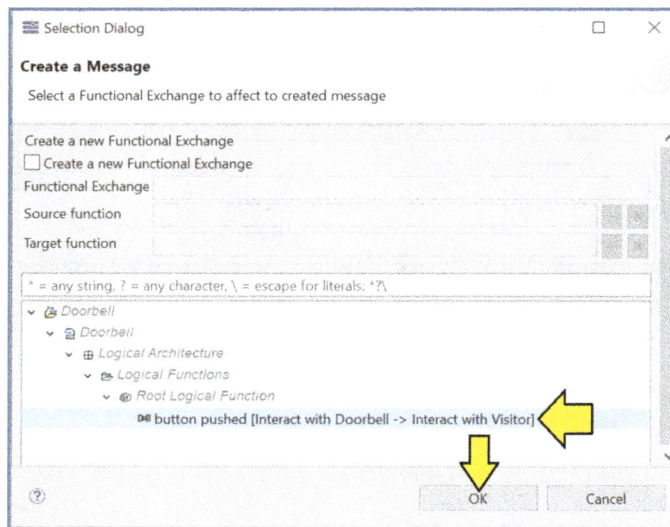

Figure 2-47 – Select the relevant functional exchange

Select the relevant functional exchange. Click: "OK".

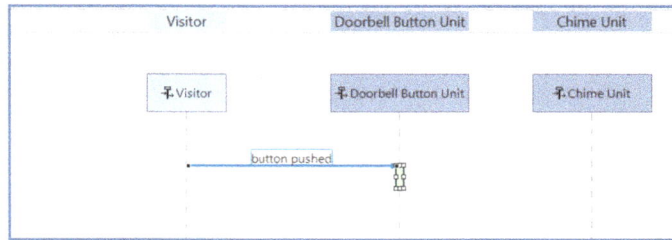

Figure 2-48 – Message created

Capella has created an arrow from the visitor to the doorbell button unit. The tool has also created a green rectangle on the lifeline of the doorbell button unit, called "Execution". We will look at this concept more carefully in the chapter on scenario diagrams.

This green box is similar to the "execution specification" concept in SysML and UML. However, *Capella* has a nice feature that is not supported in most SysML tools: we can link the green box to an actual function in the model.

Note: *Arcadia* does note really seem to have a formal name for this green box. The UML/SysML term "execution specification" is a bit arcane. We will refer to this green box as the ***execution bar.***

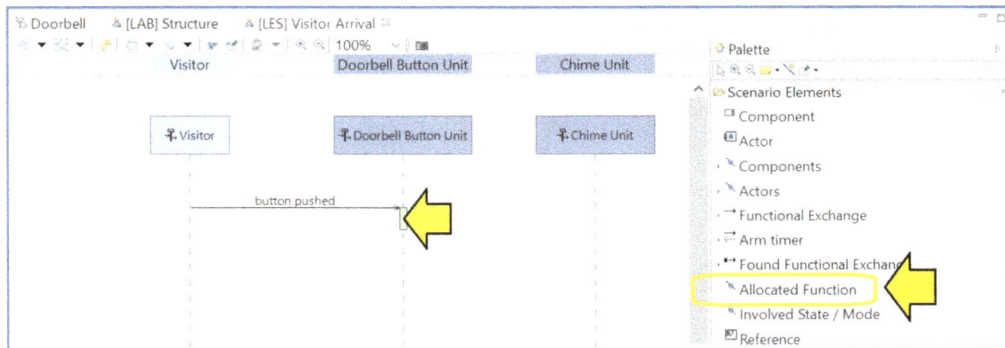

Figure 2-49 – Add allocated function

In the palette, select the "Allocated Function" tool. Click in the execution bar (green box) to add the allocated function to the lifeline.

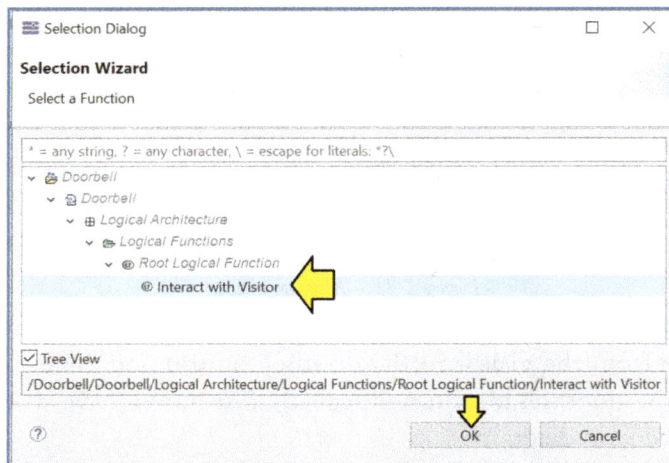

Figure 2-50 – Select a relevant allocated function

Capella will only allow functions that are allocated to the lifeline you have selected to be inserted. In this case, only the logical function: "Interact with Visitor" is available. Select that function and click: "OK".

Note that the text of the allocated function may initially appear to be truncated. In order to resize the allocated function box in the diagram, you may need to first increase the length of the (green) execution bar underneath the green allocated function box.

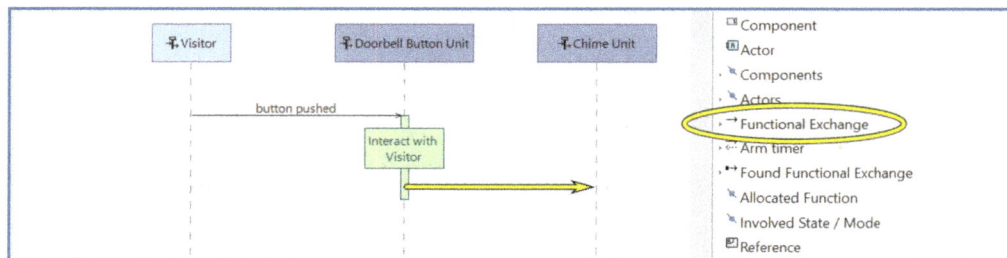

Figure 2-51 – Draw another functional exchange message

Draw a second functional exchange message from the doorbell button unit to the chime unit. Be sure to click inside the execution bar on the doorbell button unit lifeline to indicate that this new message is an output of the allocated function. Again, *Capella* will only allow you to select the existing "start chime" functional exchange.

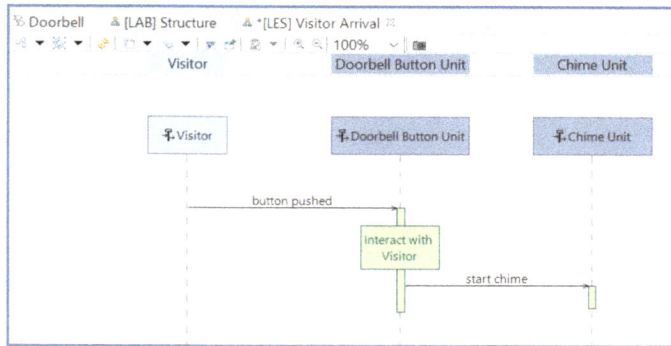

Figure 2-52 – Second message is now visible

The Second message is now visible.

Add some additional messages and functions to complete the modeling of the user story:

- Insert the "Manage Sound and Visual Indication" function on the lifeline for the chime unit.

- Add a message that references the functional exchange "start flashing" from the chime unit to the doorbell button unit.

- Add a message that references the functional exchange "button begins to flash" from the doorbell button unit to the visitor. The flashing is just a visual indication but it is a message, nonetheless.

- Add a message that references the functional exchange "stop flashing" from the chime unit to the doorbell button unit.

- Add a message that references the functional exchange "button returns to dull glow" from the doorbell button unit to the visitor. The sudden absence of flashing is itself a message and can be modeled as such.

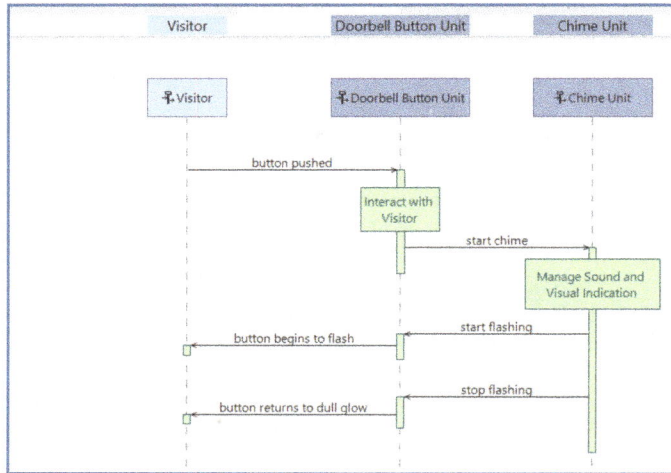

Figure 2-53 – Scenario is complete

The finished scenario should look like Figure 2-53.

This concludes our quick look at *Arcadia* and at the *Capella* tool. In the next chapter, we will cover some general topics and questions before we start looking at each of the main *Arcadia* diagram types in more detail.

Chapter 3 – Take a Deep Breath

In the previous chapter, you constructed a simple model and familiarized yourself with some of the basic functions of *Capella*. In the chapters that follow, we will look more carefully at each type of diagram and extend your familiarity with the tool and with *Arcadia* modeling.

However, before we dive into the details of all of the *Arcadia* diagram types, it is worth taking a deep breath and discussing the goals a bit. What is it that we are trying to achieve with model-based systems engineering? If you are in a hurry to get to the details, you can certainly skip this chapter and come back to it later. However, we recommend that you read through it briefly at this point before continuing to the detailed diagram chapters.

The Model-Based Systems Engineering Vision

There really isn't a universal "laws of physics" style, crisp, clear definition of just exactly what "model-based systems engineering" is (and is not). It certainly has something to do with visual diagrams and a central database containing "the model" (whatever that is). Beyond these basics, however, there is plenty of room for disagreement. If you assemble ten distinguished experts in a room, pull out the magnifying glass, and try to define exactly where the border of "model-based" and "not model-based" lies, you are likely to end up with thirteen or more conflicting opinions.

Rather than engaging in that sort of magnifying glass exercise, it will be more helpful to back into the subject by discussing some goals and examples.

Goal 1 – The Shared Vision for the System

The first goal for model-based systems engineering (henceforth "MBSE") is to enable the establishment and maintenance of the *Shared Vision for the System* for a large and complex project. We can illustrate the problem by considering the example of a project to make a "Pay as You Drive Auto Insurance" box. If you drive responsibly, your auto insurance rates are low. If you start driving like a drunken sailor, your rates instantly increase.

Clearly, we are going to mount the box in the customer's car. Therefore, we can start by modeling a car and an insurance box. We also expect this to be a high-volume product; unit costs have to be as low as possible.

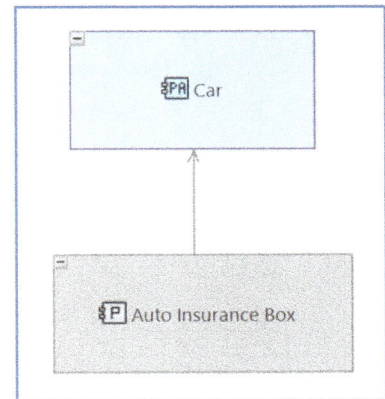

Figure 3-1 – The box mounts in the car

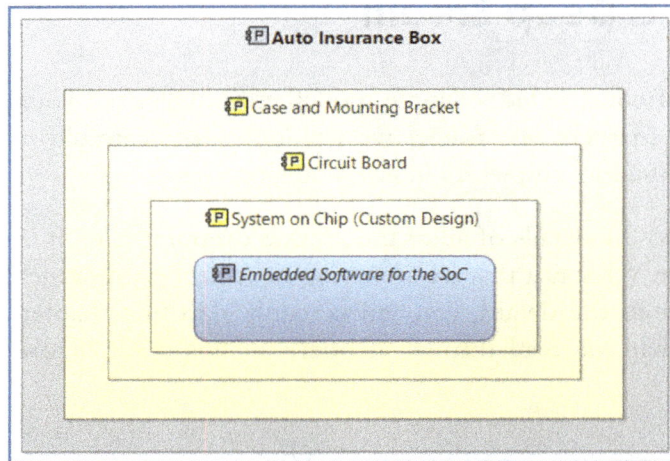

Figure 3-2 – The box contains a number of elements

The auto insurance box will have at least four major elements:

- **System on Chip –** We are going to sell millions of these. We can't pay for a disorganized collection of generic components. We will need a customized "System on Chip" (henceforth "SoC") that sweeps as many of components into one silicon die as possible.

- **Embedded Software –** The SoC will contain one or more processor cores. We will need some embedded software developed to run on the SoC.

- **Circuit Board –** We can't just attach the SoC to the dashboard with duct tape. The SoC will need to be mounted to a circuit board with other components to condition the power, watch for ignition on and off, and so on.

- **Case and Mounting Bracket –** Of course, we can't mount the circuit board in the car with duct tape any more than you can mount the unprotected SoC in the car with duct tape. We will need a properly designed case which protects the components, has the proper thermal characteristics, is built to withstand automotive shock and vibration, and meets other automotive mounting requirements.

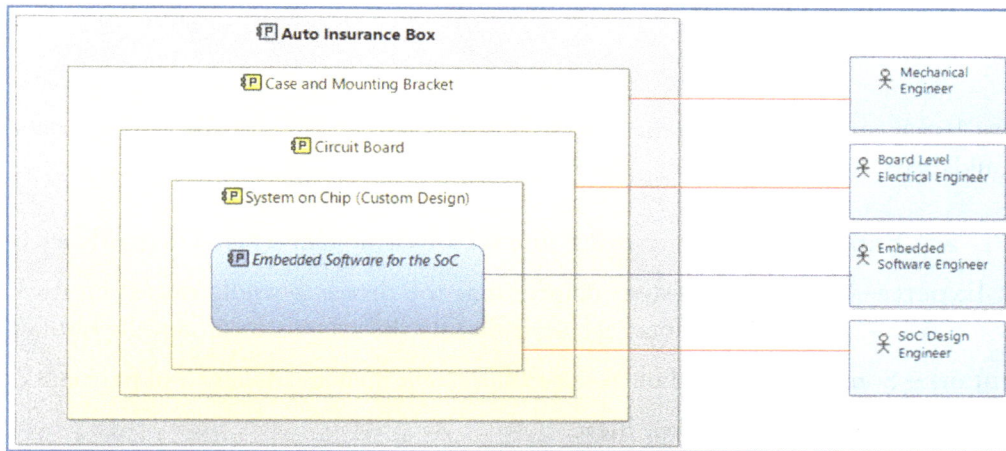

Figure 3-3 – Each element requires specialized design expertise

Each element requires its own specialized design engineer. [5]

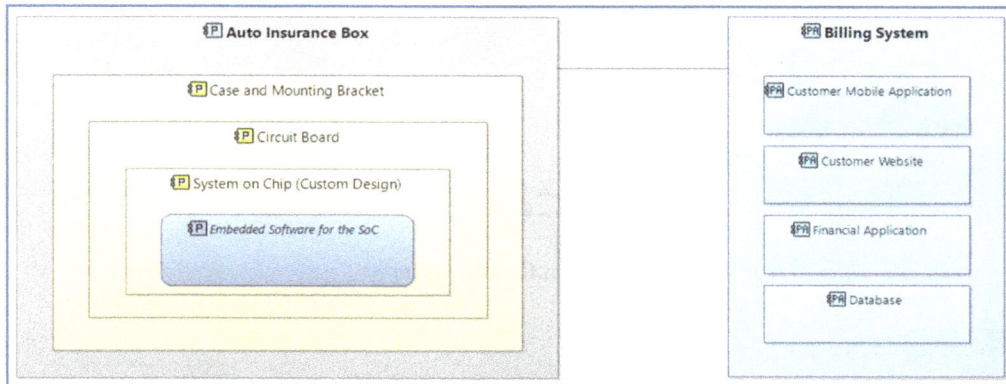

Figure 3-4 – The system will need quite a bit of software

But wait! There is more. The system will need quite a bit of software and this software will require several more development experts:

- **Cloud Application Designer** – Someone will have to design the main cloud-based software pieces.

- **Database Designer** – If the system is going to serve millions of customers, a specialist will be required to design the database.

- **Web Designer** – We need to provide a web portal to help customers sign up and to handle other administrative tasks.

---

[5] It is actually worse than this. The SoC alone may require as many as 20 different kinds of specialized design engineer. Also, we have not considered the verification and validation teams or the teams that would handle all of the automotive qualification testing. However, as we will soon see, even this modest underestimate of the number of stakeholders is difficult enough.

- **Mobile Application Designer** – No one will take us seriously if we don't also provide a mobile application.

This team is just barely sufficient to get the system bolted together. We are also going to need some other specialists for guidance:

- **Data Privacy Expert** – We need to make sure we are compliant with all data privacy laws.
- **Ethics Expert** – What if the system detects that the driver is wildly out of control and barreling directly towards a crowded elementary school? Should the system automatically notify the police?
- **Accountant** – Someone needs to make sure the system manages billing and payments properly.
- **Lawyer** – We might need several of these.
- **Marketing Expert** – What sorts of features should the system offer?

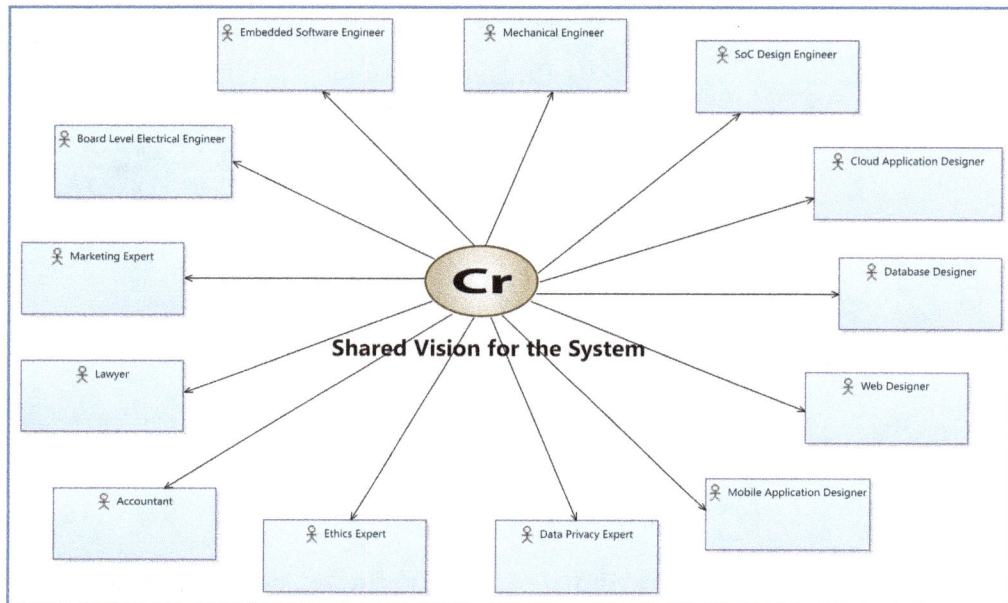

Figure 3-5 – The shared vision for the system

Eventually, all of these stakeholders need to arrive at a common understanding of what the system does and how it works. Obviously, the web designer does not need to know every detail of how the accountant is going to set up the accounting system. However, the accountant needs to know in pretty fine detail what is going to be tracked and what should determine the billing rate. Likewise, concerns from the data privacy expert will have significant impacts on all of the software pieces. Less obvious is the fact that these sorts of data privacy concerns could reach all the way into the SoC and require the chip designer to implement hardware-enforced data isolation between different blocks of memory in the chip.

Creating this common understanding is the key role of MBSE in general and *Arcadia* in particular. The goal is to provide a set of diagrams that are semantically precise, but flexible and simple enough to be understood by everyone – from the mechanical engineer to the ethics expert.

> **Arcadia does not replace the tools of each specialist**
> It is important to note that each specialist has specialty tools. The mechanical engineer has multiple computer-aided design tools. Each of the software engineers uses different specialized language and design tools. The accountant definitely has specialized tools as does the lawyer and probably even the ethics expert. *Arcadia* does not replace any of these specialized tools. The role of *Arcadia* is to create the *Shared Vision for the System* common understanding that enables each specialist to use the appropriate specialized tools to deliver his or her part of the overall system.

Goal 2 – Divide and Conquer

So far, we have shown you a goal of getting all of the diverse stakeholders to share a *Shared Vision for the System*, but we have not yet shown a compelling reason why that goal could not be achieved with PowerPoint or Visio. Goal 2 concerns the ability of MBSE to allow you – the systems engineer – to have very focused discussions with each stakeholder, one-at-a-time.

The key here is that true MBSE allows you to make simple diagrams for each stakeholder that focus exclusively on aspects of the problem of concern for that particular stakeholder – while letting the model consolidate all the different perspectives underneath the covers.

We will take a quick look at a simple industrial coffee maker as an example. We can construct the model by interviewing individual stakeholders. For each stakeholder, we can create a **system architecture blank diagram [SAB]**. The diagram will be a simple, easy-to-understand representation of that stakeholder's concerns. [6]

---

[6] The procedure to create a "[SAB]" is similar to the procedure to create a "[LAB]" shown in *Adding an Architecture Blank Diagram* on page 16 and will be covered in detail in Chapter 6 on page 123 . For the moment, don't worry too much about the details of the diagram and how to construct it. Instead, visualize the face-to-face discussion with each stakeholder and see how the contents of the diagram (not the intricate construction details) convey the concerns of the stakeholder.

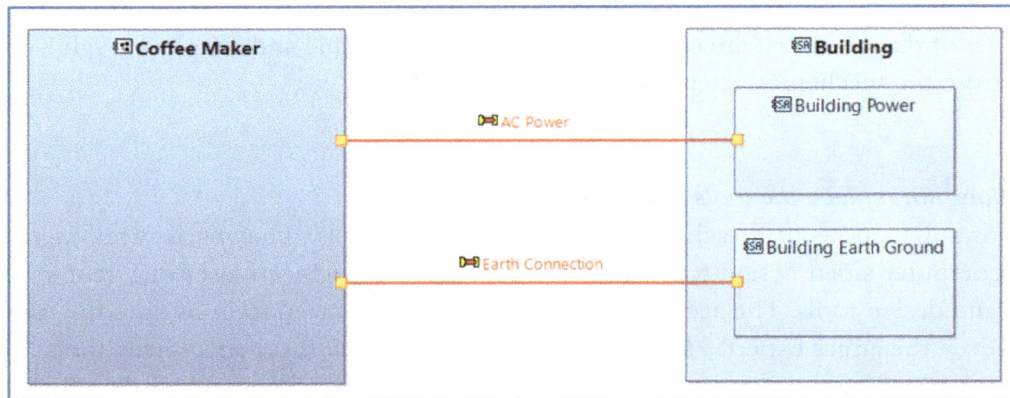

Figure 3-6 – Electrical concerns

First up is the building electrician. The coffee maker will need a connection to AC power. The coffee maker will be dealing with fluids and directly in contact with the users. It will definitely need a high-quality connection to earth ground for safety reasons. While we are interviewing the electrician, we make a diagram for the electrical concerns.

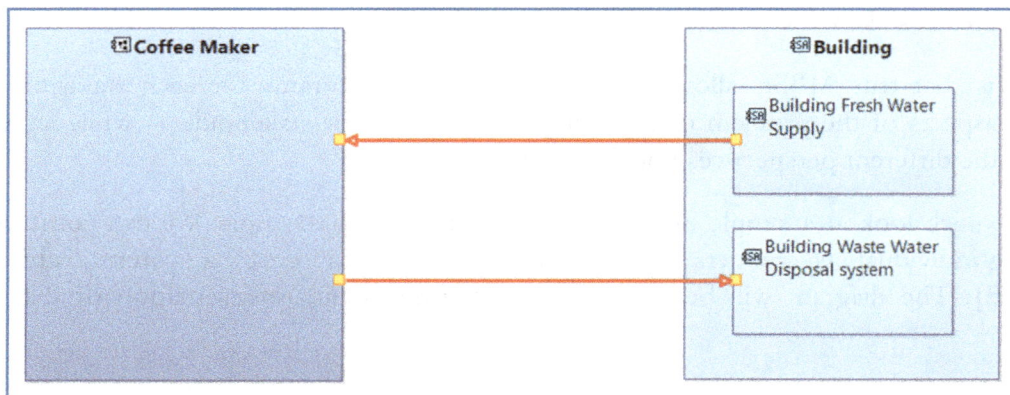

Figure 3-7 – Plumbing concerns

The coffee maker should be automatic. Obviously, it will need a source of fresh water. It will also have a spill tray. The spill tray needs to be connected to the drain. We make another diagram for the plumbing concerns.

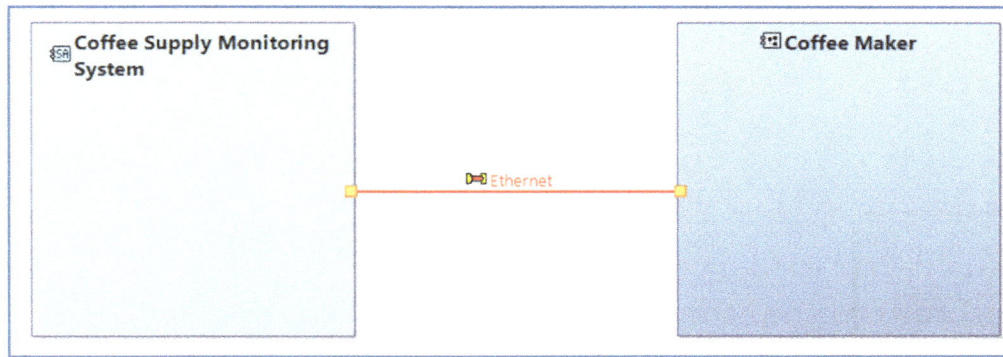

Figure 3-8 – Catering concerns

As we interview the catering department, we find that they expect to have a large number of these machines in operation. They need a central monitoring system to check when specific machines are running out of coffee. At this point, we don't really know the details of how this monitoring will work, but we specify an Ethernet connection for this purpose.

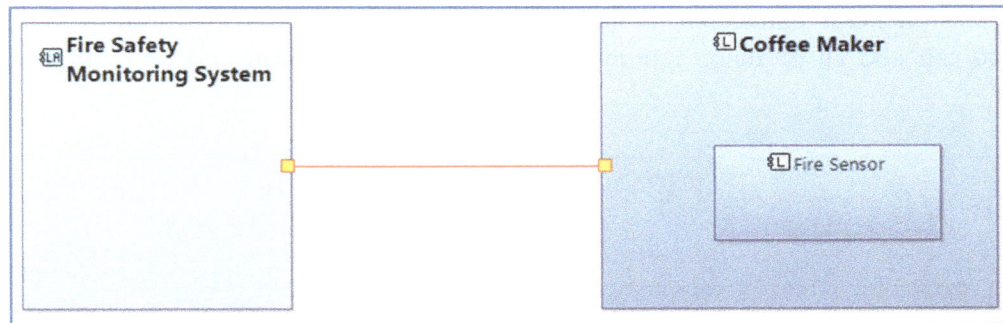

Figure 3-9 – Fire safety concerns

Finally, the fire safety department wants the coffee maker to have a fire sensor and a connection to the central fire safety monitoring system. Initially, we have offered to let them share the Ethernet connection with the catering department, but they are not so enthusiastic about that idea. For the moment, we have modeled an unspecified connection to the fire safety monitoring system.

So far, we have created a diagram for each stakeholder – in line with our MBSE strategy. We can confirm that the model has consolidated all of the stakeholder input by creating a System Architecture Blank [SAB] to explore the overall Coffee Maker system.

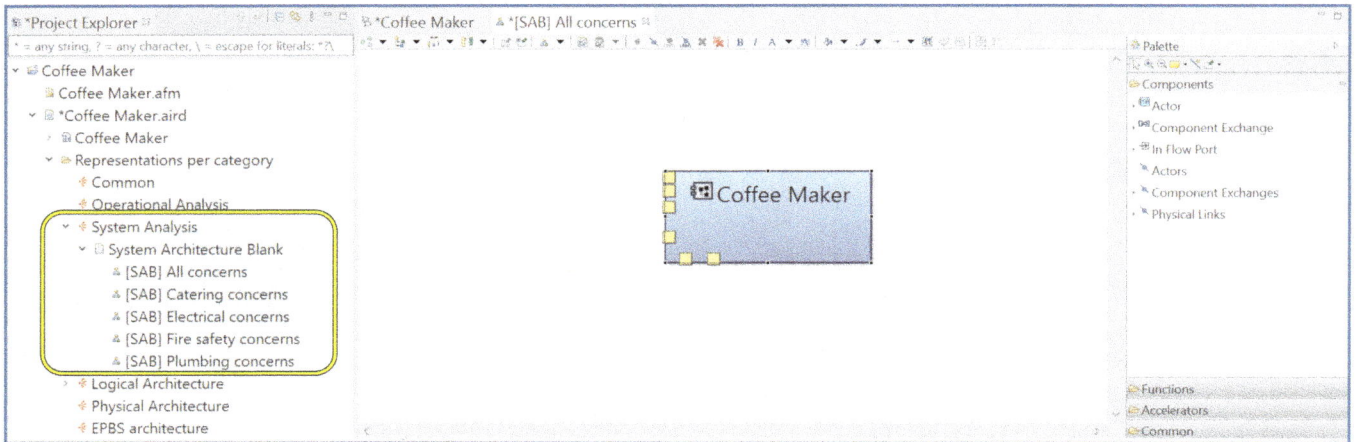

Figure 3-10 – Create another system architecture blank diagram

Create another system architecture blank diagram. By default, *Capella* automatically adds the coffee maker system component and all of its ports (created in each previous partial diagram) to the new SAB.

All of the previous elements connected to the coffee maker were modeled as (non-human!) system-level actors. Next, we can add all of these system-level actors to the diagram and let *Capella* fill in the links automatically.

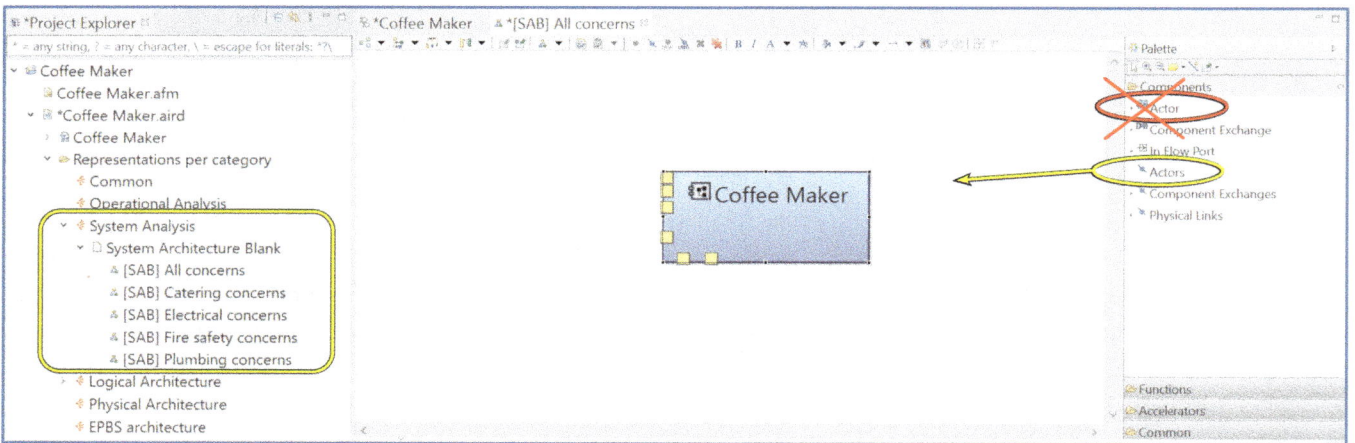

Figure 3-11 – Insert actors from the palette

In the palette select the "Actors" tool – not the "Actor" tool. Drag it into the diagram.

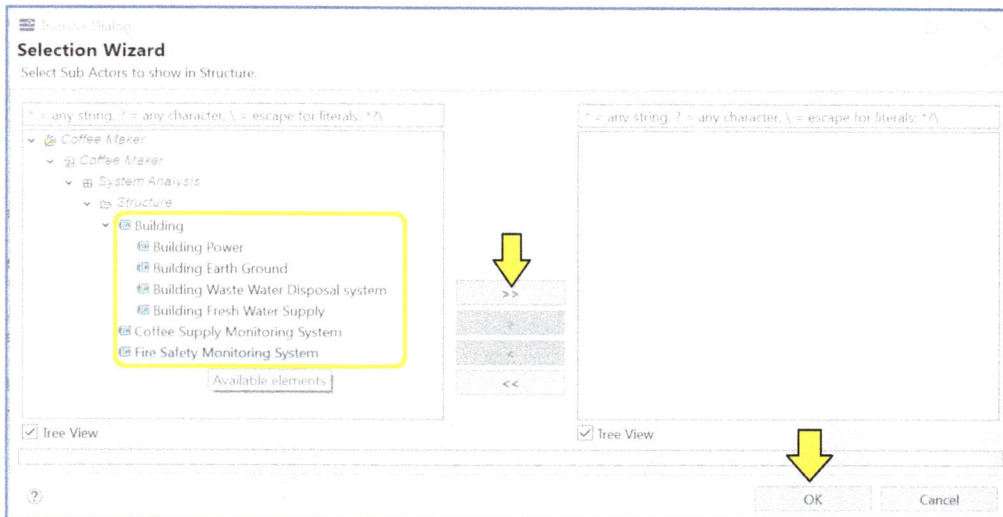

Figure 3-12 – Select all related elements, click transfer, and click OK

A transfer dialog will appear. In this dialog box we can see that the model has captured and consolidated all the information from the different stakeholder diagrams. Select all of the system-level actors. Click on the double arrow (>>) to move them into the diagram. Click "OK".

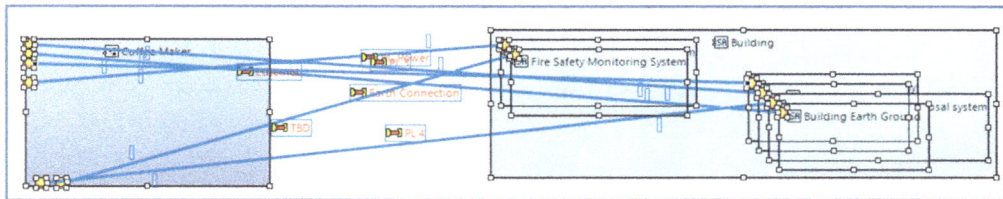

Figure 3-13 – Diagram shows all related elements

The diagram now shows all of the related elements.

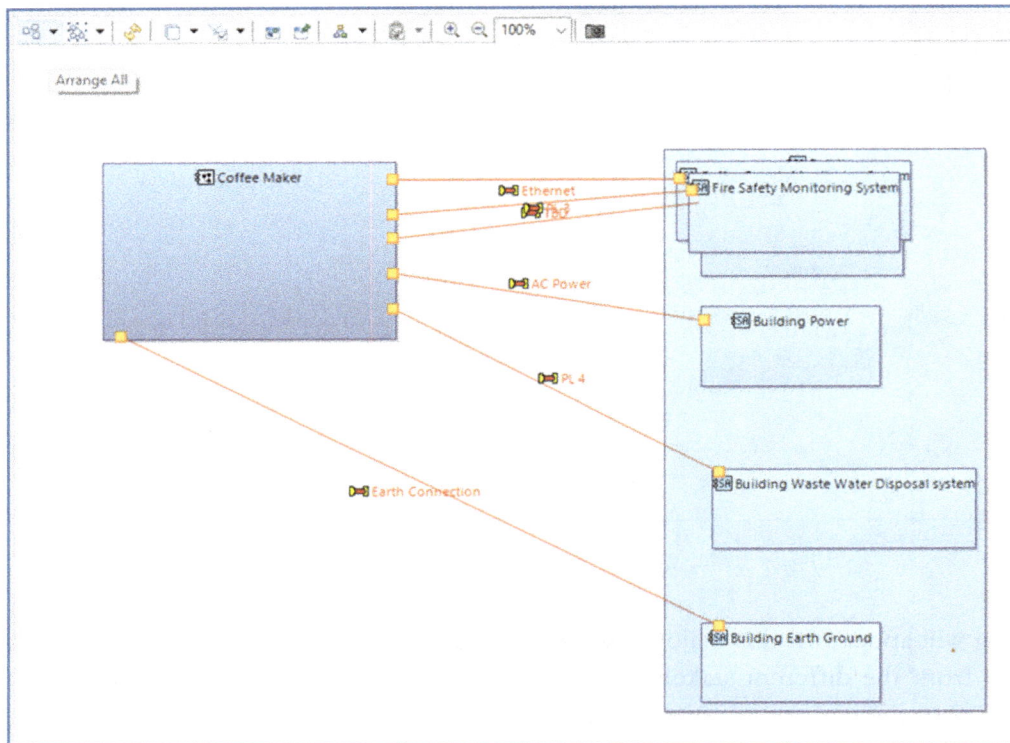

Figure 3-14 – Arrange all cleans up the diagram

We just need to click on the "Arrange All" button on the top left of the diagram toolbar the layout so that the diagram is more readable.

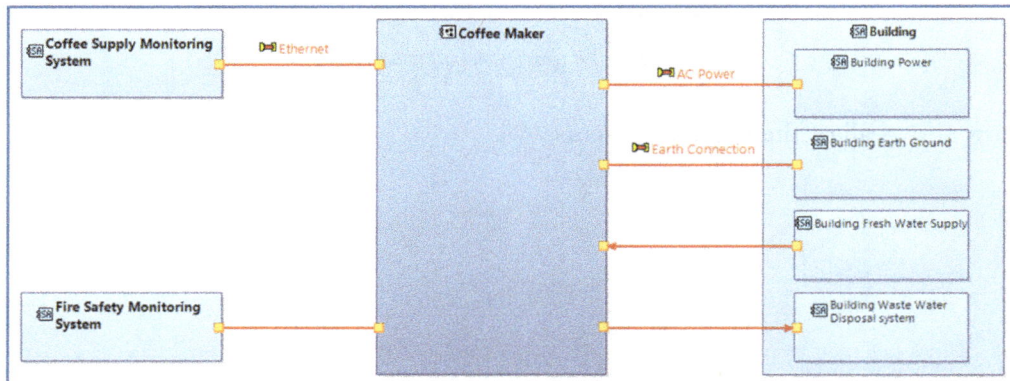

Figure 3-15 – Additional manual polishing

Another minute or so of additional manual polishing will give the diagram a nice polished appearance.

The Tool/Model Gets Progressively More Helpful

The ability of the tool to automatically insert related elements gets more and more helpful as your skill improves and you elaborate your model. You will quickly learn how to make new views by dragging certain elements to a new diagram and letting the tool automatically add selected relevant relationships and elements.

Goal 3 – Common Understanding of Diagrams

Goal 3 sounds similar to Goal 1, but Goal 3 is important enough to be worth noting individually. While it is important that all stakeholders buy into the *Shared Vision for the System* it is even more important that the stakeholders **understand** that vision and all understand it the same way. [7]

Goal 4 – Simplification

MBSE originated in the aerospace industry. Legend has it that one day, the airplane makers woke up and realized that their specification documents weighed more than their airplanes. Goal 4 of MBSE and *Arcadia* is to reduce the dependence on complicated text artifacts in large system development projects.

The ancient saying goes:

> *A picture is worth a thousand words.*

Arcadia will not completely eliminate the need for text descriptions of things. However, a good *Arcadia* model can reduce the need for long blocks of written prose. Not only will the amount of text needed drop, but well-designed *Arcadia* diagrams should be easier to understand than the blocks of dense prose they replace.

Goal 5 – Knowledge Management

Used properly and consistently, creation of *Arcadia* models can be a great way to extract and codify the tangled "tribal knowledge" embedded in large systems development organizations.

David Hetherington once talked with a technical fellow at one of the largest defense contractors in the United States. He explained that his company was systematically going through its base of legacy software and modeling the behaviors simply as an intellectual property best practice. That is, code on its own was very difficult to scan and consider for reuse. Good graphical models made it much easier to understand the function and capabilities of the legacy software blocks at a high-level. This improved understandability in turn helped the company manage this legacy software as a library of reusable intellectual property assets.

---

[7] This goal – making diagrams that everyone can understand – sounds rather obvious. However, it is easy to get carried away and produce diagrams that your stakeholders don't actually understand.

Goal 6 – Avoiding High-Level Mistakes

Goal 6 is that MBSE should help a team avoid high-level mistakes. Recent studies of companies that have deployed MBSE confirm that the largest perceived benefits are early in the project.

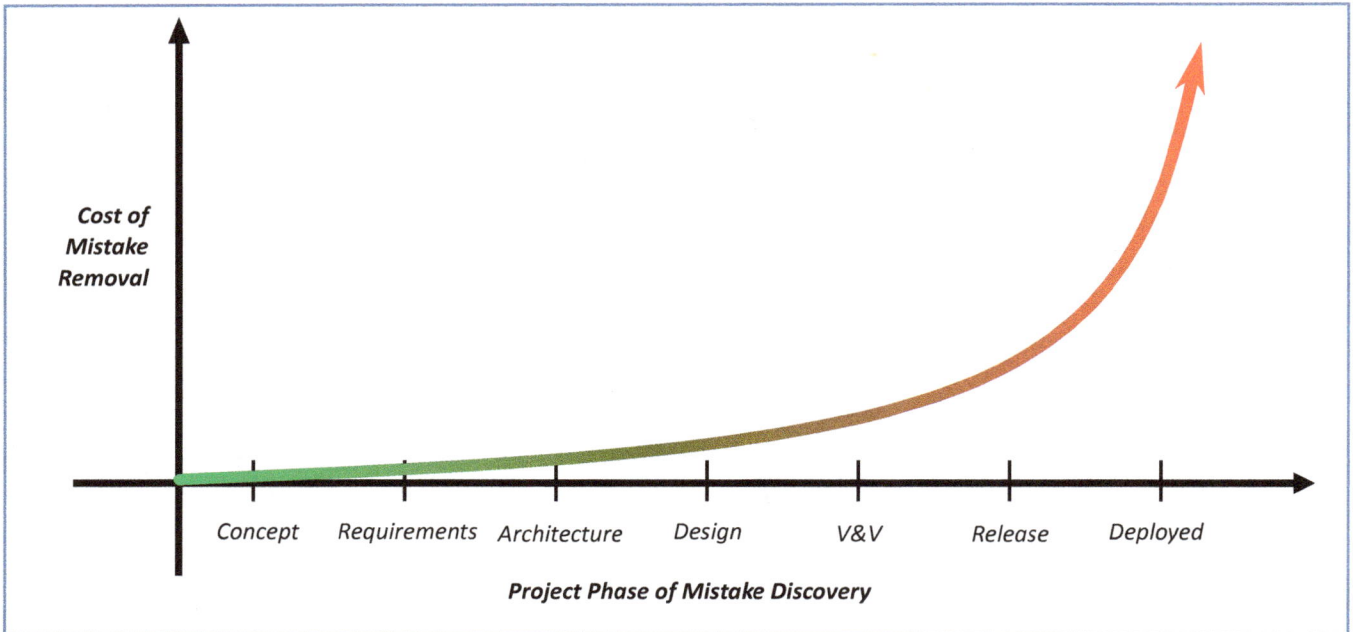

Figure 3-16 – Mistakes are cheaper to remove early in the project

In the software development world, the benefits of removing bugs early in the project are well understood. A bug detected at the requirements or architecture stage is several orders of magnitude less costly to remove than a bug detected after a product has been deployed to thousands of customers. MBSE and *Arcadia* are very helpful for moving the detection of fundamental mistakes, misunderstandings, goal conflicts, and other problems towards the left side of the curve shown in Figure 3-16.

Why Arcadia?

In the previous section, we laid out a series of six goals for model-based systems engineering. In this section, we will go back and review how well *Arcadia* meets those goals. We will also look at a few other aspects of *Arcadia* along the way.

Does Arcadia Meet the Goals?

The first question is whether *Arcadia* meets the goals. The first three goals are easy to assess.

- **Goal 1 – The Shared Vision for the System**
- **Goal 2 – Divide and Conquer**
- **Goal 3 – Common Understanding of Diagrams**

Pascal uses *Arcadia* on a daily basis and loves using *Arcadia* meet these three goals. He runs circles around colleagues who are still struggling to express themselves in PowerPoint and Visio.

The next three goals require individual answers:

- **Goal 4 – Simplification**
  David Hetherington recently completed a safety manual for an automotive semiconductor product written entirely around an integrated *Arcadia* model. The total number of pages in the manual was similar to previously written manuals for similar products. However, the number of diagrams increased by more than a factor of five – with a corresponding dramatic reduction in the number and size of long blocks of dense text. If you focus on this goal, you definitely can use *Arcadia* to take some of the dense text bloat out of your product documents.

- **Goal 5 – Knowledge Management**
  Knowledge management can take many forms. Many projects involve reuse and variants in models and requirements. These are a form of knowledge management and *Arcadia* models are part of the team strategy.

- **Goal 6 – Avoiding High-Level Mistakes**
  This goal is more difficult to assess. We certainly believe that *Arcadia* can be very helpful in helping teams avoid high-level mistakes. However, proving the absence of mistakes is challenging. SysML was first released in 2006. The *Capella* implementation of *Arcadia* was first publicly released in 2014. Model-based systems engineering is still in its very early stages of development compared to other disciplines such as mechanical or electrical engineering. So far, we do not yet have enough data as an industry to make solid statistical comparisons. Nevertheless, we are personally convinced. We have seen MBSE methods flush misunderstandings between teams to the surface. MBSE can definitely help complex teams avoid high-level mistakes.

Flexibility

Arcadia is quite flexible and types can be used to model a wide variety of different systems. *Arcadia* can be used to model airplanes, ships, automobiles, medical devices, railway networks, semiconductor architecture, and just about any other system that you can imagine.

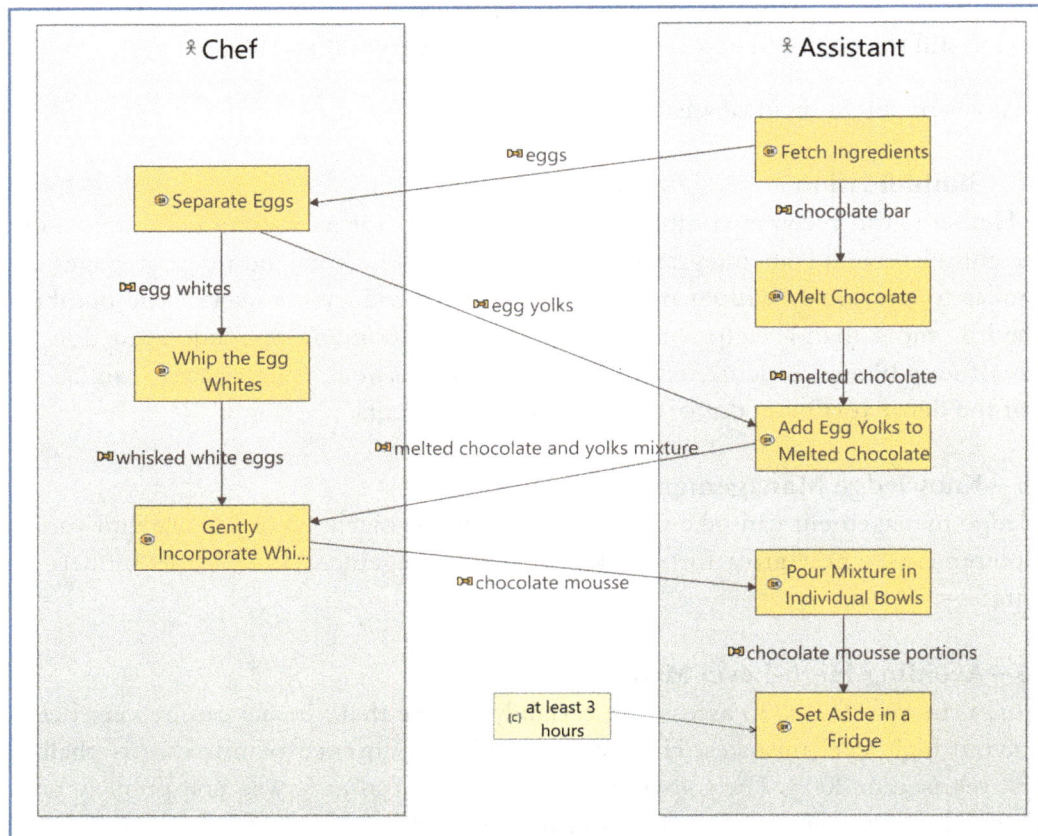

Figure 3-17 – Arcadia model of the process for preparing a Chocolate Mousse

Arcadia can even be used to model the process for preparing a Chocolate Mousse! [8]

Why Capella?

Capella itself has two main strengths:

1)   Business Model
2)   Built-In Methodology

Business Model

In 2014 Thales released the *Capella* MBSE workbench as an open source via the Eclipse Foundation with the help of Obeo. At that time, *Capella* was already successfully deployed across various Thales business units worldwide. The objective of Thales in making *Capella* open source was to extend its adoption by allowing other organizations to use it. There were other important objectives too:

---

[8] Chocolate Mousse recipe provided by Pascal Roques; David Hetherington has no idea how to make a Chocolate Mousse.

- Improve quality by getting more feedback from a wider variety of users.

- Share costs with other companies.

- Extend the life of the tool, as open-source projects tend to live longer than commercial tools.

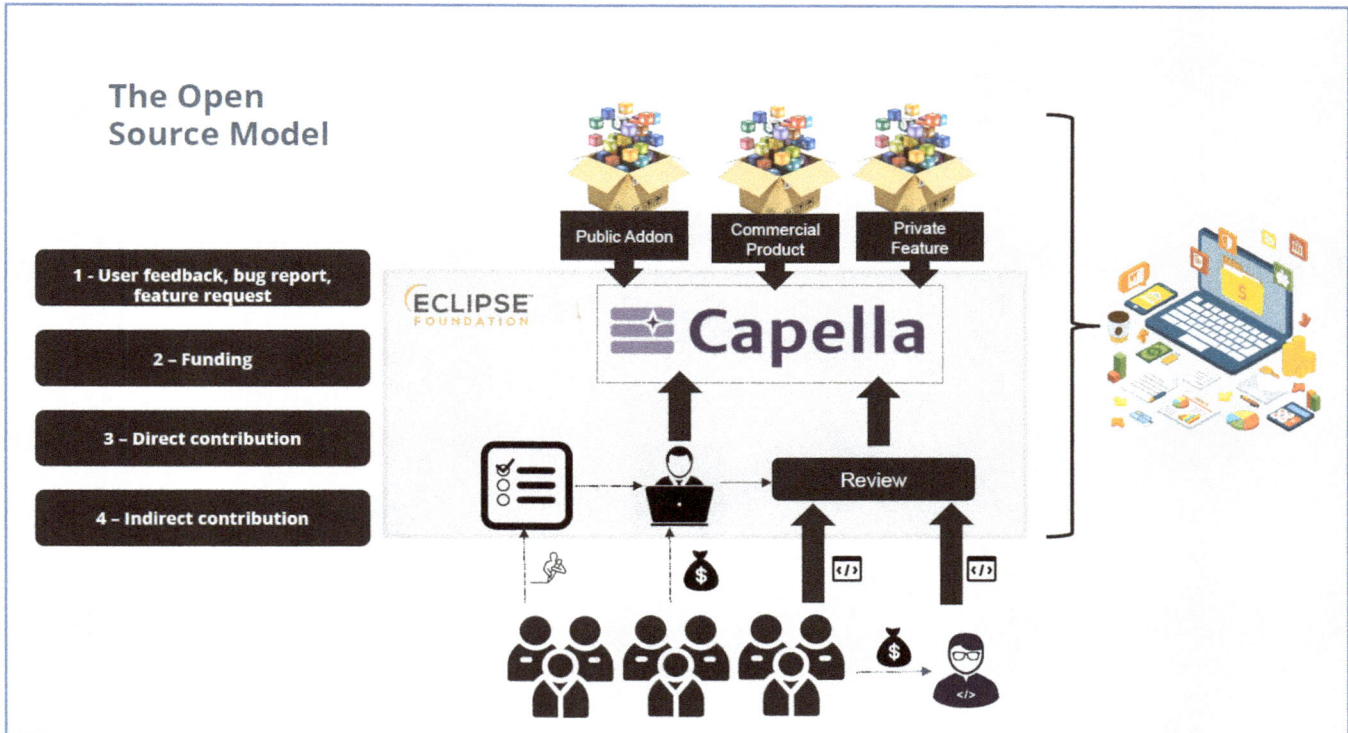

Figure 3-18 – The open source model of Capella

Figure 3-18 shows the open source business model behind *Capella*. [9] In 2022, with more than 600 identified organizations using *Capella* over 55 countries. There are also more than 50 organizations officially endorsing the tool on the *Capella* Adopters page. [10] *Capella* has definitely reached its goal of becoming a reference solution, used worldwide in multiple and various domains.

Built-In Methodology

The *Capella* workbench is an Eclipse application that provides:

- An implementation of the *Arcadia* methodology.

- A graphical modeling language that supports the methodology.

---

[9] Reproduced with permission from an Obeo presentation at **SESE tour 2021:** *Capella (3/3), Focus on Deployments & Community*

[10] https://www.eclipse.org/capella/adopters.html

- An integrated tool that supports both the workflow of the method, the diagram editing, and the model creation and management for the methodology.

The embedded methodology browser conveniently provides in-context *Arcadia* methodology prompts and guidance – there is no need to refer to a separate publication for guidance on the methodology steps.

Chapter 4 – Mission Capability Blank Diagrams

In this chapter, we will take a closer look at the **mission capabilities blank diagram**, as well as the important concepts **mission**, **capability**, and **actor**.

Mission capabilities blank diagrams are particularly useful at the system analysis level. These diagrams can be used to highlight the relations between system missions, capabilities and actors. The example used in the next few chapters is a simple clock radio such as the one shown in Figure 4-1.

Figure 4-1 – Clock Radio

A Word About the Arcadia Levels

In *Our First Glimpse of the Five Arcadia Levels* on page 14, we briefly introduced the five *Arcadia* levels. Inexperienced engineers might assume that there is some sort of mandate to proceed linearly from top to bottom. Nothing could be farther from the truth. In real life, you start with the best information you have about the project and model that. This initial best information is likely to be scattered between the levels. Modeling this scattered information is OK. You don't need everything to connect perfectly and be 100% consistent at the beginning. As you work your way into the problem, you continue to elaborate and adjust the content of the model, cleaning up inconsistencies and tying the model into a coherent overall picture.

It is also not the case that you need all five levels for every model. What you need depends on the circumstances of what sort of system you are designing and which areas need additional clarity. In many cases, there might not be much business value in fleshing out a physical level. In others, the operational analysis level might not matter much.

In the following chapters and sections, we will be visiting all of the levels, but not in rigid sequence. In fact, we are going to start with system analysis, then back up briefly to operational analysis, and continue from there.

Missions

Mission – A system **mission** is a high-level goal to which the system should contribute. To be fulfilled, a mission should use a number of system functions, regrouped within one or more system capabilities.

We will start by creating a model called "Clock Radio" using the procedure described in *Creating a Model* on page 7.

In our example, the main mission of the Clock Radio is to "Wake Up User on Time" with the sound of his/her preferred radio station (or buzzer).

To create a **mission** with *Capella*, we can either do it directly from the project explorer, or through a specific diagram called "Mission (Capabilities) Blank".

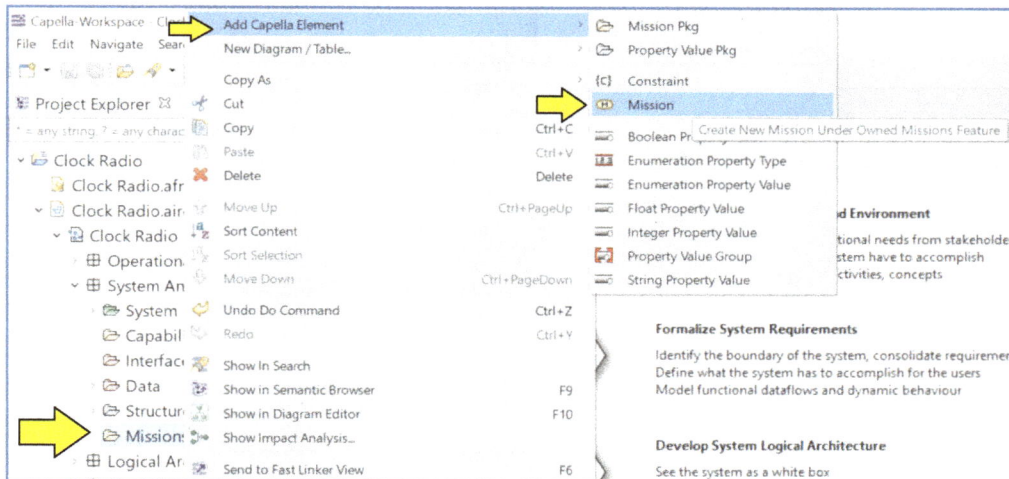

Figure 4-2 – Adding a mission from the project explorer

The procedure for creating a mission from the project explorer is as follows:

1) Expand "Clock Radio.aird".

2) Expand "Clock Radio".

3) Expand "System Analysis".

4) From inside "System Analysis", right-click on the "Missions".

5) Select "Add Capella Element".

6) Select "Mission".

In practice, however, it is easier to add a mission by creating a diagram for it in the activity explorer.

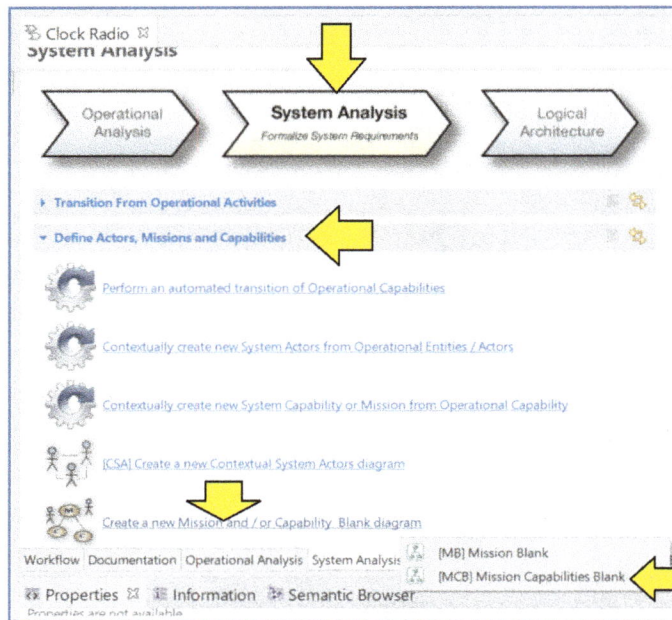

Figure 4-3 – Add a Mission Capabilities Blank (MCB)

1) Select "System Analysis".

2) Expand "Define Actors, Missions and Capabilities".

3) Select "Create a new Mission and/or Capability Blank diagram".

4) A small, two-line menu will appear offering a choice of "[MB] Mission Blank" or "[MCB] Mission Capabilities Blank". The "[MCB] Mission Capabilities Blank" diagram will allow us to conveniently create the capabilities that will be needed for the mission. If we choose the "[MB] Mission Blank", we will not be able to create capabilities: only missions and actors.

5) Select "[MCB] Mission Capabilities Blank".

6) Name the diagram "[MCB] Clock Radio Mission & Capabilities".

7) Click "OK".

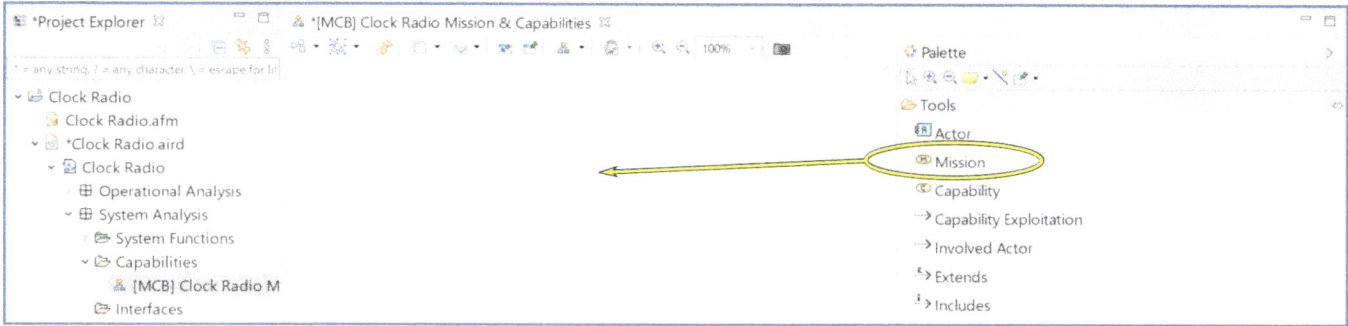

Figure 4-4 – Add a mission

In the palette, select the "Mission" tool and drag it to the diagram to create a mission. Name the mission "Wake Up User on Time".

Capabilities

Capability – A system **capability** is the system's expected ability to supply a service contributing to fulfilling one or more missions. A system capability represents a system usage context. It is characterized by a set of functional chains and scenarios that it references, and which more precisely describe the conditions for performing the system functions that contribute to it.

In our example, the main capabilities of the Clock Radio are:

- **Manage Current Time –** Help the user to always have the correct current time shown by the Clock Radio.

- **Manage Alarm Time –** Enable the user to set/clear the desired alarm time and to modify it.

- **Sound Alarm –** Wake up the user with the sound of his/her preferred radio station (or buzzer).

We can create a **capability** with *Capella* directly from the project explorer, or through a specific diagram called "Mission Capabilities Blank".

As we have already started our "Mission Capabilities Blank" diagram, it will be quicker to create the three capabilities with the palette in the diagram.

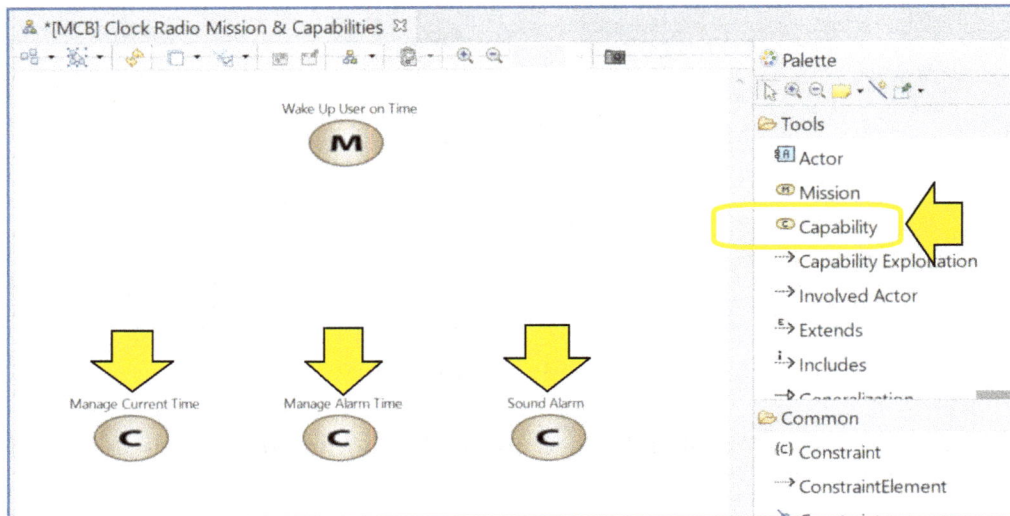

Figure 4-5 – Add capabilities

In the palette, select the "Capability" tool and drag three capabilities to the diagram. Name the capabilities "Manage Current Time", "Manage Alarm Time", and "Sound Alarm".

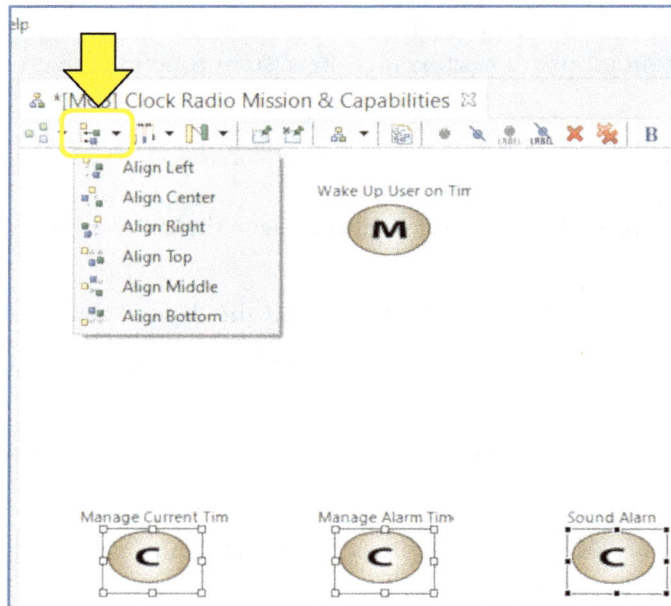

Figure 4-6 – Align the capabilities

Notice that we can easily align our three capabilities with the "Align" button of the diagram toolbar.

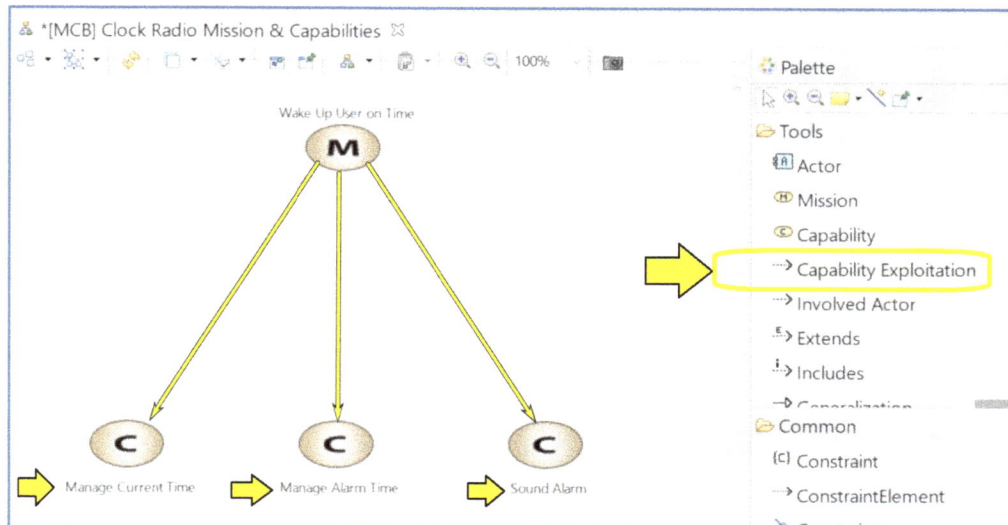

Figure 4-7 – Link the mission to the capabilities

The names of the capabilities (and missions) can only be located left, right, top or bottom, with respect to the rectangle containing the oval shape.

1) Move the titles of the three capabilities out of the way by dragging them to a spot below their respective icons.

2) In the palette, select the "Capability Exploitation" tool.

3) Draw exploitation relationships from the mission to each of the capabilities.

Actors

Actor – A system **Actor** is an entity that is external to the system (human or not), interacting with it, especially via its interfaces.

In our example, the main actors of the Clock Radio are:

• "User": the person using the Clock Radio to be waken up.

• "Radio Station Transmitter": the organization/ physical device emitting radio waves.

To create an actor with *Capella*, we can either do it directly from the project explorer, or through several diagrams, such as "Contextual System Actors (CSA)", "System Architecture Blank (SAB)", or the one we already created: "Mission Capabilities Blank (MCB)". We will continue with the "Mission Capabilities Blank (MCB)" diagram.

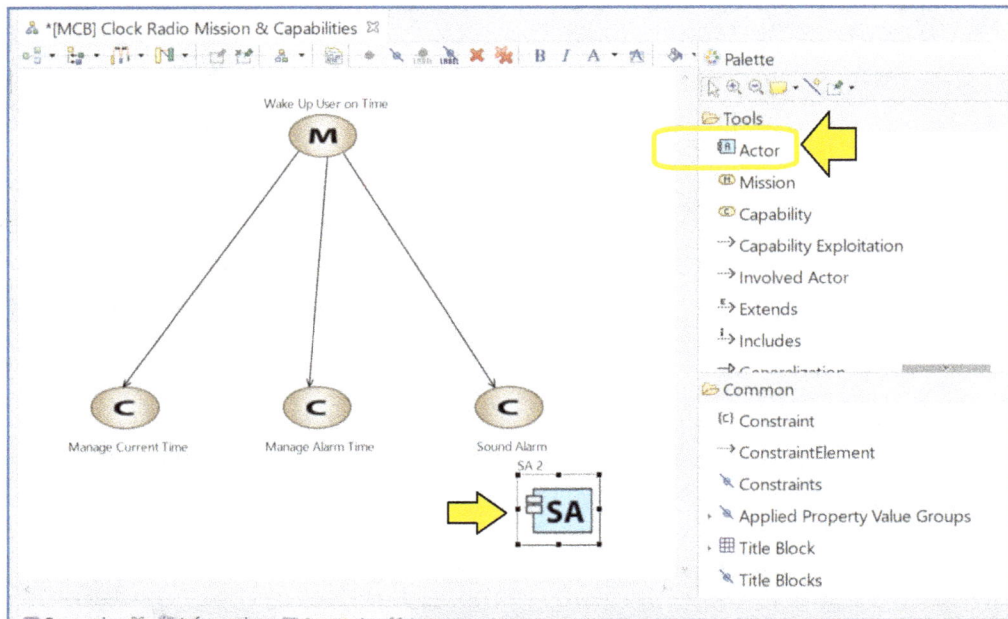

Figure 4-8 – Create the first actor

In the palette, select the "Actor" tool. Click in the lower part of the diagram to create an actor. [11] Double-click on the actor, name the actor "Radio Station Transmitter", and click "Finish".

Actors can be either systems or humans. By default, actors are shown as boxes which is intuitive for external systems, but less intuitive for human actors.

---

[11] Like most other MBSE tools, *Capella* allows for either "select and click" or "drag and drop" operation when creating new elements.

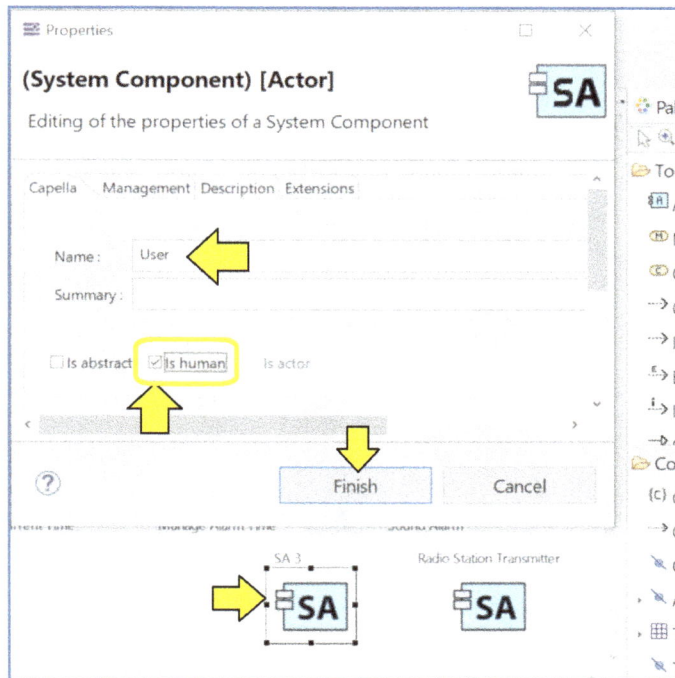

Figure 4-9 – Create the second actor

Create a second actor as before and name it "User". Before closing the properties panel, select "Is human".

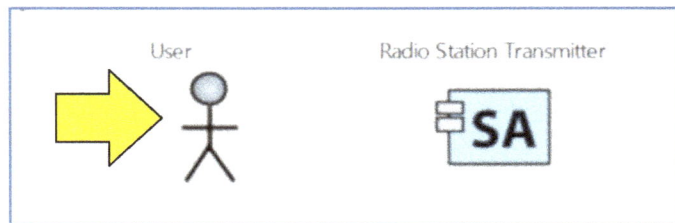

Figure 4-10 – The user actor looks like a human

After you click "Finish", the icon will change.

The last thing we need to do is connect each capability to the relevant actors. The "Radio Station Transmitter" actor is only involved in the "Sound Alarm" capability, whereas the "User" actor is involved in all of the capabilities.

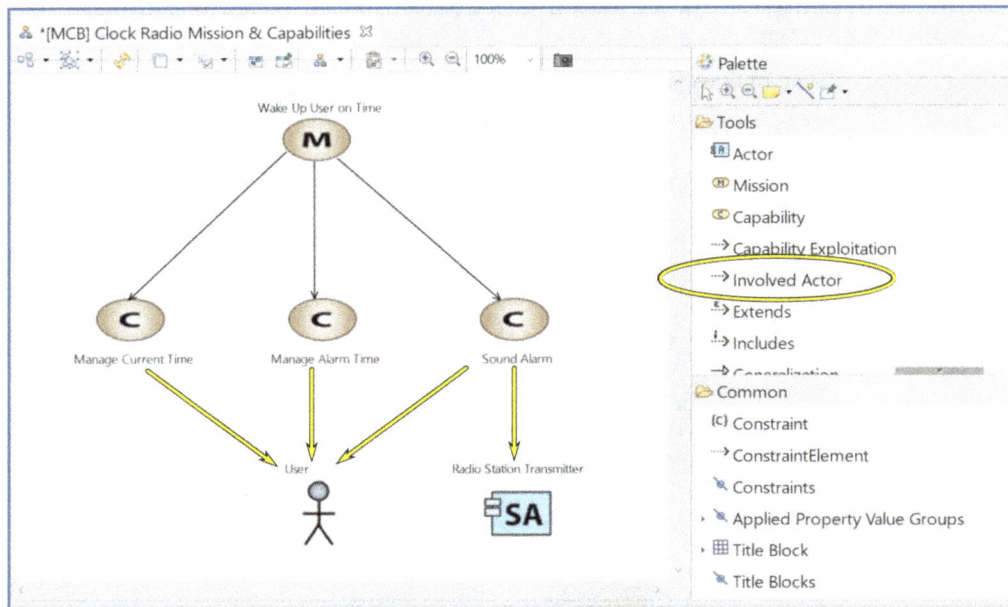

Figure 4-11 – Connect the capabilities to the actors

1) In the palette, select the "Actor" tool.

2) Drag a relationship from the "Sound Alarm" capability to the "Radio Station Transmitter" actor. Note that the direction of the arrow is important.

3) Drag relationships from each of the three capabilities to the "User" actor.

Operational Analysis

Next, we will return from the system analysis level to the top operational analysis level of the model.

Missions and capabilities have counterparts at operational analysis level. Capabilities, declared at system analysis level, can be traced to logical and physical architecture levels. The operational analysis level uses concepts similar to those described previously. The main difference in *Capella* is that the concept of "Operational Mission" is not available. [12]

> *Operational Capability* – An **Operational Capability** is an ability to provide a service on behalf of the system. An operational capability represents a context for a set of operational activities that support the capability.

---

[12] This is one of the few odd cases in which *Capella* does not fully support the *Arcadia* architecture. Operational missions are described in [Voirin] – Chapter 17, section 15.

Operational Entity – An **Operational Entity** is a real-world entity (a physical element, a group or organization, an existing system, etc.), carrying out operational activities to which the system is likely to contribute, or which can influence the system. An operational entity can be broken down into sub entities or actors.

Operational Actor – An **Operational Actor** is a particular type of operational entity, which is human and cannot be broken down.

Operational analysis involves carrying out domain modeling, sometimes referred to as "Business Modeling". In domain modeling, we are not (yet) concerned with the actual system. Rather, we are focused on the needs of the stakeholders.

Why do we need a clock radio? We need a clock radio in order to wake up in the morning. The "Business" context of the clock radio system is the one of a person at home (typically) needing to wake up at a predefined time. We capture this context by creating an **operational actor** called "Person", inside an **operational entity** called "Home", and connected to an **operational capability** "Wake Up on Time".

In *Capella*, we can do all of this using an "Operational Capabilities Blank (OCB)" diagram.

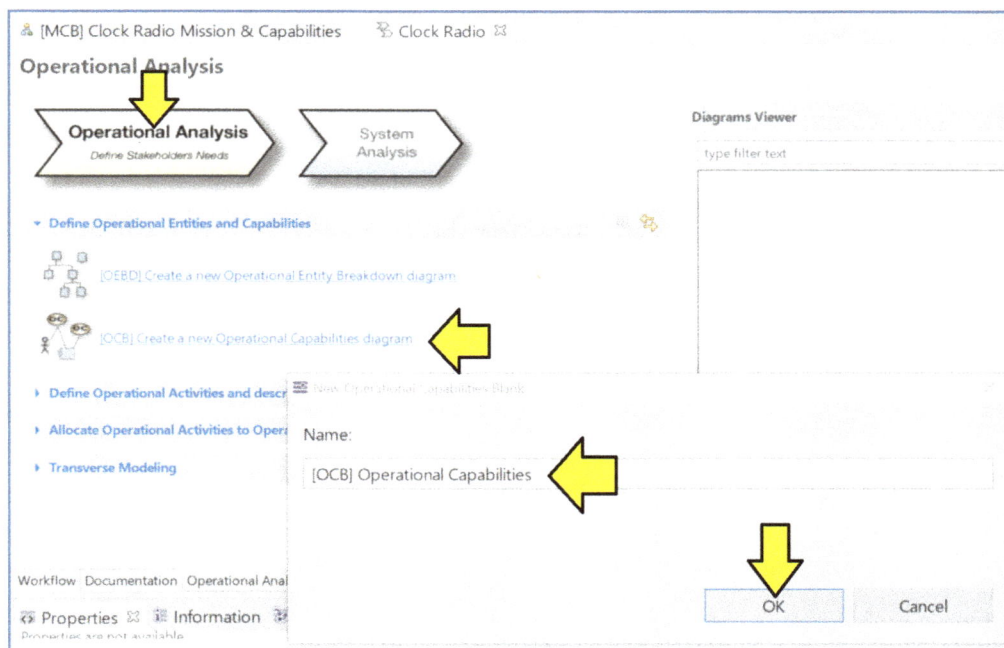

Figure 4-12 – The Operational Capabilities Blank (OCB) diagram

1) Return to the activity explorer and select "Operational Analysis".

2) Expand "Define Operational Entities and Capabilities".

3) Select "[OCB] Create a new Operational Capabilities Diagram".

4)   Accept the default diagram name of "[OCB] Operational Capabilities" and click "OK".

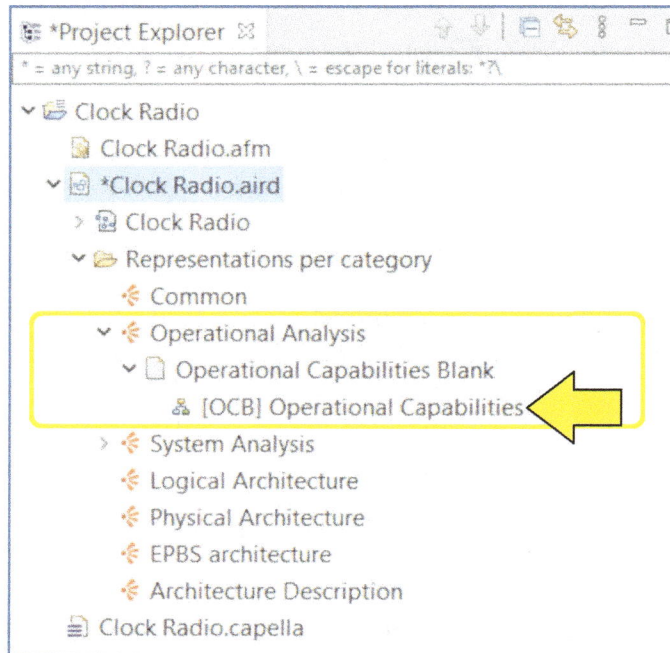

Figure 4-13 – The new diagram is automatically placed in the right place

Notice that *Capella* has automatically placed the new diagram in the correct place in the project explorer. We don't have to do anything in particular to decide where the diagram should go. *Capella* finds the right location for us.

The OCB diagram is empty by default. It allows for the creation of operational capabilities, but also operational entities and actors, as well as all the possible relationship between them.

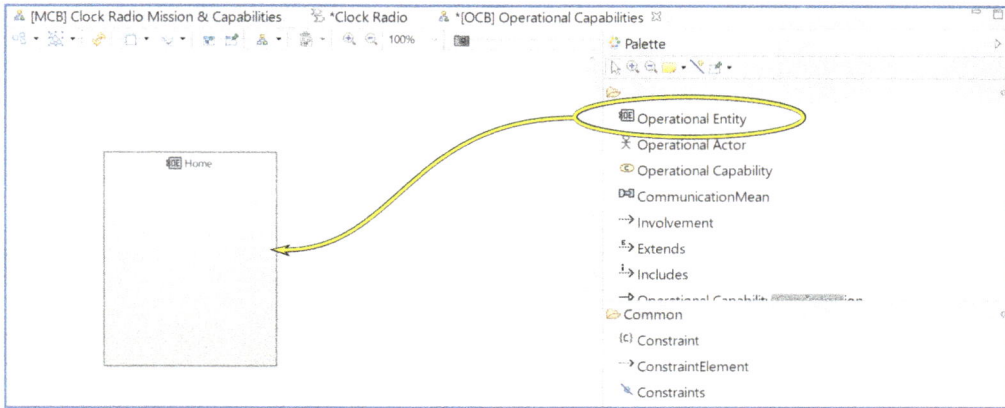

Figure 4-14 – Create an operational entity called Home

In the palette, select the "Operational Entity" tool and drag it to the diagram to create an operational entity. Name the new operational entity "Home".

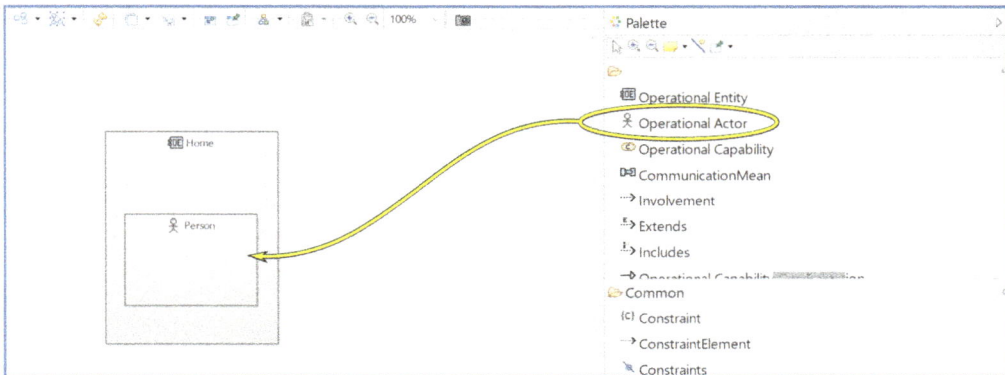

Figure 4-15 – Create an operational actor called Person

In the palette, select the "Operational Actor" tool and click inside the "Home" operational entity to create an operational actor inside of the home. Name the operational actor "Person".

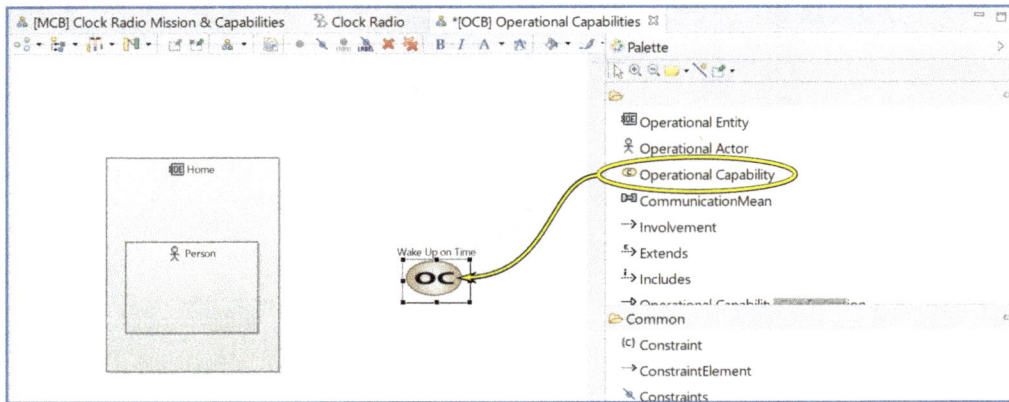

Figure 4-16 – Create an operational capability called Wake on Time

Then we can create an operational capability called "Wake Up on Time".

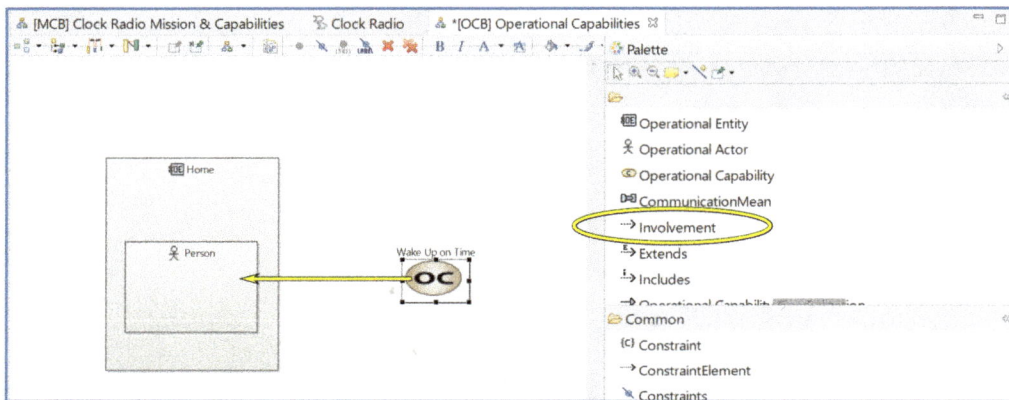

Figure 4-17 – Link the capability to the actor

And finally, we can link the operational capability to the operational actor using "Involvement" relationship. Notice that we can only draw relationship from the operational capability to the operational actor. *Capella* will not allow us to draw the relationship in the "wrong" direction.

When we have created System level capabilities, it is good practice to create traceability links between System capabilities and Operational capabilities, by means of **Realization** links.

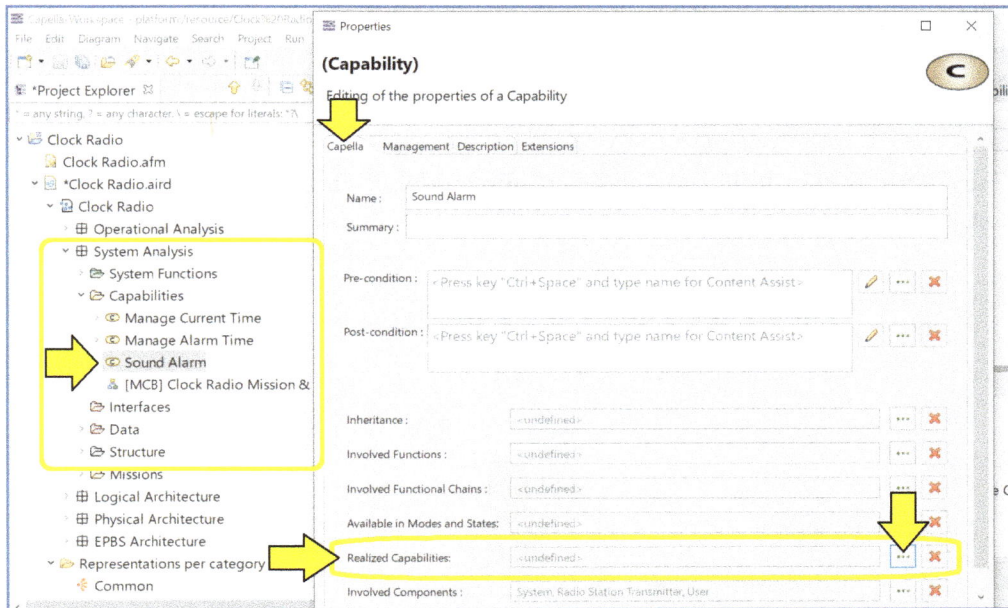

Figure 4-18 – Open the realized capabilities transfer panel

1) Remembering that we are in the operational analysis level and that the system analysis level is "below" us, return to the system analysis level in the project explorer and expand the "Capabilities" section.

2) Double-click on the "Sound Alarm" capability to open its properties panel.

3) In the main "Capella" tab, locate the "Realized Capabilities:" entry and click on its three-dots icon to open the transfer panel.

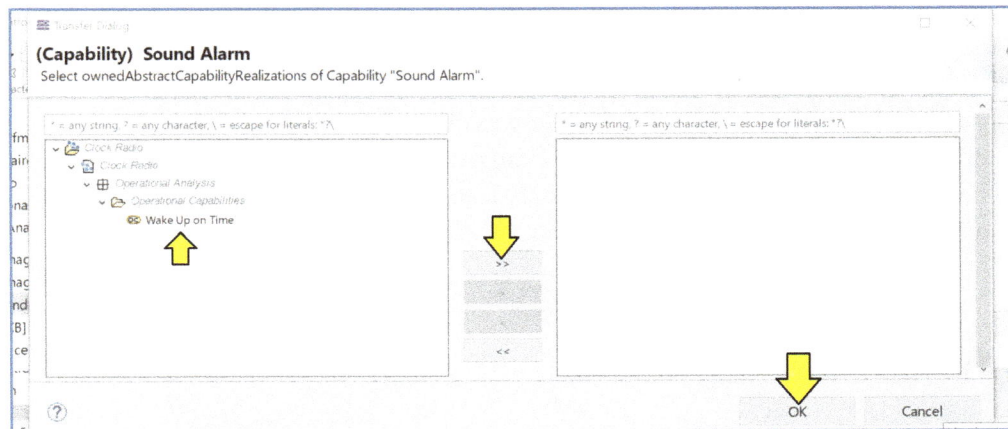

Figure 4-19 – Fill in the realized capabilities

Select "Wake Up on Time" and click the double arrow icon to transfer it. Click "OK". Click "Finish" to exit the properties panel of the "Sound Alarm" system capability.

With the sequence of steps above, we have modeled the concept that the "Sound Alarm" is a more concrete (but not yet fully concrete) system level capability that "realizes" the more abstract "Wake Up on Time" operational level capability. As we continue to model, we will continue down the levels linking them with these sorts of relationships.

Actually, all three of the system level capabilities in our model contribute to the one operational capability. Go ahead and establish realization links for the other two capabilities.

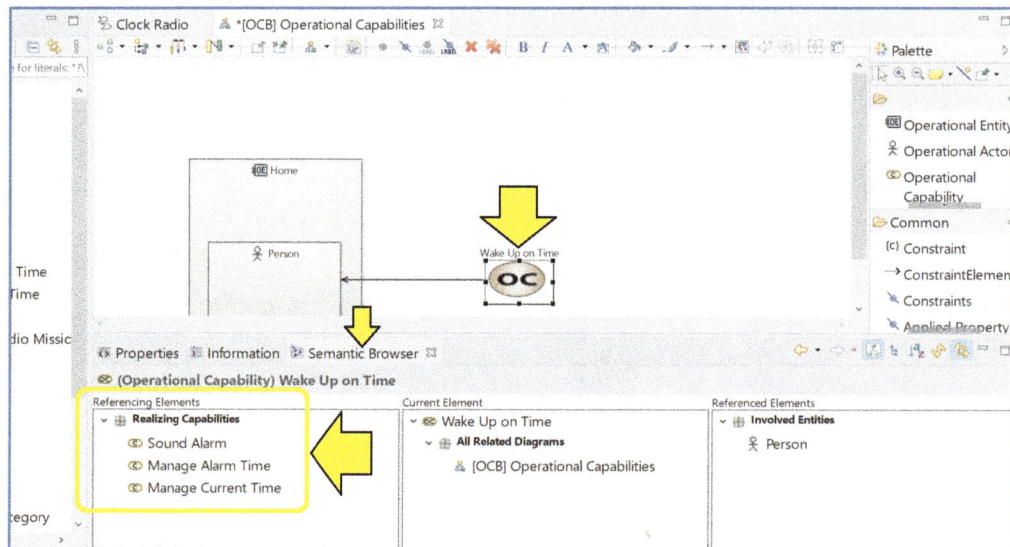

Figure 4-20 – Links in the semantic browser

If you open the semantic browser and select the "Wake Up on Time" operational capability, you will see that there are now three elements that realize this capability.

> ***Logical and Physical Architecture Levels***
> In *Arcadia*, system missions and capabilities are really defined at system analysis level. As the design progresses, the modeler adds components at the logical and physical levels to realize them. This sort of progressive linking of specification down to design is a key focus of the *Arcadia* method.

Further Study

Here are a few topics for more in-depth study:

- **Actors in *Capella*** – In this video, you will find a in-depth explanation of the new way to manage actors in Capella, one of the most important new features of the tool since it was first designed.
  - https://url4ap.net/CH-4-MCB-System
- **Practical MCB example** – At the beginning of this short video, you will find a real example of a mission capability blank diagram. The diagram is even enhanced with requirements and properties. (Slides 24-37 starting at 14:38 in the video)
  - https://url4ap.net/CH-4-MCB-Actors

Chapter 5 – Data Flow Blank Diagrams

In this chapter, we will take a closer look at the Data Flow Blank Diagram, as well as the important concepts of Function, Functional Exchange and Functional Chain.

Data Flow Blank Diagrams are useful at all *Arcadia* levels. They allow us to precisely define the information or matter dependencies between the functions. These diagrams provide a diverse set of mechanisms for managing complexity. Simplified links can be calculated between the high-level functions, exchanges can be categorized, and functional chains can be represented as highlighted paths.

The example used in this chapter is again a simple Clock Radio for domestic usage, started in the preceding chapter.

Functions

> *Function* – A **function** is a behavior or service provided by any structural element: system, actor, logical component, physical component. Examples: detect a threat, acquire temperature, etc. A function owns function ports that allow it to communicate with the other functions.

To distinguish functions defined at different *Arcadia* levels, we refine the term further:

- **System Function** – function defined at System Analysis level.
- **Logical Function** – function defined at Logical Architecture level.
- **Physical Function** – function defined at Physical Architecture level.

Let's return to our Clock Radio example, started in Chapter 4. We had identified three main system capabilities of the Clock Radio:

- **Manage Current Time** – helping the user to always have the right current time displayed by the Clock Radio, including automatic adjustment for daylight savings time.
- **Manage Alarm Time** – enabling the user to set/unset the desired alarm time and to modify it.
- **Sound Alarm** – waking up the user with the sound of his/her preferred radio station (or buzzer).

It is best to elaborate these capabilities one-at-a-time. We will want to identify the necessary system level functions to describe the capabilities precisely, both the functions of the system itself and also the functions that we expect external actors to perform when interacting with the system.

For the first capability, for instance, we need at least:

- One system level function inside the user actor to modify/read the current time.

- One system level function inside the Clock Radio to manage the current time. This function needs to increment the displayed time for each minute and hour and handle daylight savings time and/or other adjustments input by the user.

One of the main themes of *Arcadia* is the fully traceable allocation of functions and components from level-to-level in the model. We will cover this allocation in more generally and in more detail in *Allocation Relationship* on page 123. In this chapter, we will focus more narrowly on elements related to functions.

Capella provides various features to visualize this sequence of allocation. For example, we will see that functions are green rectangles by default but become light blue as soon as they are allocated to an external actor.

Let's start by creating a **System Data Flow Blank (SDFB)** diagram from the activity explorer.

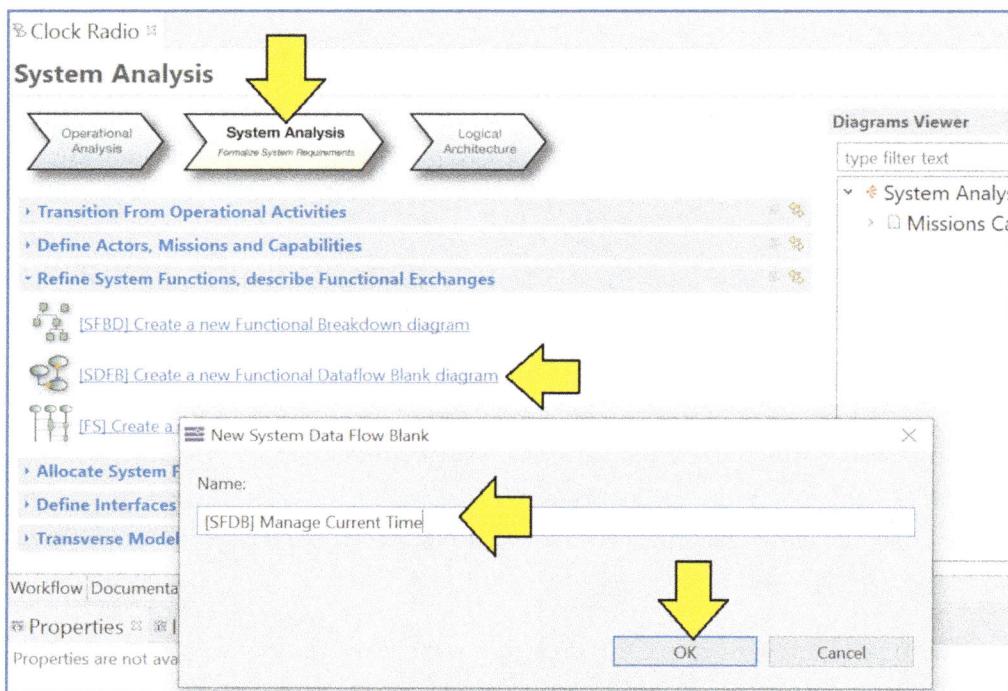

Figure 5-1 – Create and name a System Data Flow Blank

Name the diagram to match the first system capability: "[SFDB] Manage Current Time". [13]

---

[13] It is also possible to create Data Flow Blank diagrams directly from the Project Explorer, by right-clicking on a capability, and then selecting "New Diagram - Data Flow Blank Diagram". The name of the diagram would then be automatically set with the name of the capability, and the diagram would be saved under the capability in the "Capabilities" package.

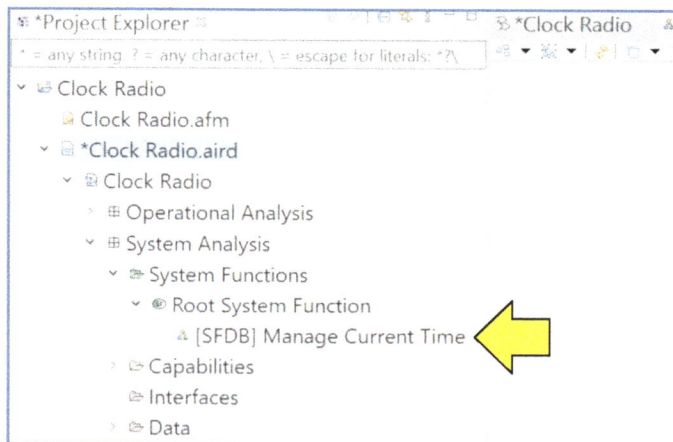

Figure 5-2 – The SDFB diagram is automatically saved

The SDFB diagram is automatically saved under "Root System Function" in the "System Functions" package.

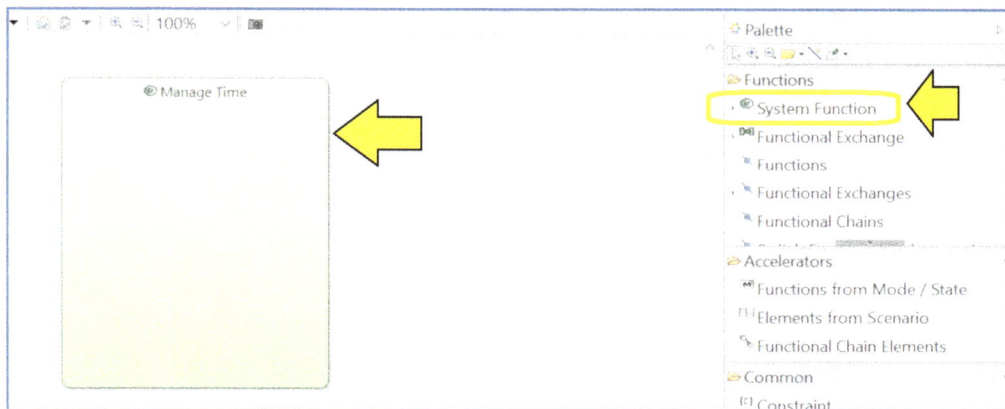

Figure 5-3 – Create a function

In the palette, select the "System Function" tool. Click in the diagram to create a function. Name it "Manage Time". Drag to resize it to be a bit taller.

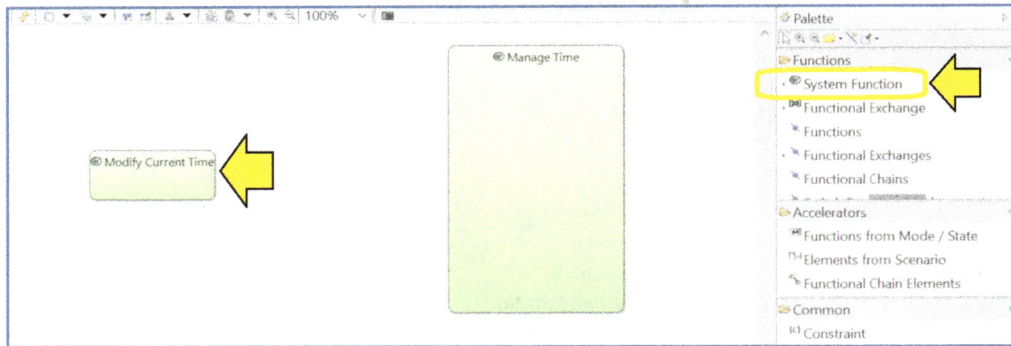

Figure 5-4 – Create another function with default size

Create another function called "Modify Current Time". Don't resize this one for the moment. Notice that while both functions are green for the moment, the first function will be allocated to the system, but this new function will be allocated to the user.

Let's make both rectangles the same size.

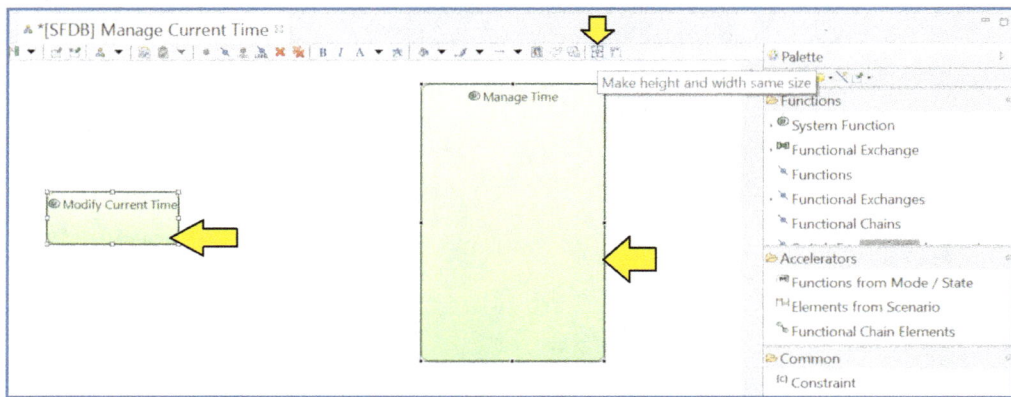

Figure 5-5 – Use the same size tool

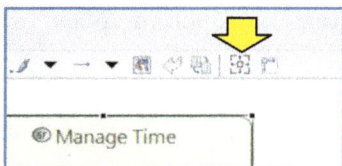

Figure 5-6 – Closer look at the icon

Select both functions (hold down the shift key while selecting). Notice that as soon as you have two functions selected, several new tool icons will appear above the diagram. Use the **same size** tool to make both rectangles the same size.

Both rectangles are now the same size. Next, let's align the rectangles.

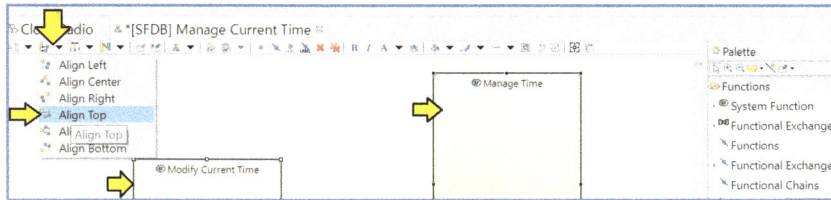

Figure 5-7 – Use the alignment tool

Select both rectangles again. Use the **alignment tool** at the top of the diagram. Notice that the tool works either as a button or a pull-down menu. If you simply click on it, it will perform the same alignment operation as the last time you clicked on it which may not be the one you intend to use. For now, pull down the menu and select "Align Top".

> *Graphical Anchor*
>
> Notice that the last graphical object selected always plays the role of the **graphical anchor** or reference for the operation. In the case of the **same size** tool, all selected elements are resized to match the size of the element selected last. In the case of the **alignment** tool, all selected elements are moved into positions relative to the last element selected.

Functional Exchanges

> *Functional Exchange* – A **Functional Exchange** is a unidirectional exchange of information or of matter between two functions, linking one output function port of the source function to one input function port of the target function.

Creating Functional Exchanges

In our example, we can now create our first functional exchange called "new current time" between "Modify Current Time" and "Manage Time".

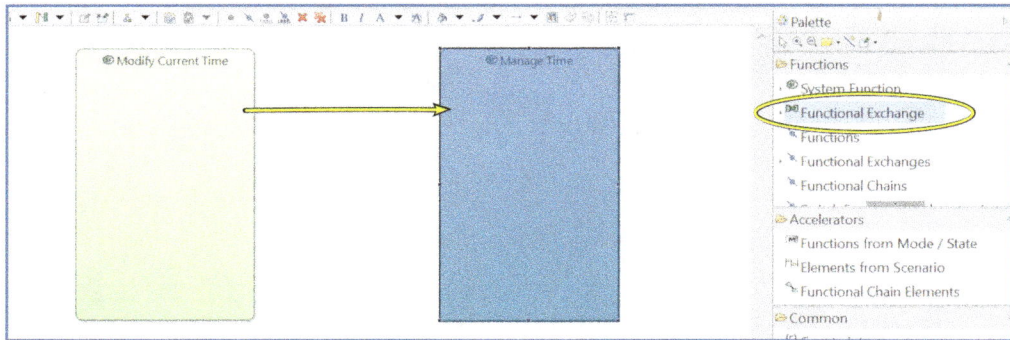

Figure 5-8 – Draw functional exchange

In the palette, select the "Functional Exchange" tool and drag an arrow from "Modify Current Time" to "Manage Time".

Notice that as you hover over "Modify Current Time" it turns dark blue. Once you click and begin to drag to create the connection, "Modify Current Time" returns to its previous color. When you reach "Manage Time" it turns dark blue and stays that way until you click once to create the connection.

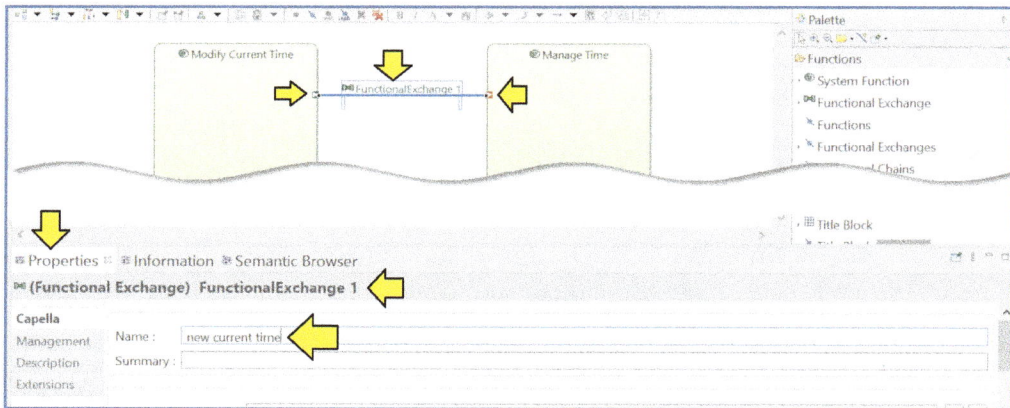

Figure 5-9 – Name the functional exchange

Capella automatically creates a **functional exchange** with a default name of "FunctionalExchange 1"). *Capella* automatically creates the output and input ports on the functions for the functional exchange.

Using the "Properties" tab below the diagram, name the functional exchange "new current time". You can also double-click on the functional exchange itself to open its **property sheet** and name it there.

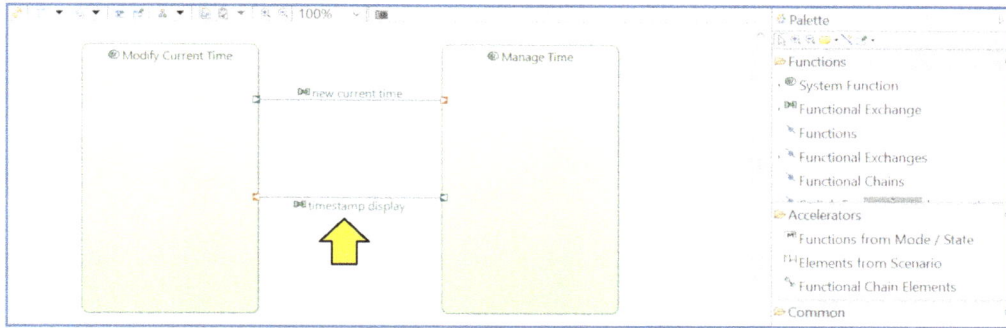

Figure 5-10 – Create a second functional exchange

We can now create a second functional exchange called "timestamp display" between "Manage Time" and "Modify Current Time", in the same way.

Naming Functions and Functional Exchanges

Each function is defined by its name. The name of a function should generally be a verb or command form such as "Display the Time". Don't be afraid to add additional words such as "the" to clarify the intent. "Display Time" is ambiguous and could mean:

- "Display the Time" – verb (command).

- "The time that is currently being displayed" – noun.

- "The time of day that the display is expected" – noun.

Don't be lazy. Use enough words to make the meaning unambiguous.

As for function inputs, function outputs, and functional exchanges, nouns are expected. "What is it that comes out of or goes into a function?" The functional exchanges in a model show the flow of these outputs and inputs in the form of a graph – commonly called "data flow".

Subfunctions

Defining simple functions is a great first step in the analysis process. However, in most systems the team will want to decompose the main functions, adding **subfunctions** to provide more granular detail about the intended operation of the system. In systems engineering, we think of this sort of decomposition as a "tree" and functions that do not contain any subfunctions are referred to as **leaf** functions.

Let's create two subfunctions inside "Modify Current Time". We will separate the modification of the time from the simple reading of the current time. [14]

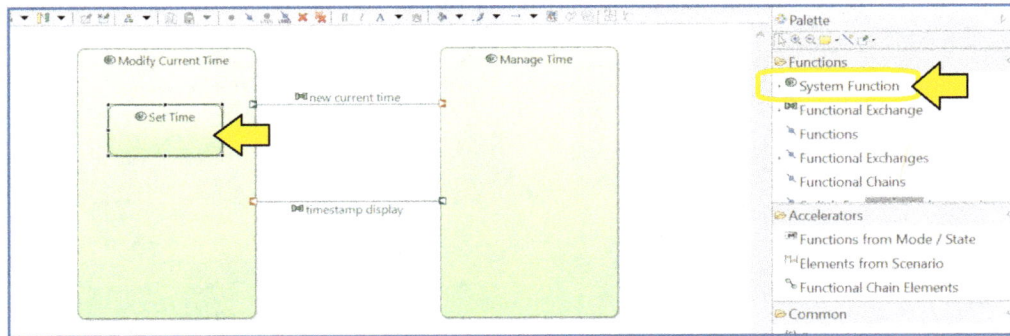

Figure 5-11 – Create a subfunction

In the palette, select the "System Function" tool (again). Instead of clicking in the background of the diagram in an empty spot, click inside the function we want to decompose and name the subfunction "Set Time".

Create a second subfunction called "Get Time".

Next, let's take a look at the *Capella* **validation** feature.

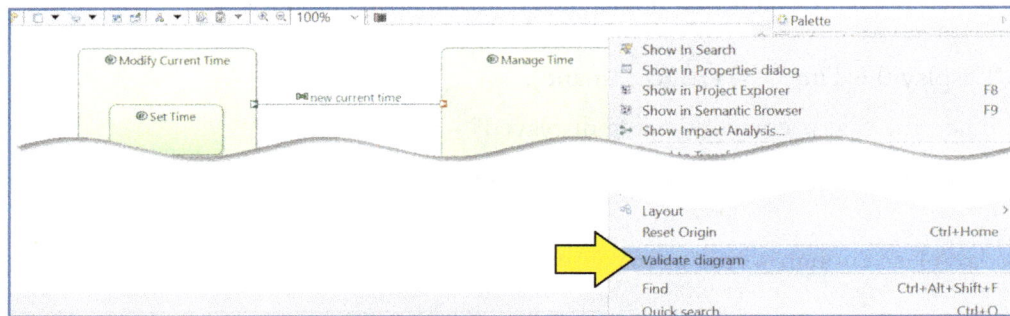

Figure 5-12 – Validate the diagram

Right-click in the background of the diagram and select "Validate diagram".

---

[14] Some modelers might not bother with this sort of trivial decomposition; we are just using this simple example to demonstrate the *Arcadia* subfunction decomposition concept.

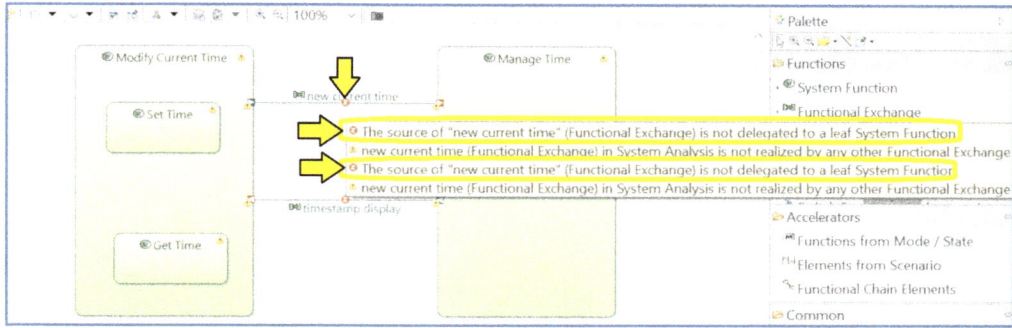

Figure 5-13 – Leaf function port warning

If you look closely, in the center of the "new current time" functional exchange, there is now a small red circle with an "x" in it. Move the mouse cursor to hover over that small red circle. A yellow box will appear with four warnings in it. Two of the warnings indicated that some systems analysis model elements are not yet realized by elements at the logical level. That information is not a surprise. However, the other two warnings tell us that we are in violation of an important *Arcadia* rule which states that "only leaf functions can have ports".

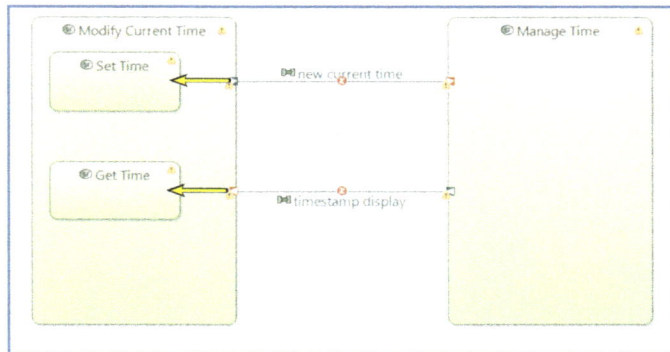

Figure 5-14 – Move ports to the leaf function

In order to correct the problem that is causing the leaf node port the warnings, simply grab the ports with the mouse and drag them to the subfunctions.

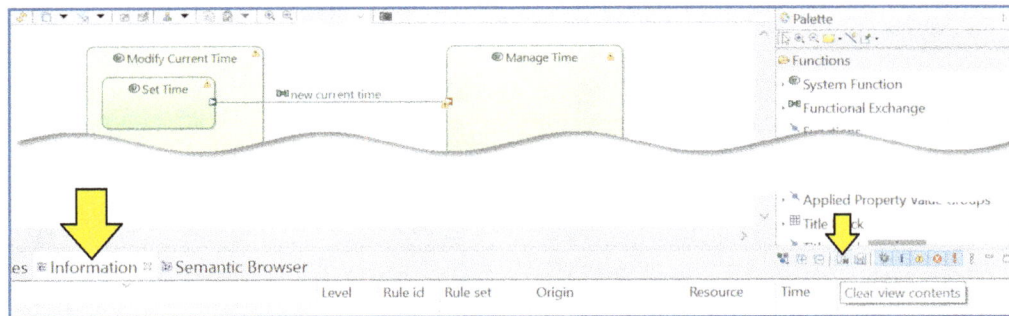

Figure 5-15 – How to clear warnings

By the way, if the boss suddenly drops by for a visit and you want to quickly remove the warnings from your current diagram, select the "Information" tab below your diagram and then click on the "Clear view contents" tool icon.

Hiding Subfunctions

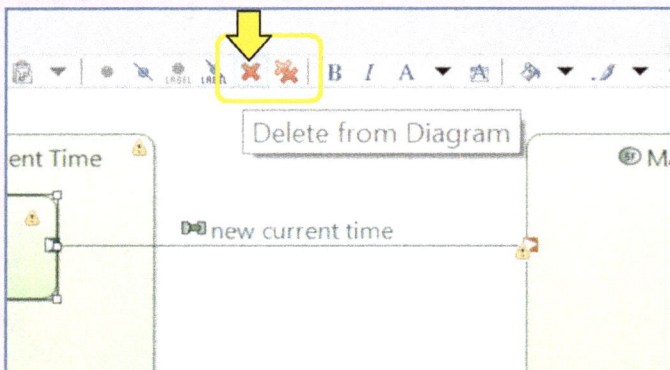

Hiding versus Deleting Elements

The toolbar at the top of the *Capella* diagram area has two tool icons for hiding or deleting elements.

- **X** – This tool deletes the selected model element(s) from the diagram, but not from the model. In other words, this tool effectively "hides" the selected element(s).

- **XX** – This tool deletes the selected model element(s) from the diagram *and also* from the model. Once you use this tool to delete an element, it is gone. Period.

As your model becomes more complicated, you will often want to define subfunctions, but not show them on a certain diagram. For example, suppose we wanted to retain the definition of the "Set Time" function, the "Get Time" function, and the functional exchanges we defined previously but not show the detail of the subfunctions in the diagram?

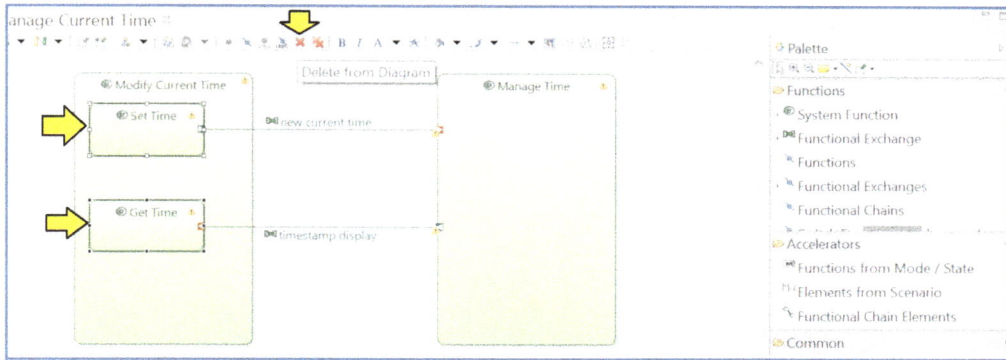

Figure 5-16 – Be careful of hide or delete

We might want to use the "Delete From Diagram" tool. [15] However, if we merely select the two subfunctions and click on the "Deleted from Diagram" tool, the results will not be quite what we expected. Yes, the subfunctions disappear from the diagram. However, the functional exchanges disappear as well!

This is the standard behavior of *Capella*: as soon as either the source or target function is not displayed in a diagram, the functional exchanges between them are not displayed.

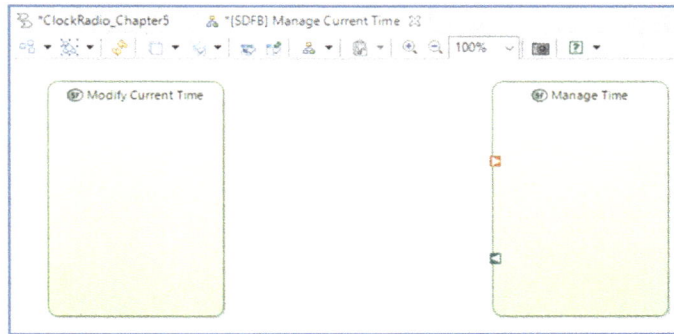

Figure 5-17 – Function ports not connected

In this case, the function ports belonging to "Manage Time" are still visible on the diagram, but not connected.

Fortunately, there is a way to make *Capella* show the data flow connections, even if the subfunctions are hidden.

---

[15]We definitely would **NOT** want to use the "Delete From Model" tool!

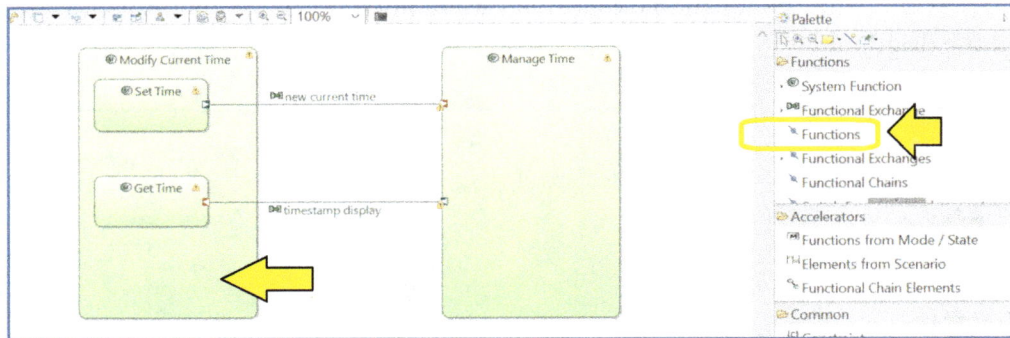

Figure 5-18 – Use the insert/remove function

In the palette, select the "Functions" tool and then click inside the parent function, in this case the "Modify Current Time" function.

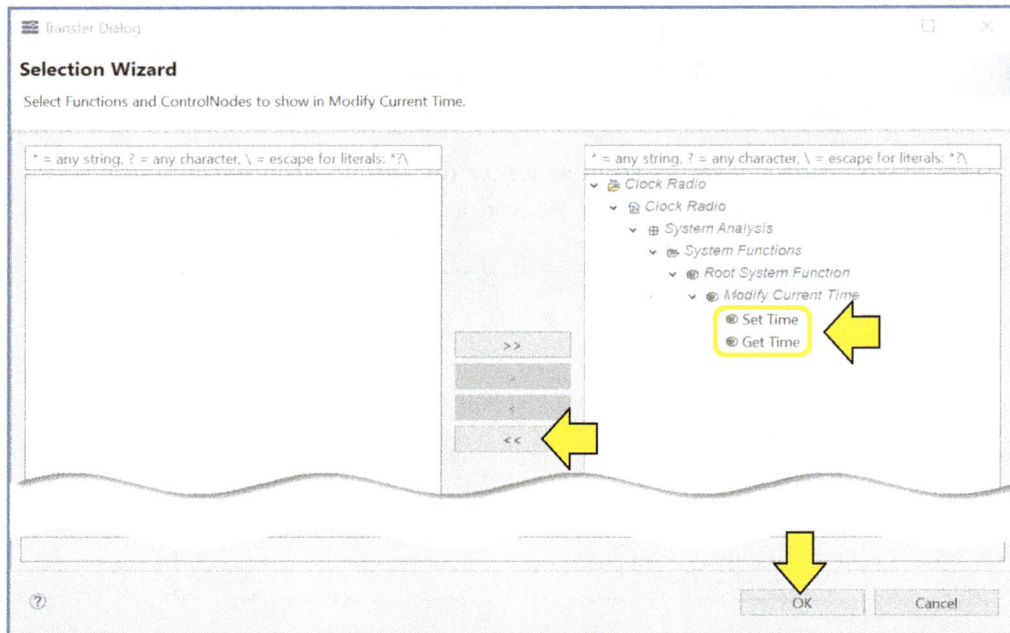

Figure 5-19 – Use the transfer dialog to remove the subfunctions from the diagram

The "Transfer Dialog" appears. It has detected the two subfunctions "Set Time" and "Get Time" within the parent function "Modify Current Time". Clicking the "«" button removes the two subfunctions from the diagram.

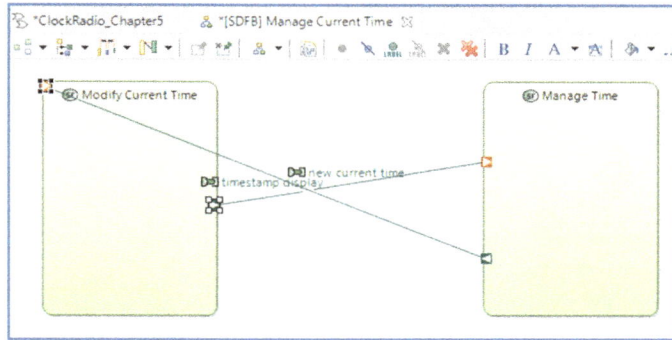

Figure 5-20 – The tool reconnects function ports

In this case, however, *Capella* notices that the data flows to and from the (hidden) subfunctions cross the border of the parent function and displays function ports on the border of the parent function to support the data flows.

Of course, we can drag the ports around to make the diagram look like Figure 5-10 *Create a second functional exchange* on page 81. That is, we can make the diagram look like it looked before we inserted the subfunctions in the first place.

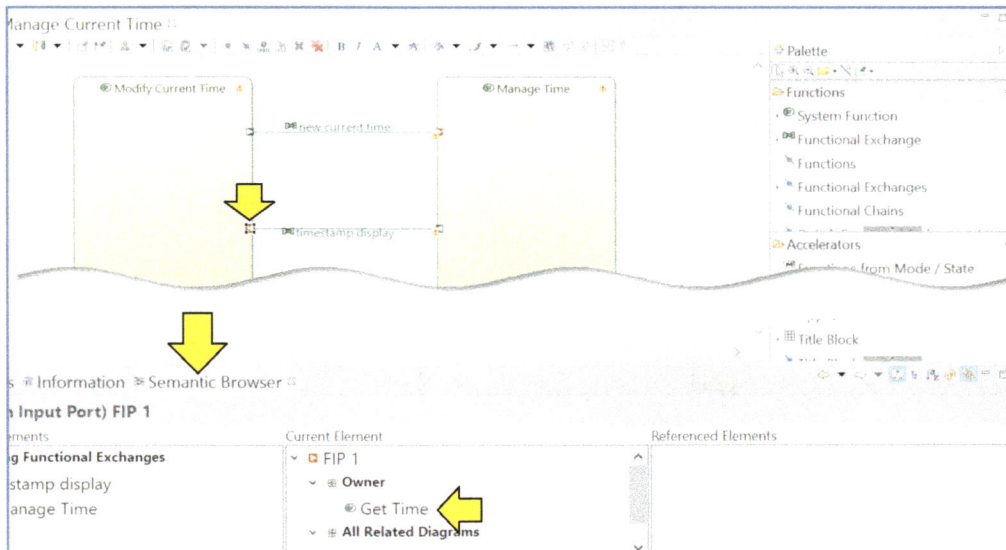

Figure 5-21 – Port actually still owned by the subfunction

However, if we select one of the seemingly restored ports in the diagram and select the "Semantic Browser" tab below the diagram, we will see that the port is actually still owned by the hidden subfunction.

This ability to collapse down the amount of functional detail shown while still retaining display of the data flow is a unique feature of *Capella*.

Next, we will create another data flow blank diagram and create subfunctions in the same style for the second capability "Manage Alarm Time".

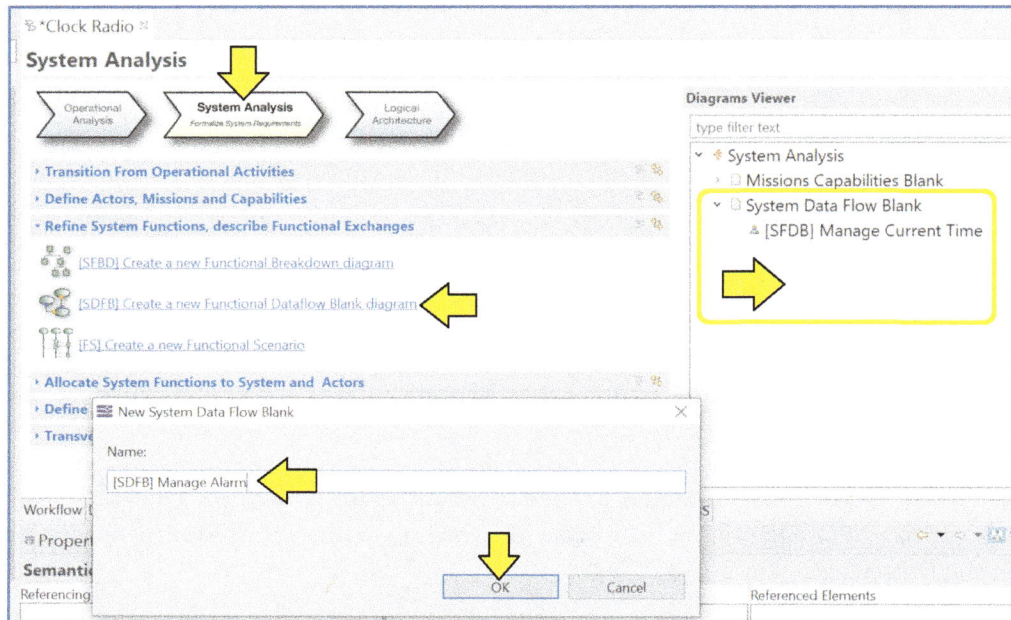

Figure 5-22 – Create functional data flow blank

Create a second system data flow blank (SDFB) diagram from the activity explorer. Name the diagram "[SDFB] Manage Alarm".

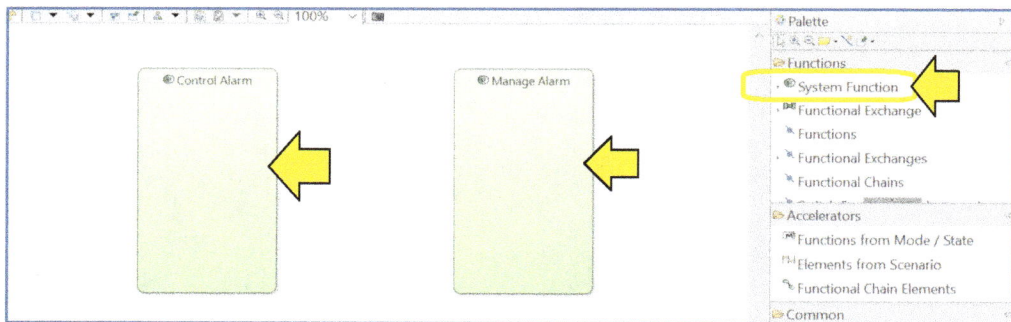

Figure 5-23 – Create two system functions

Create two new system functions called "Manage Alarm" and "Control Alarm".

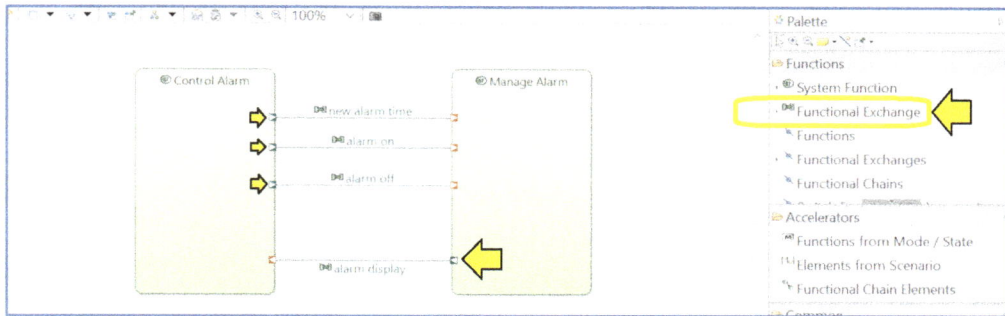

Figure 5-24 – Create New functional exchanges

Create three functional exchanges from "Control Time" to "Manage Time":

- new alarm time
- alarm on
- alarm off

Create one functional exchange from "Manage Time" to "Control Time":

- alarm display

After further consideration, we realize that the "Manage Alarm" function will need the current timestamp as an input from "Manage Time" to check whether it corresponds to the alarm time. We will want to insert the existing "Manage Time" function in this second SDFB.

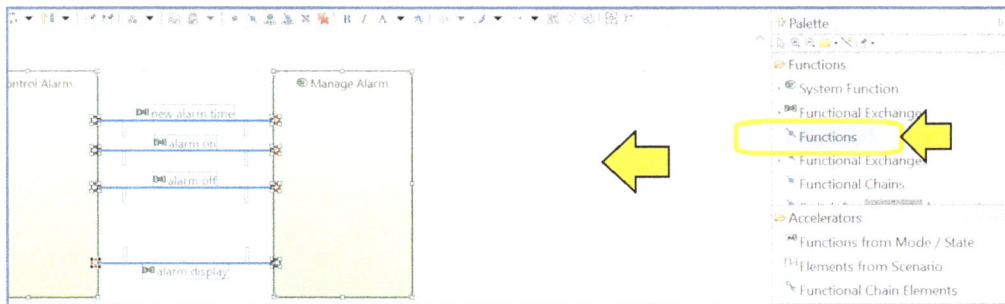

Figure 5-25 – Insert existing function

Select the two functions in the diagram and drag them to the left to make some space. In the palette, select the "Functions" tool and click in the blank space in the diagram.

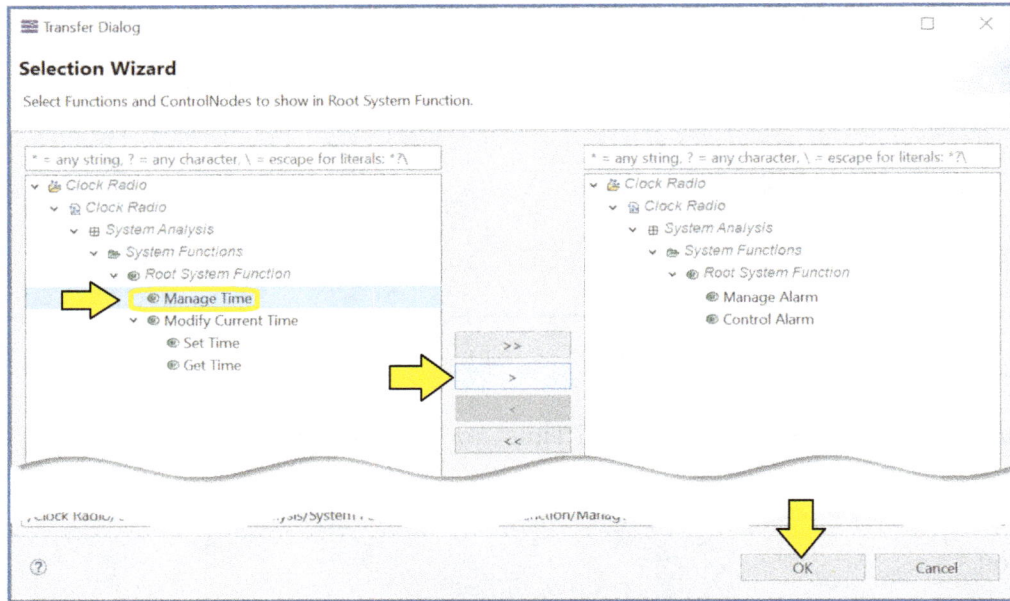

Figure 5-26 – Add the Manage Time function

The transfer dialog will appear. On the left, select "Manage Time" and click on the ">" tool. Click "OK". The existing function "Manage Time" will be added to the diagram.

Automatic Filtering

As we have seen in the previous section, it is possible to hide specific elements on diagrams one-by-one. However, *Capella* also has some features to automatically **filter** out certain categories of content.

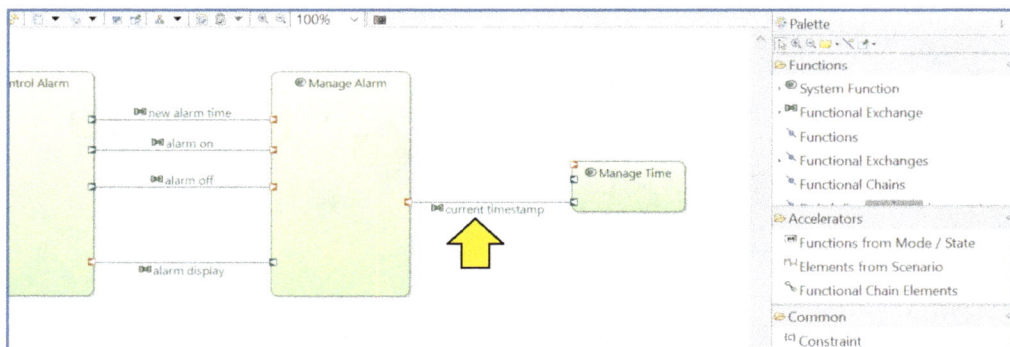

Figure 5-27 – Create another functional exchange

Create a new functional exchange from "Manage Time" to "Manage Alarm" and name it "current time-stamp".

When we added "Manage Time" to this diagram, *Capella* automatically added its existing ports. These existing ports are not relevant in this particular diagram. For model-based systems engineering in general, diagrams should never contain model elements that are not relevant to the concept that the particular diagram is supposed to illustrate. Such extraneous model elements are visual noise and impede the reader's understanding of the model.

In order to hide these extra ports, we could select them one-by-one and hide them. That is a tedious approach, especially for large, complicated models where elements can have a large number of ports. Instead, we can apply a filter to the diagram. A filter is a dynamic diagram property that we want *Capella* to remember and apply automatically whenever the diagram is updated.

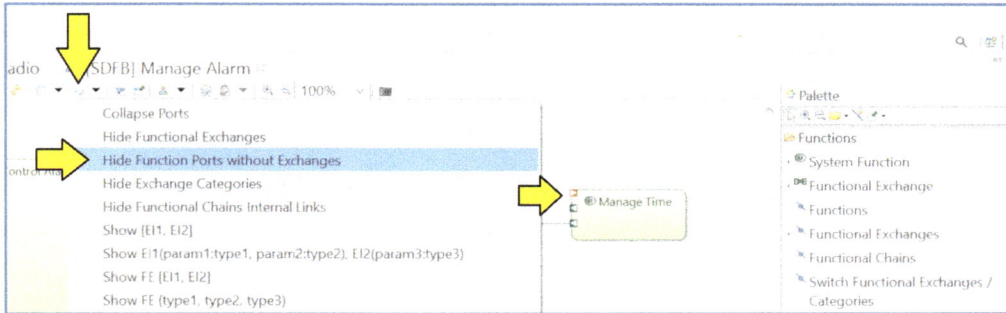

Figure 5-28 – Hide function ports without exchanges

From the top toolbar, pull down the filter tool menu. Select "Hide Function Ports without Exchanges".

All ports that are not connected are automatically hidden. If we add other existing functions to the diagram, their unconnected ports will also be hidden.

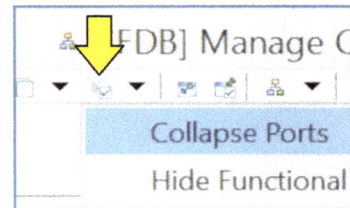

Figure 5-29 – Closer look at the filter tool icon

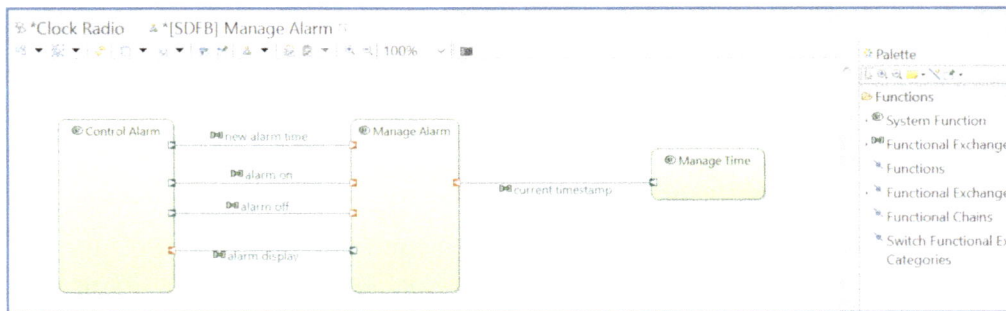

Figure 5-30 – Ports without exchanges are now hidden

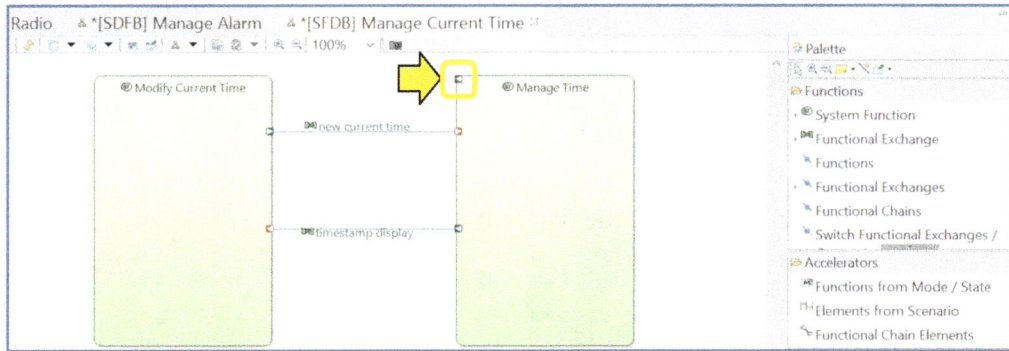

Figure 5-31 – New port shows on previous diagram

Remember we said that when a function is shown on a data flow blank diagram, all of its ports are automatically displayed. As we just added a new output port to "Manage Time", if we now go back to the first SDFB, we will see that it now appears automatically on this diagram too.

Let's review by creating a third system data flow blank for the last capability, using all of the techniques we have just introduced.

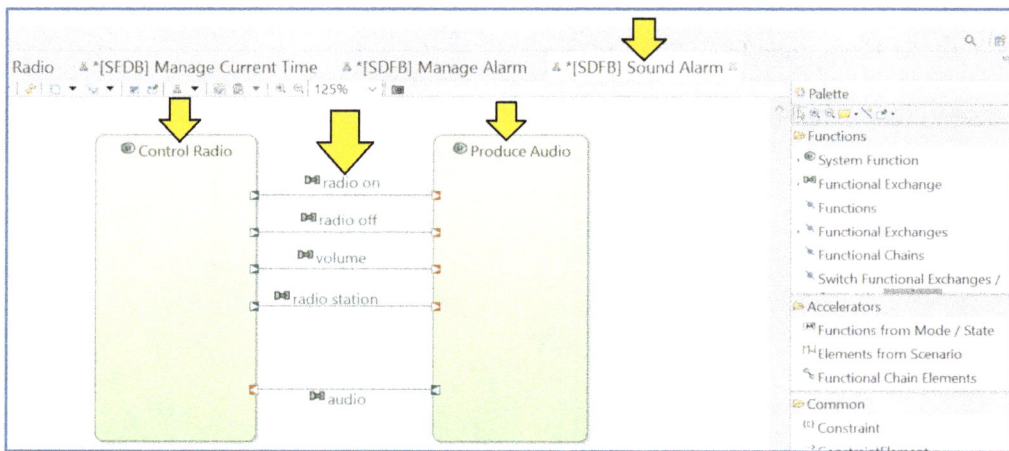

Figure 5-32 – Create new functions and exchanges

Create a new functional data flow blank called "[SDFB] Sound Alarm". Add two new system functions "Control Radio" and "Produce Audio". Create four functional exchanges from "Control Radio" to "Produce Audio":

- radio on
- radio off
- volume
- radio station

Create one functional exchange from "Produce Audio" to "Control Radio":

• audio

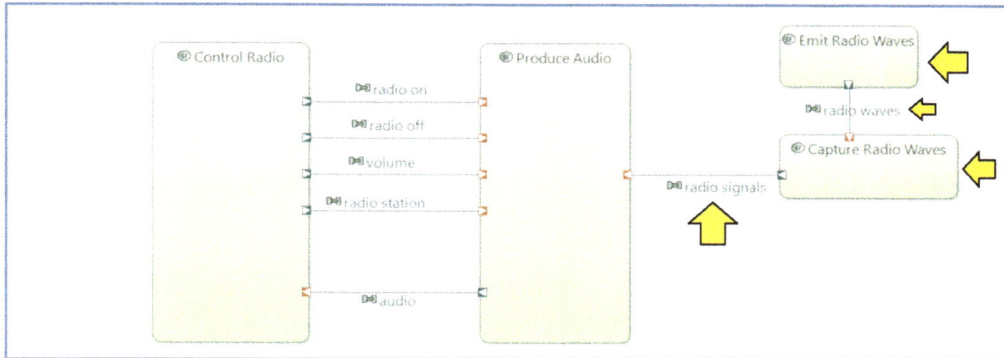

Figure 5-33 – Create functions for radio wave emit and capture

Introduce two additional functions for the emission of the radio waves by the radio station transmitters and capture of them by the clock radio.

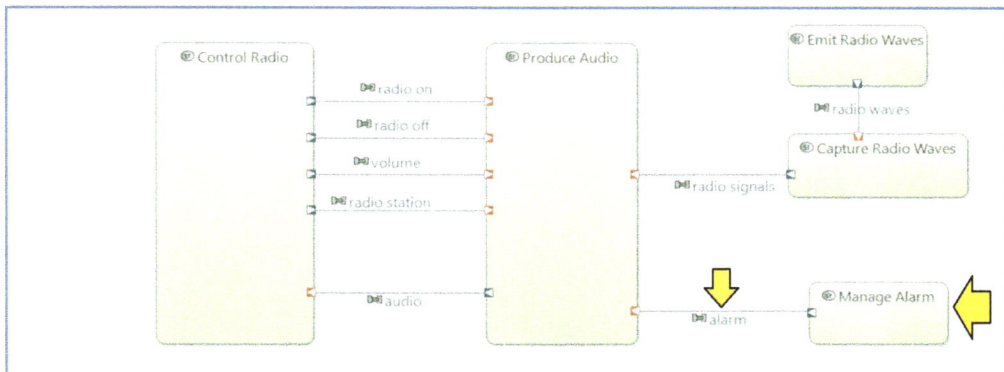

Figure 5-34 – Add existing function to diagram

As for the second capability, we realize that our new "Produce Audio" function needs an input from the existing "Manage Alarm" function telling it that the alarm has been triggered, and that radio sound should be emitted.

1) Add the existing "Manage Alarm" to the diagram.

2) Add a functional exchange called "alarm" from "Manage Alarm" to "Produce Audio".

3) Hide the previous ports that are not connected to functional exchanges.

Functional Chains

Functional Chain – A **functional chain** is a model element that highlights a specific flow of data through the system. The functional chain can contain both functions and functional exchanges. Functional chains are particularly useful for assigning constraints such as latency and criticality as well as organizing tests.

In our simple example, we will create one functional chain for each capability. In a complex model, there are usually several functional chains for each capability.

To create a functional chain, you cannot just pick the command in the palette. You will only find the "Insert/Remove Functional Chains" command which manages the display of existing functional chains in the diagram. You will not find a command to create a new functional chain.

Instead of simply clicking to create a functional chain, you first have to select the model elements you want to be included in the functional chain, and then create the functional chain from a right-click context menu. You will not need to select functions and functional exchanges. As soon as you select a functional exchange, *Capella* will understand implicitly want also the source and target functions of that functional exchange to be part of the functional chain.

Restoring Hidden Subfunctions

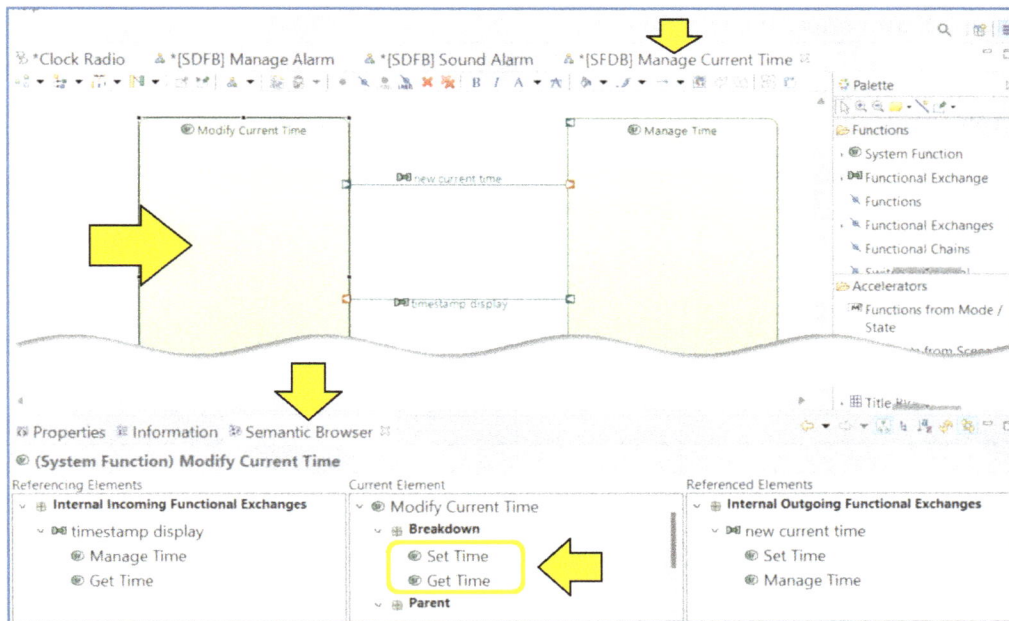

Figure 5-35 – Return to the Manage Current Time SFDB

Return to the "Manage Current Time" system data flow blank (SFDB) diagram. We will want to make use of the "Set Time" and "Get Time" functions. Selecting the "Modify Current Time" system function and checking the semantic browser we can see that they are there. However, we just finished removing them from the diagram in Figure 5-16 on page 85. We will want to restore them to the diagram.

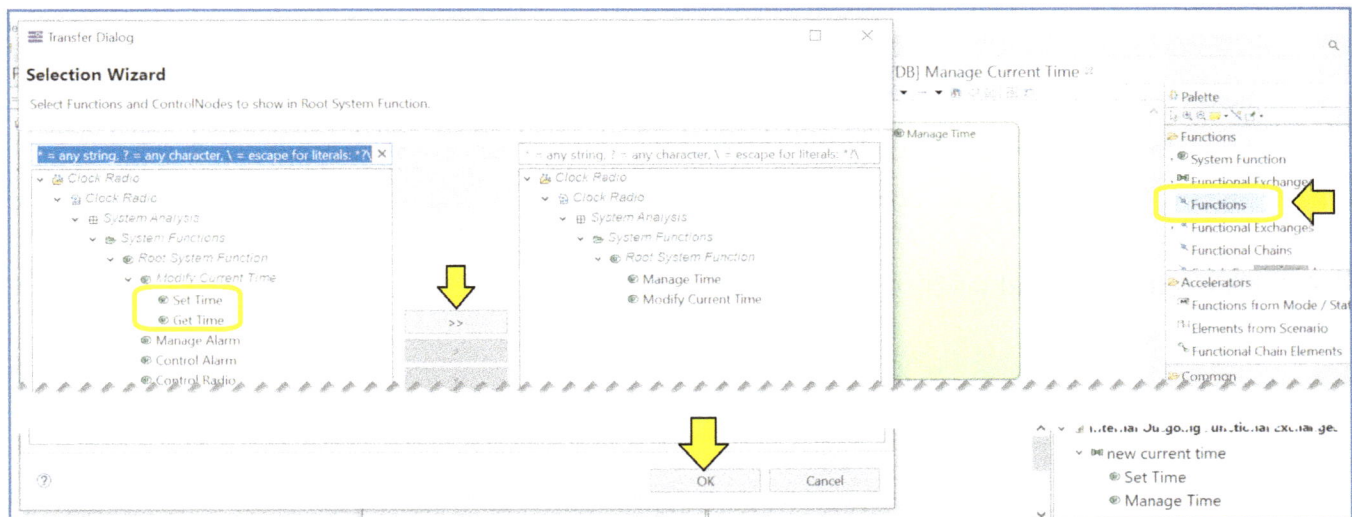

Figure 5-36 – Restore the subfunctions to the diagram

1)   In the palette, select "Functions".

2) Click inside the diagram.

3) The transfer dialog will open.

4) In the left column select the two functions.

5) Click on the single arrow (>) to move them into the diagram.

6) Click "OK".

7) After the transfer dialog closes move the ports on the functions around as needed to make the functional exchange connections look neat.

Simple Functional Chain

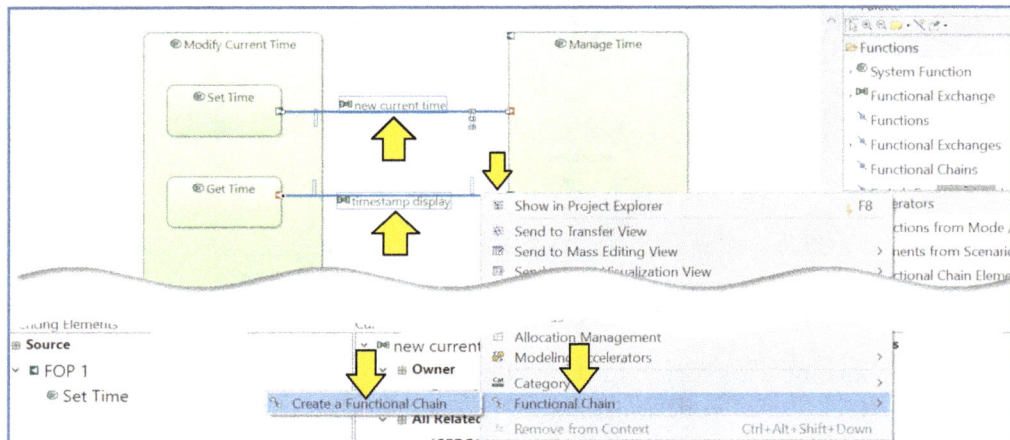

Figure 5-37 – Select two functional exchanges and create a functional chain

Now we are ready to create our first functional chain.

1) Select the "new current time" functional exchange.

2) Hold down the shift key and select the "timestamp display" functional exchange.

3) Still holding the shift key down, right-click on the last functional exchange and select "Functional Chain".

4) Slide left slightly and select "Create a Functional Chain".

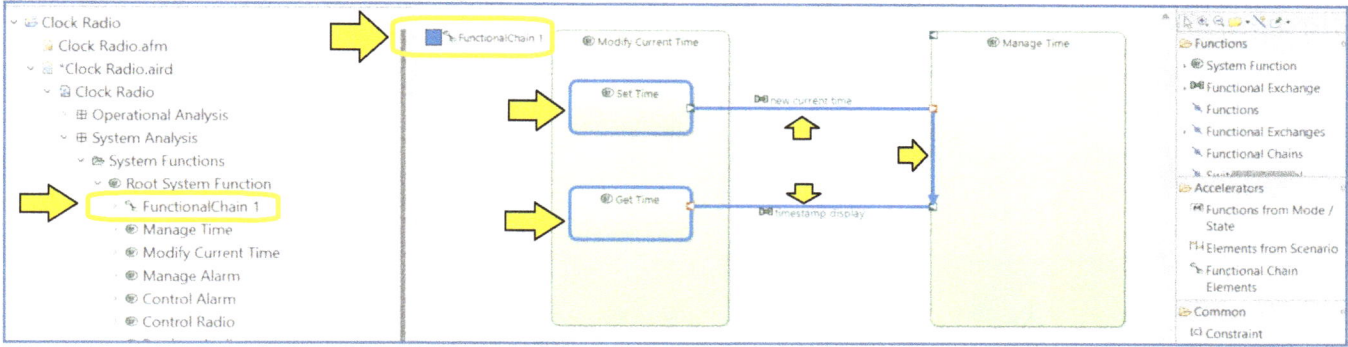

Figure 5-38 – Functional chain created

Capella creates a small blue square representing the functional chain, with a default name: "FunctionalChain 1". It also adds blue color highlights on the source and target functions and an arrow linking function ports of the intermediary function. We can see the direction of the functional chain from the highlighted arrows. The model element itself appears in the project explorer under the "System Functions" package, just above the functions and functional exchanges.

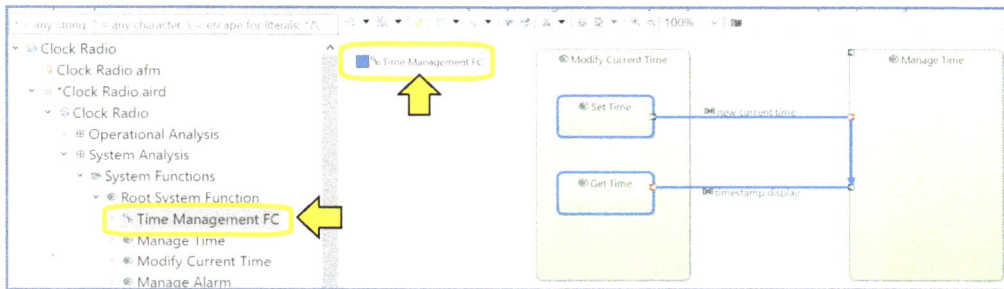

Figure 5-39 – Rename the functional chain

Select the functional chain in the project explorer and press F2 to rename it. Rename the functional chain "Time Management FC". Of course, it is not mandatory to add the "FC" suffix, as *Capella* knows it is a functional chain, and shows it with a specific icon. However, we recommend this notation style to easily distinguish this model element from capabilities, logical components, or physical components that might end up having the same name.

Functional Chain with Loops

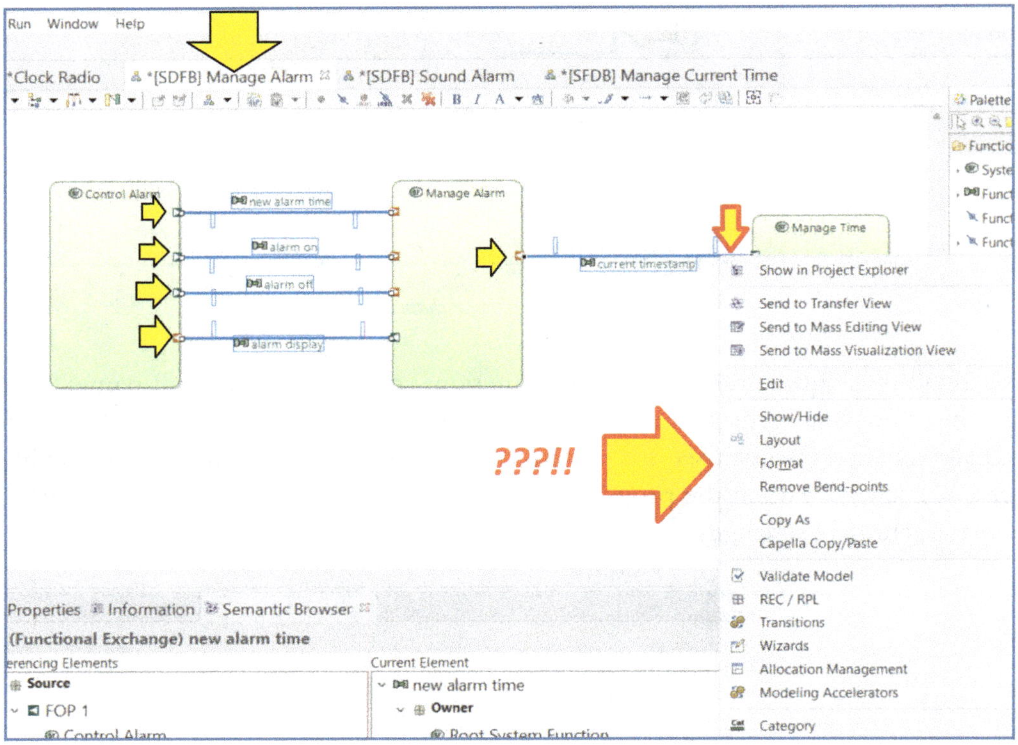

Figure 5-40 – The command does not appear

Let's make a more complex functional chain.

1) Return to the "Manage Alarm" SFDB diagram.

2) Starting with "new alarm time" hold down the shift key and select all of the functional exchanges.

3) Still holding the shift key down, right-click and look for "Functional Chain".

The "Functional Chain" command is no longer in the menu!

What is going on? *Capella* contains a sort of tree constructor that looks at the directions of all the selected functional exchanges. This logic does not allow the creation of the functional chain if there is a potential loop in the selected functional exchanges.

Looking back at the functional exchanges we have selected in the diagram, "alarm display" has exactly the potential to cause such a loop.

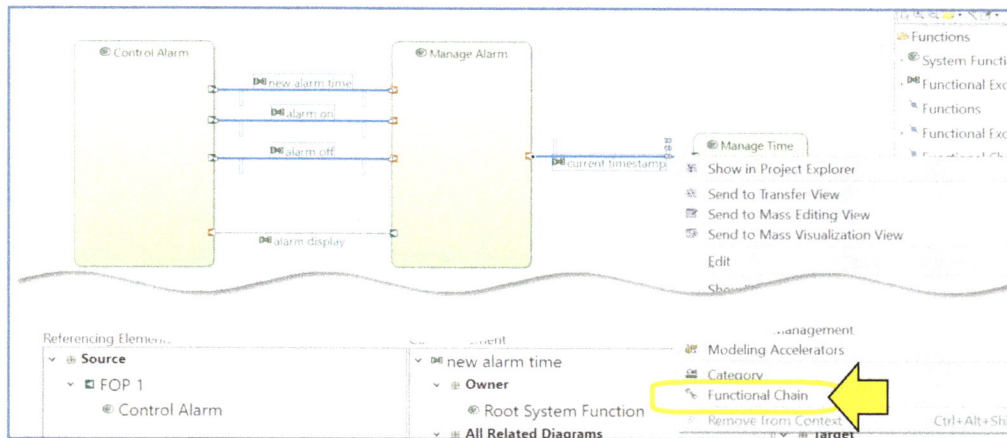

Figure 5-41 – Create a functional chain without a loop

Try again – this time skipping "alarm display". Now the command will appear. Go ahead and complete the functional chain.

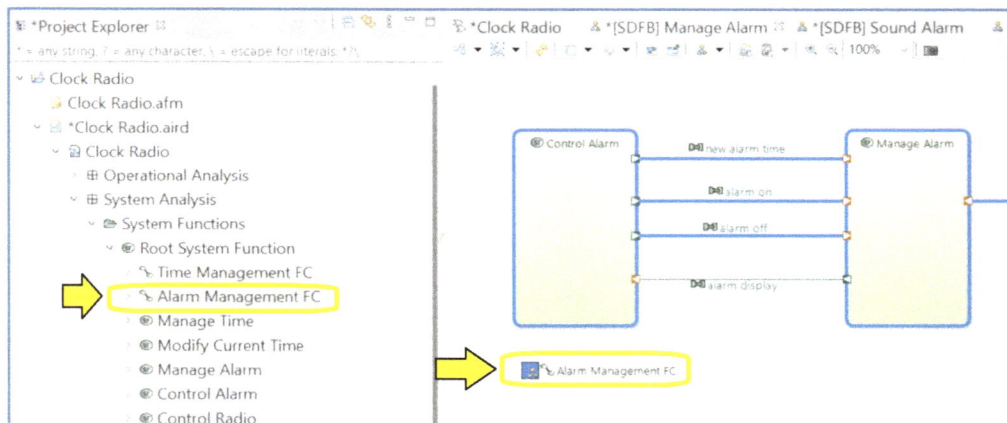

Figure 5-42 – The functional chain has blue notation again

Capella again creates a small blue square representing the functional chain, with a default name: "Functional-Chain 2". It also adds blue color blue highlights to the source and target functions as well as the selected functional exchanges.

Rename the second functional chain: "Alarm Management FC".

What if we actually need the functional exchange that causes a loop in our functional chain? *Capella* does support functional chains containing loops, but the functional exchanges that cause the loops have to be added manually after the functional chain is created.

Let's modify the functional chain to add the missing "alarm display" functional exchange.

The only way to modify a functional chain is to create a contextual diagram of type "Functional Chain Description" from the context menu in a diagram or in the project explorer.

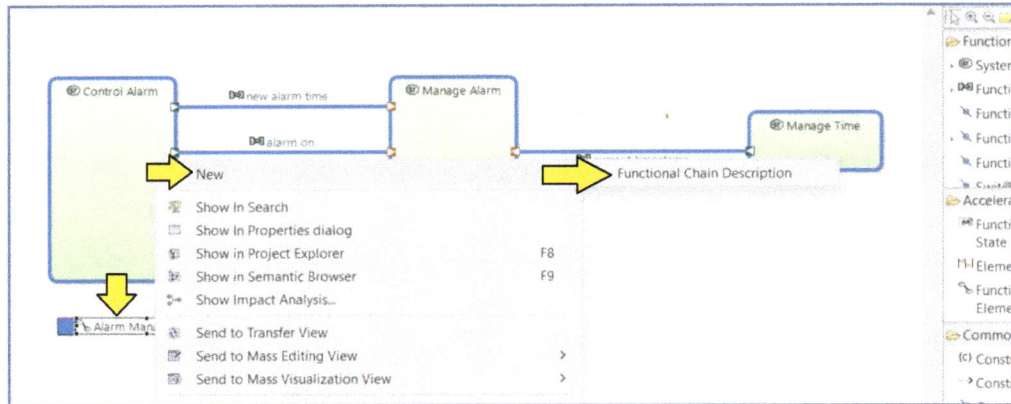

Figure 5-43 – Create functional chain description

Right-click on the functional chain icon or its name in the diagram. Select "New" and then "Functional Chain Description". [16]

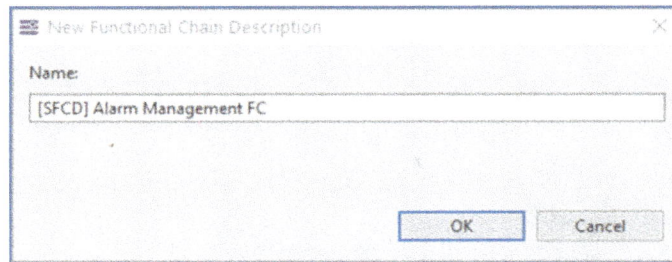

Figure 5-44 – The diagram is named automatically

Capella automatically gives the name of the selected functional chain to the contextual diagram.

---

[16] As of version 6.1, *Capella* allows you to skip this menu step. Simply double-click on the blue box. *Capella* will create a "Functional Chain Description" and bring you to the same name panel.

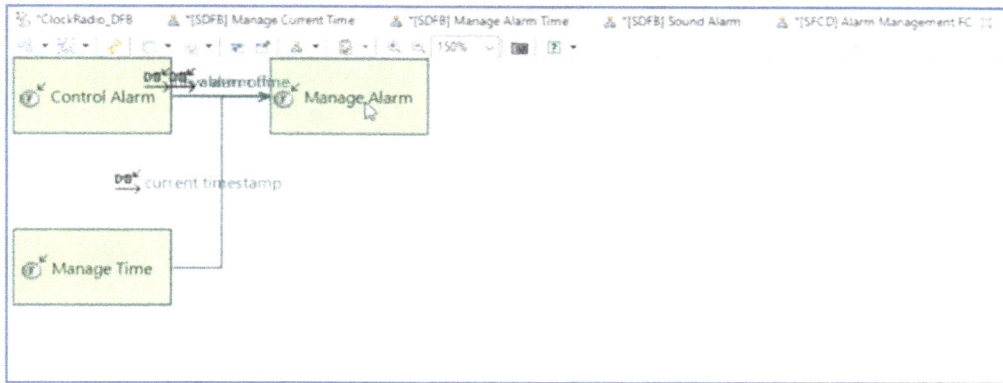

Figure 5-45 – This is a scratch pad diagram

This diagram should be seen as a "scratch pad" diagram, containing the current functional chain relationships. Without this diagram, *Capella* would need to implement a rather complicated user interface for modifying functional chains. We can think of this diagram as a sort of "short cut" for the tool designers. As such, we should not expect to keep this diagram around for the long term. We also should avoid wasting a lot of time formatting it to look pretty.

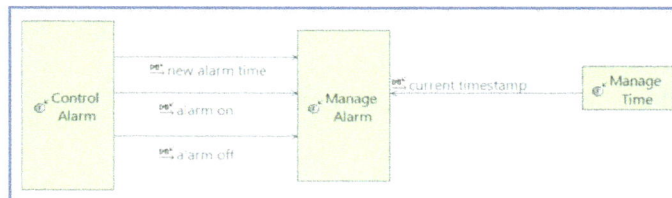

Figure 5-46 – Clean up the diagram a little

While we don't want to obsess about making this diagram pretty, it will be helpful to drag things around a little to make the relationships easier to work with.

Hint: As you drag the arrows around, the labels won't move with the arrows. In order to figure out which label belongs to which arrow, select the arrow and look in the right column of the semantic browser.

In the semantic browser, you will have noticed that the relationships are identified as **involvements**. [17]

---

[17] David Hetherington: My apologies for the odd usage of the word "involvement" here. This term really should have been "relationship". Unfortunately, the authors of *Capella* have hard-coded this odd English usage into the user interface for the tool. As such, we will continue with the word: "involvement".

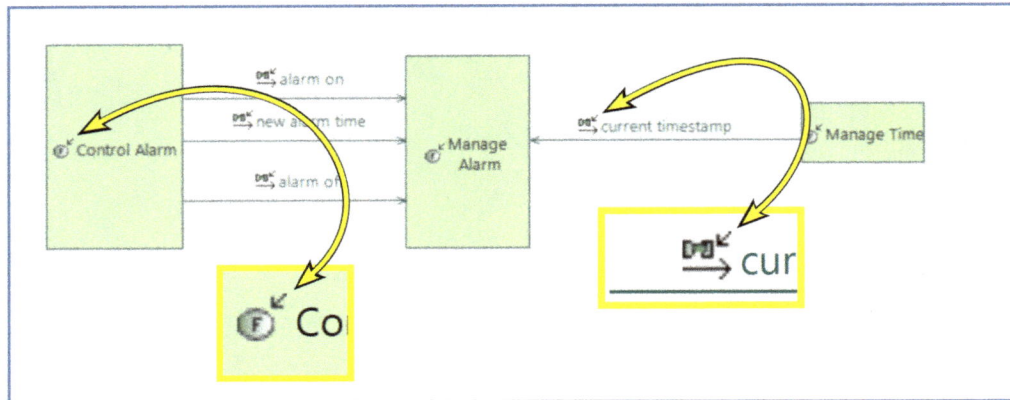

Figure 5-47 – Notice the small arrows on the icons

Examining Figure 5-47, we can see that the icons for the function and for the functional exchange in the functional chain description diagram look a little different from the icons we have seen elsewhere – they include small arrows that are not in the other icons.

This difference is important. It means that the elements shown on the diagram (which are the contents of the functional chain) are references to the other elements, *not* the elements themselves! As parts of the functional chain, these references are elements in and of themselves and are part of the model, but they are not the same elements as the original functions and exchanges. As such, we can remove any of the reference elements in the functional chain by selecting them and clicking on the "Delete From Model" button in the diagram toolbar.

With the palette, it is easy to add new relationships to the functional chain. There are two possible approaches to adding a relationship to a functional chain:

1) First add the missing function with the "Function" button and then the missing functional exchange with the "Exchange" button.

2) Add both with the smart "Exchange and Function" button.

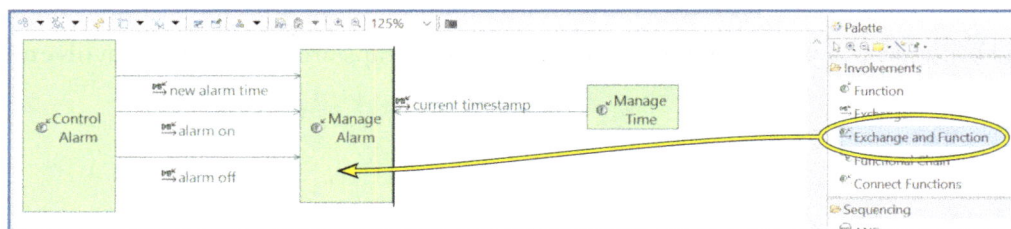

Figure 5-48 – Drag exchange and function to diagram

Select "Exchange and Function" and drag it to the "Manage Alarm" function.

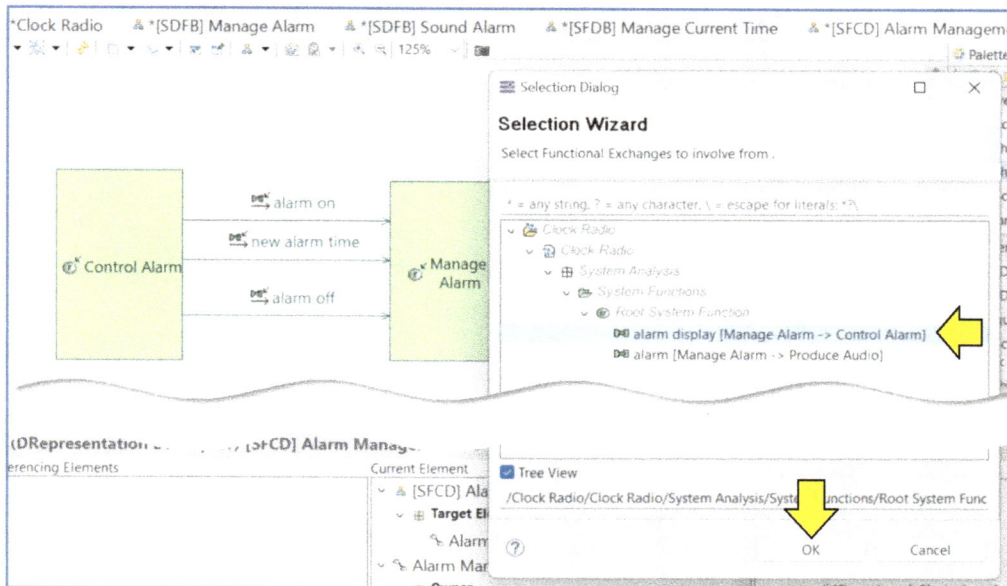

Figure 5-49 – A selection wizard will open

In the latter case, *Capella* opens a selection wizard showing all functional exchanges linked to the selected function. Select "alarm display" and click "OK".

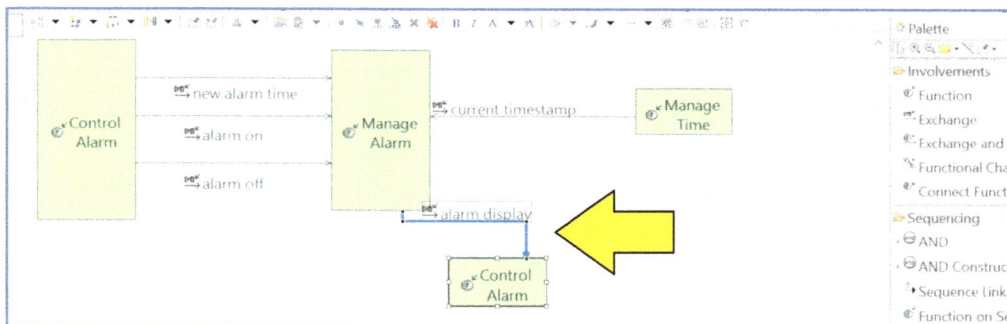

Figure 5-50 – Functional exchange added

Capella adds both the functional exchange and its source or target functions.

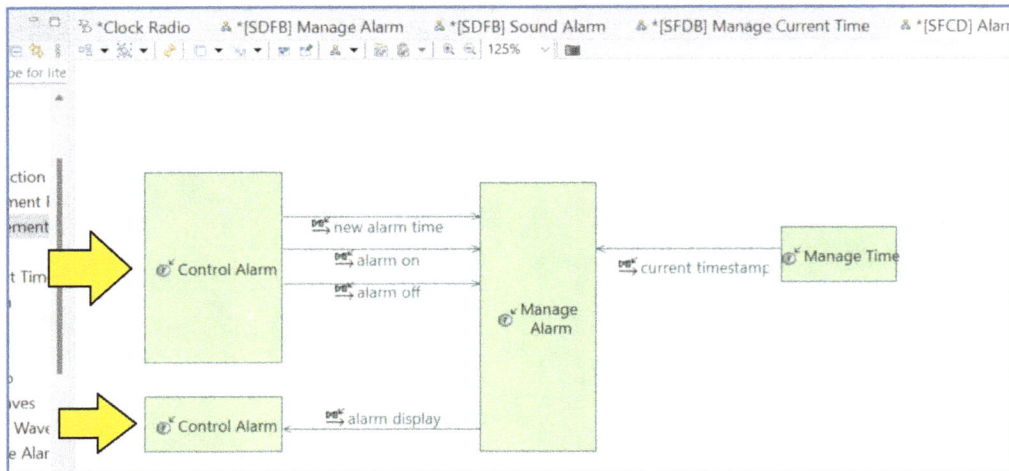

Figure 5-51 – Rearrange the diagram

We have rearranged the diagram a little in order to see clearly the loop on the "Control Alarm" function on the left. Notice that it is mandatory to have the "Control Alarm" function displayed twice on the diagram because the two rectangles are considered as different steps in the functional chain.

Note that the green rectangles are not actually functions. Rather, each green rectangle represents a sort of reference to a function in the functional chain. You will notice that there is a specific icon with an arrow on top of the "F" icon, to indicate the concept of reference.

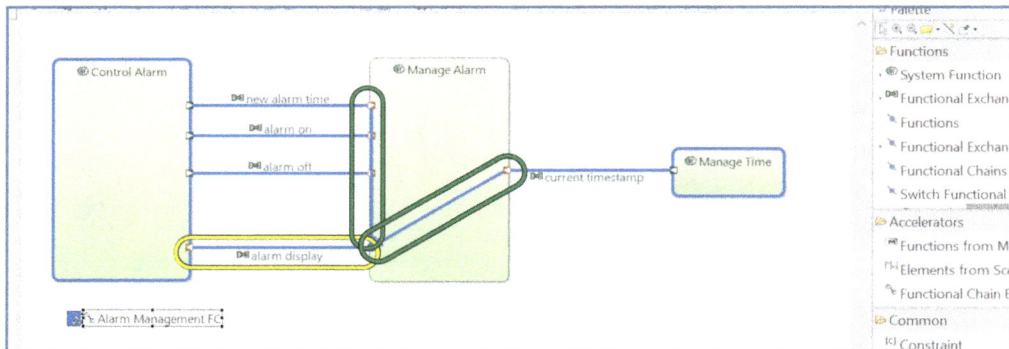

Figure 5-52 – The SDFB has been updated

When we modify the SFCD "scratch diagram", we modify the functional chain model element itself. When we open the SDFB again, we can see that the functional chain has been updated. However, there is some subtlety to the updates:

- *Capella* has added the "alarm display" functional exchange as expected. This "involvement" is shown in yellow in Figure 5-52.

- *Capella* has also used the internal tree logic mentioned previously to deduce and add some additional links in the functional chain. These are shown in green in Figure 5-52.

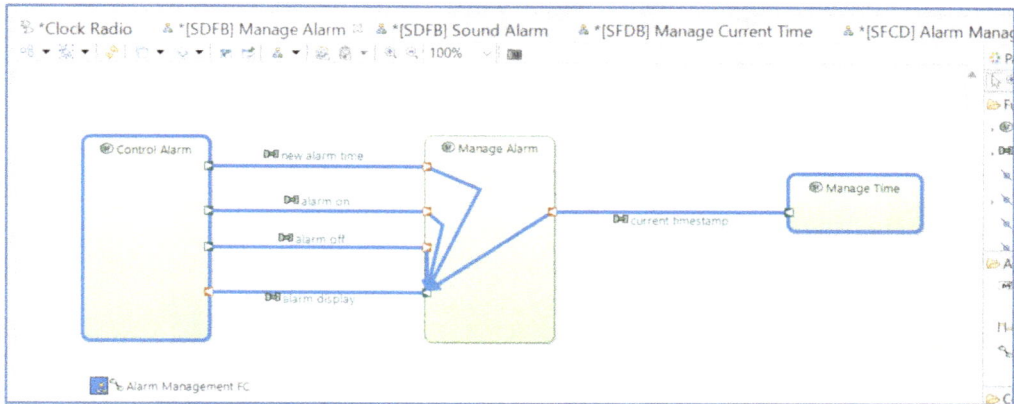

Figure 5-53 – Drag the arrows to improve readability

Some of the blue arrows ended up on top of each other. We can simply click on and drag the arrows to improve readability.

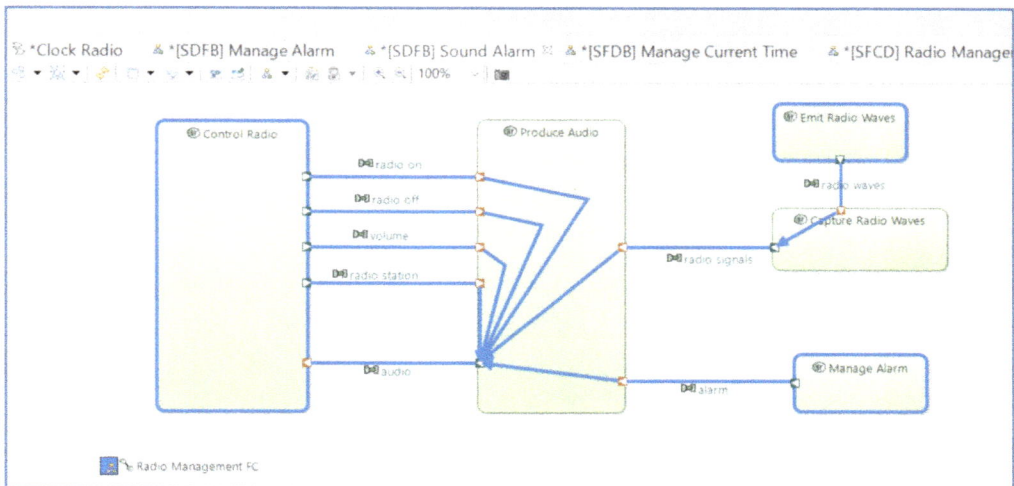

Figure 5-54 – Create a third functional chain

Finally, we can create a third functional chain for the third capability called "Radio Management FC". The steps are basically identical to the steps for the second functional chain.

Functional Chains in Project Explorer and Semantic Browser

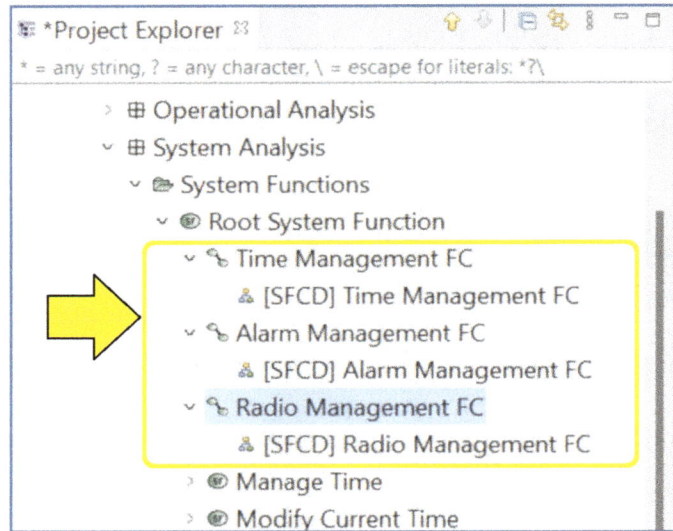

Figure 5-55 – The three functional chains appear in the Project Explorer

We can check that the three functional chains appear in the Project Explorer, before the functions. And the contextual SFCD diagrams are automatically stored under the corresponding functional chain.

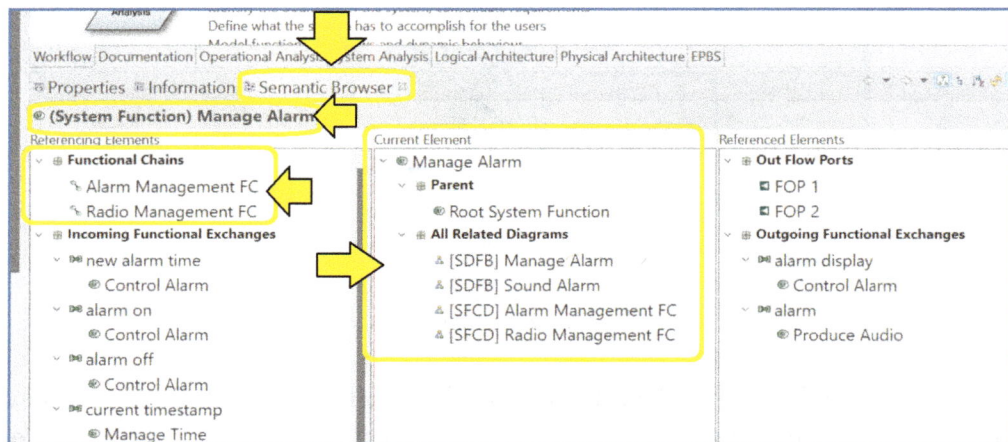

Figure 5-56 – The semantic browser provides information

Even more interestingly, if we select now the "Manage Alarm" function either in a SDFB or in the Project explorer, the Semantic Browser provides a lot of information on it, and in particular that this function is involved in both "Alarm Management" and "Radio Management" functional chains.

Global Data Flow Diagram

Next, we will create a global data flow diagram containing all of the leaf functions. This diagram will allow us to see all functions and functional exchanges at the same time. We only need the leaf functions. According to *Arcadia*, non-leaf functions should not own function ports and are hence not relevant to the data flow.

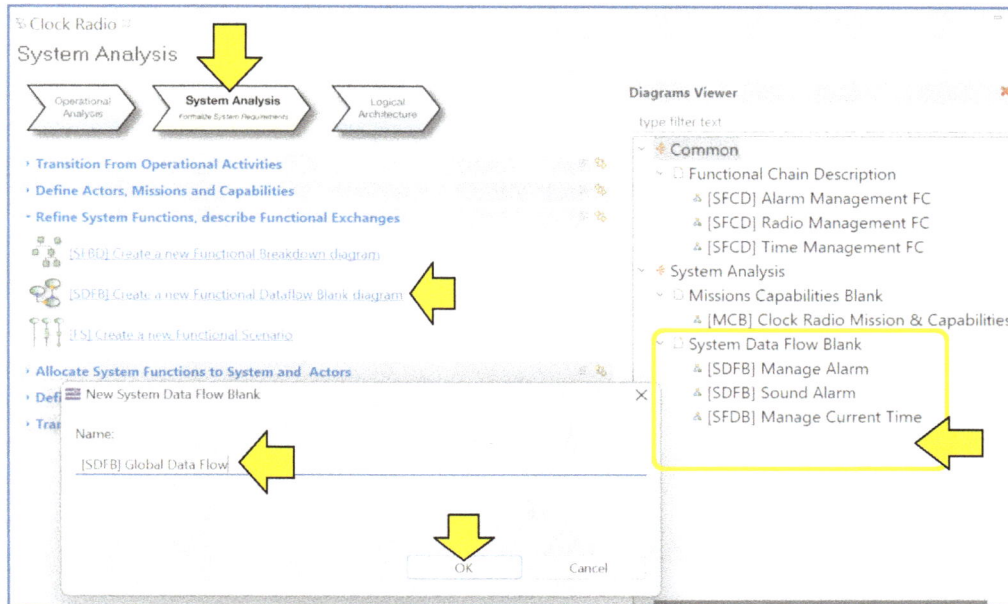

Figure 5-57 – Create another system data flow blank for the global data flow

Return to the activity explorer and create another system data flow blank (SDFB). As usual for any of the "blank" diagrams, the diagram starts out empty.

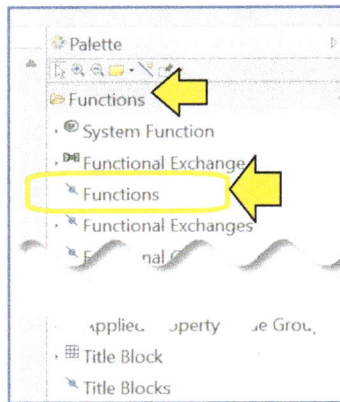

Figure 5-58 – Display all leaf functions

In order to add all of the functions to diagram, we can use the "Functions" tool within the similarly named "Functions" of the palette. Select the tool and click inside the diagram to open the selection wizard.

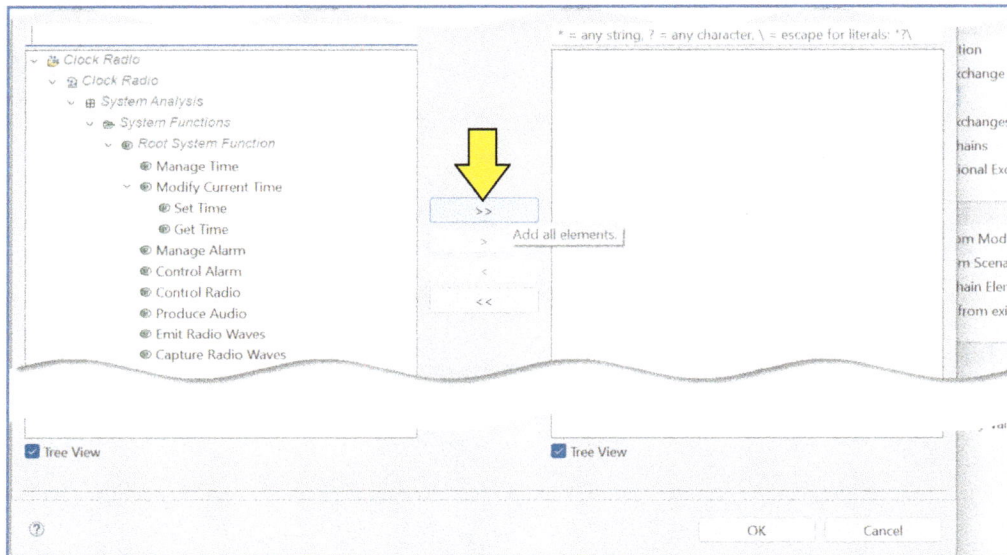

Figure 5-59 – Add all elements

We could add the leaf functions one-by-one. However, since there is really only one non-leaf function we want to skip, we will add the leaf functions in two steps. First, click on the double arrow ">>" add tool to add all the functions to the diagram. Do not click "OK" yet.

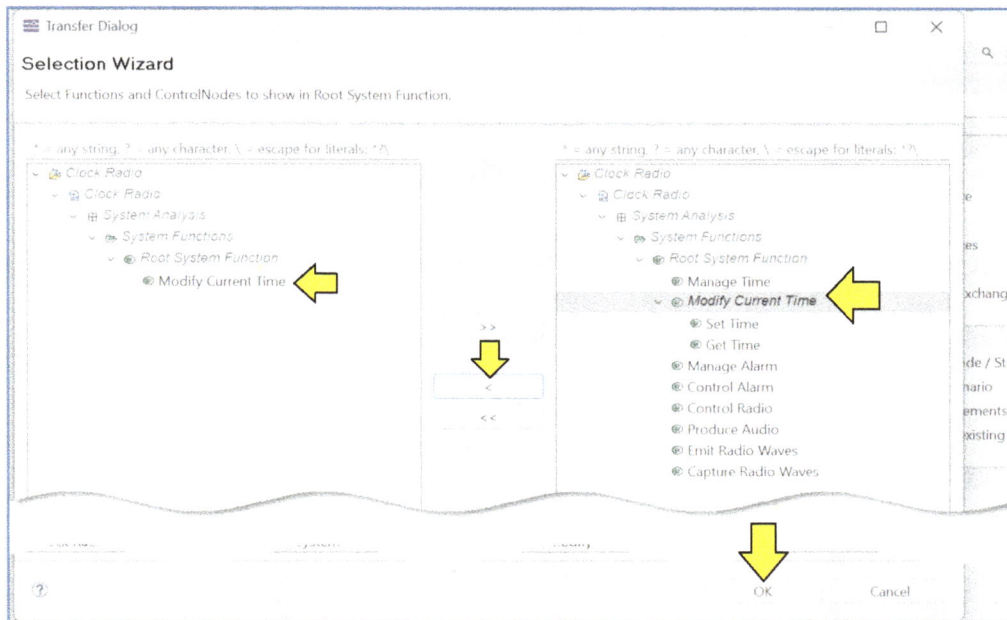

Figure 5-60 – Remove Modify Current Time and click OK

Select "Modify Current Time" and click on the single arrow "<" remove tool. "Modify Current Time" reappears in the left panel. It is still in the right panel as it has two subfunctions that we want in the diagram. However, "Modify Current Time" is now shown in italics. Click "OK".

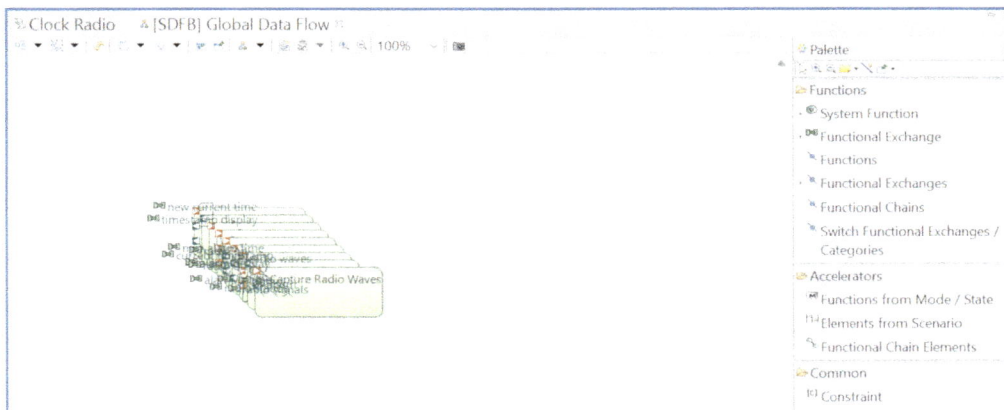

Figure 5-61 – The display is rather cluttered

Capella displays all selected functions with their ports and the functional exchanges. However, the layout is rather cluttered. We will want to find a trick to avoid spending ten minutes moving each rectangle and each line one-by-one.

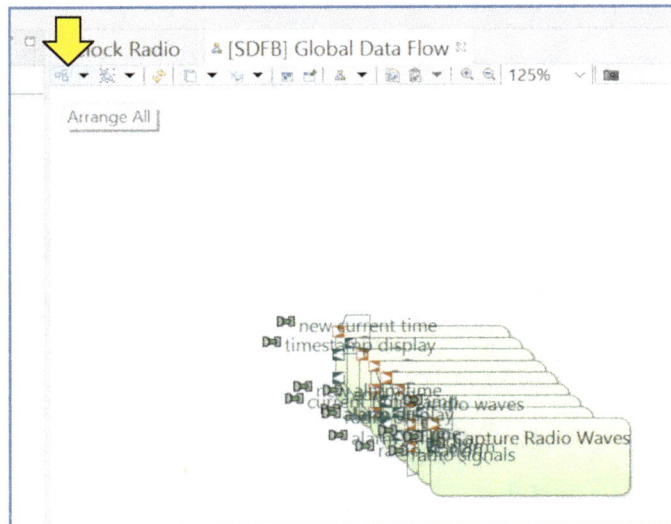

Figure 5-62 – Use the arrange selection tool

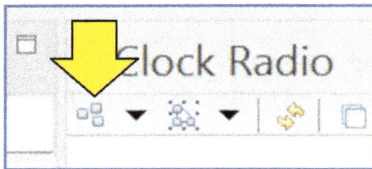

Figure 5-63 – Closer look at the icon

The little trick which saves time here is first not to click anywhere so that all graphical objects remain selected and go directly to the top left part of the diagram toolbar and click on the "Arrange Selection" button.

Capella instantly spreads the rectangles nicely in the diagram. Now, we just have to spend one or two minutes improving the layout as desired.

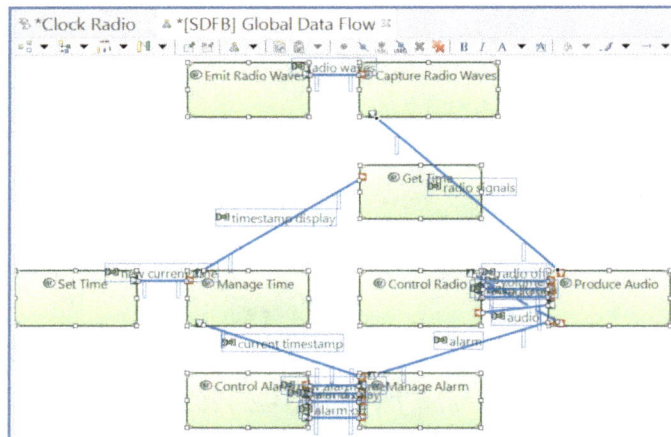

Figure 5-64 – The tool automatically spaces the rectangles

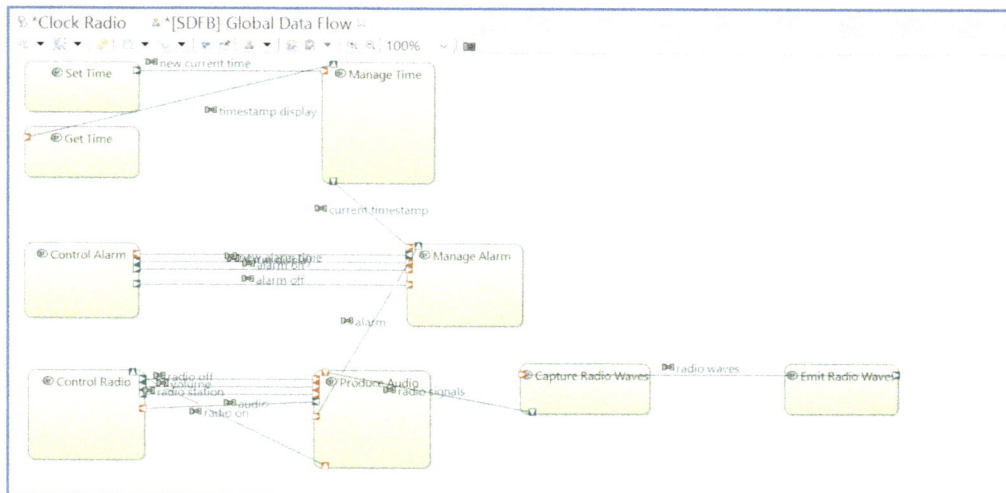

Figure 5-65 – Adjust the location, size, and spacing of the functions

First, we will want to rearrange the location of the functions to make a bit more sense. We will also want to make some of them larger so that it is easier to see multiple ports on one side. Finally, we recommend spacing them out a bit so that some of the longer functional exchange names can fit between them. Note: you will find it helpful to hide the palette to make more space in the diagram.

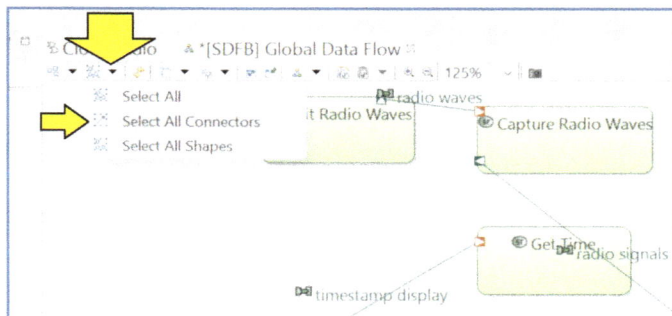

Figure 5-66 – Select all connectors

Next, we can pull down the "Select All" menu and select "Select All Connectors".

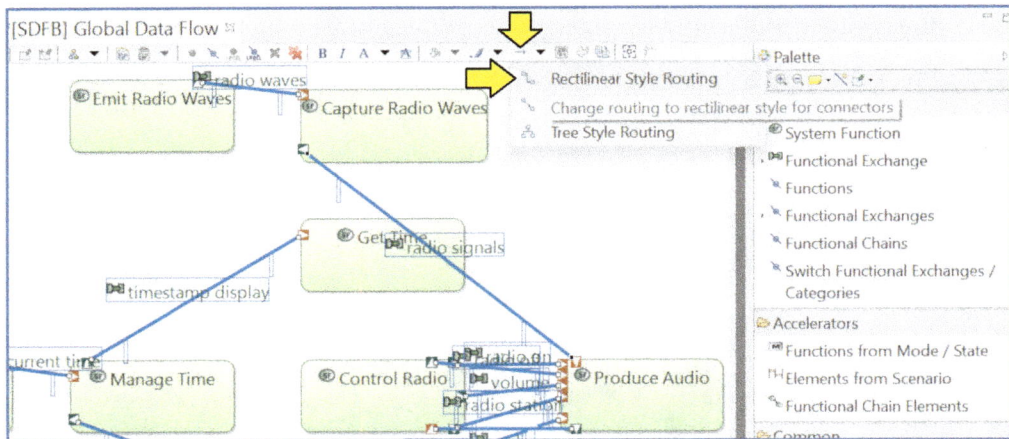

Figure 5-67 – Select rectilinear style routing

Next, we need to pull down the "Line Style" menu and select "Rectilinear Style Routing".

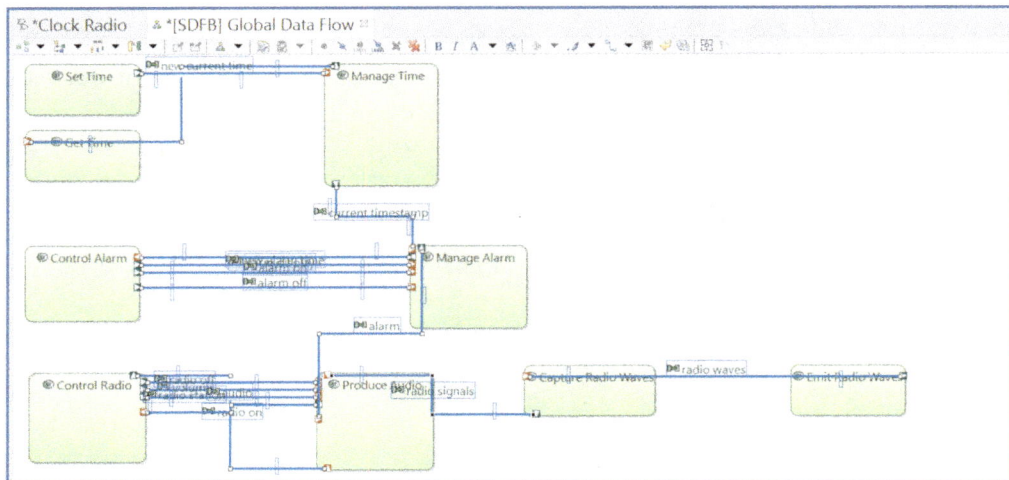

Figure 5-68 – All functional exchanges are now rectilinear

All functional exchanges are now rectilinear, and easier to move so that the diagram rendering is neat. By default, in all blank diagrams the lines are oblique.

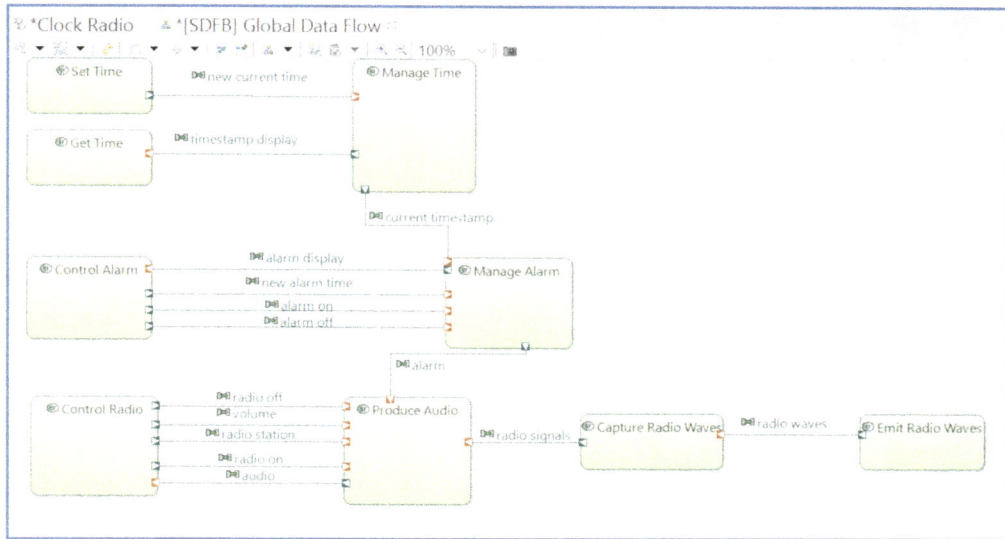

Figure 5-69 – The diagram is cleaned up

Finally, we can move the ports around to make the lines straight and space them more evenly. The diagram is diagram is more readable and shows all of the functional chains.

Functional Chains and Allocations

In the next chapter, we will cover the allocation relationship. All system level functions will be allocated either to the system as a black box (and remain green), or to an external actor (and become light blue). Here is a quick preview of what that will look like.

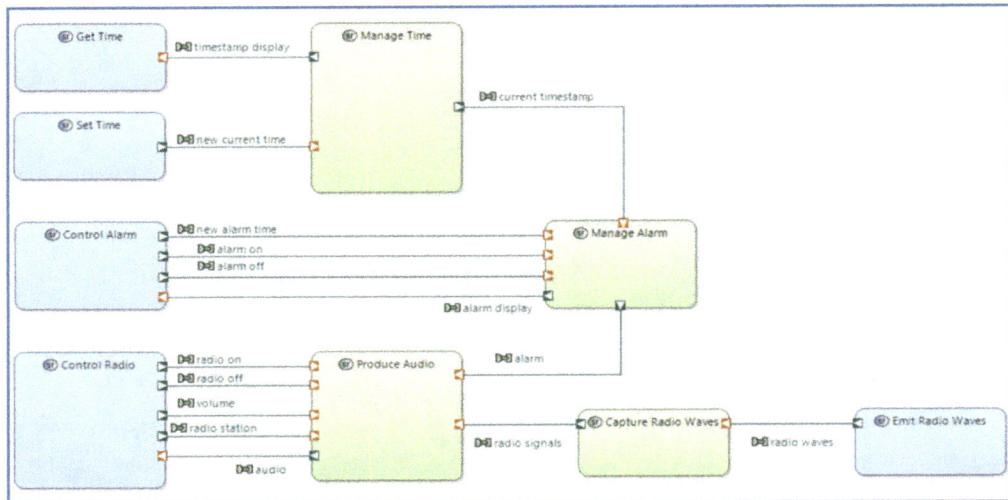

Figure 5-70 – Glimpse of what allocation will look like later

The light blue functions on the left will be allocated to the human actor called "User", and the light blue function on the right will be allocated to the actor called "Radio Station Transmitter". All remaining green functions will be allocated to the Clock radio system.

What if we would like to add the functional chains into our current diagram? Adding them is quite straightforward.

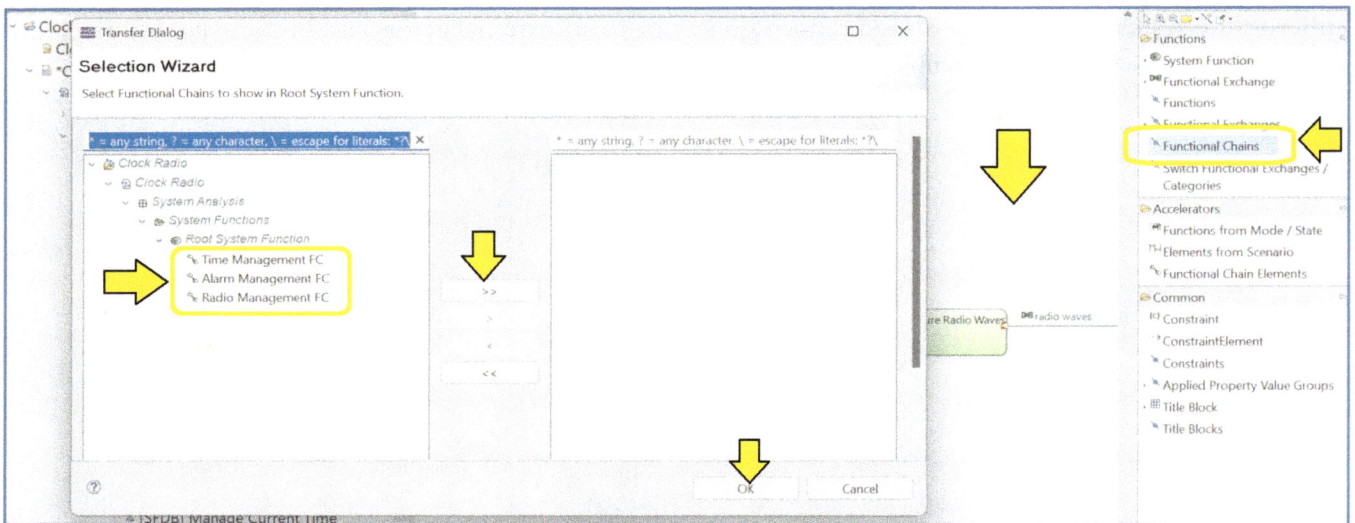

Figure 5-71 – Insert functional chains

Return to the "Global Data Flow" SDFB without allocations, as in Figure 5-69. As the functional chains already exist, we just need to insert them in this last SDFB with the "Functional Chains" tool in the palette.

1)   Select the "Functional Chains" tool in the palette.

2) Click anywhere in the diagram.

3) The transfer dialog will open.

4) The existing functional chains will be visible in the left pane of the transfer dialog. You do *not* need to select them.

5) Instead, simply click on the double arrow "**>>**" add tool to add all of the functional chains to the diagram.

6) Click "OK".

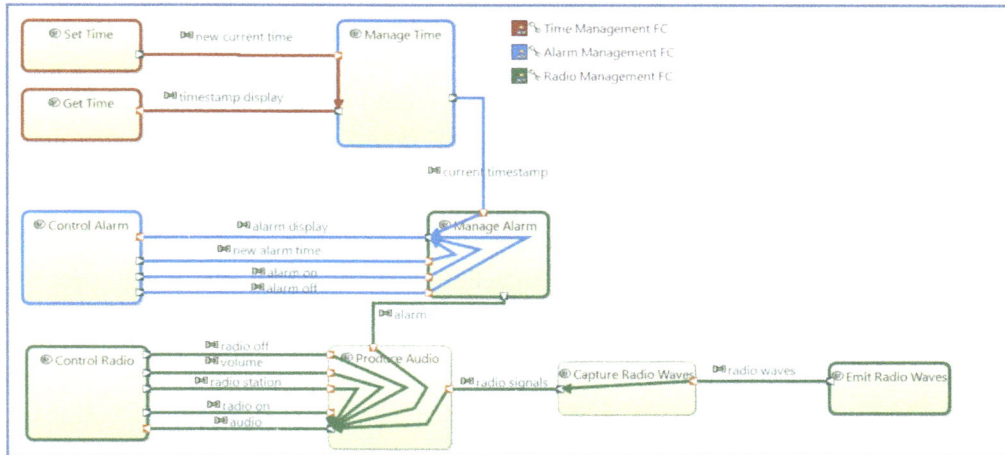

Figure 5-72 – All three functional chains are displayed

All three functional chains are automatically displayed by *Capella* both as small colored rectangles and graphical highlights on functions and functional exchanges which are part of the functional chain. By default, *Capella* uses the blue color for the first displayed functional chain, the red color for the second one, the green color for the third, and so on. You may want to further adjust the spacing of lines and the size of the functions to make the resulting diagram easier to read.

Functional Chain Colors

What if you want to change the color of a functional chain? For instance, the red color could mean "safety-critical" in your project context. The functional chain that *Capella* has shown in red is not critical at all. It is definitely possible to change the color of a functional chain. However, the procedure may not be obvious to all readers.

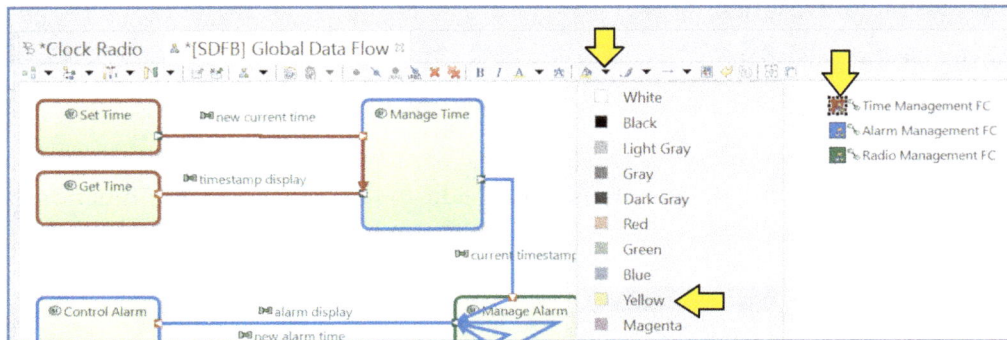

Figure 5-73 – Select yellow

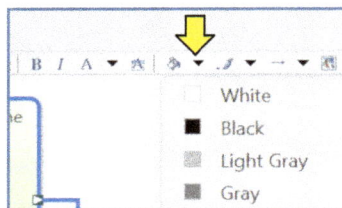

Figure 5-74 – Closer look at the icon

Select the rectangle representing the "Time Management FC" functional chain in our diagram. Pull down the "Fill Color" menu in the diagram toolbar. [18] We can either select a predefined color or click on "More Colors…". Choose the yellow color.

The rectangle is now yellow, but all graphical highlights are still red.

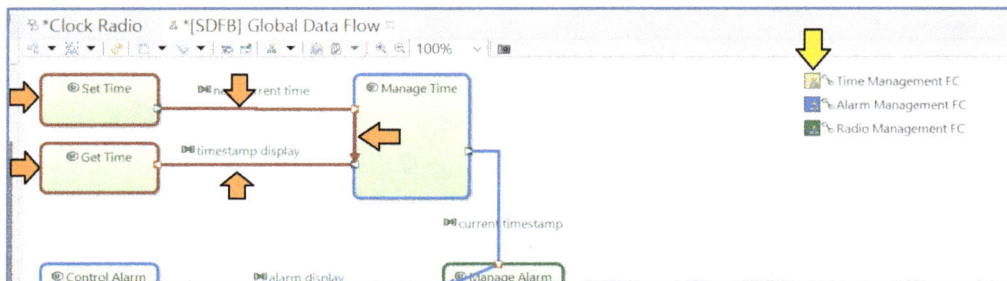

Figure 5-75 – The rectangle is now yellow

---

[18] Note that the "Text Color" menu (the one next to the pitcher pouring icon) and the "Fill Color" menu (the one next to the letter "A") are right next to each other in the toolbar. We want the "Fill Color" menu.

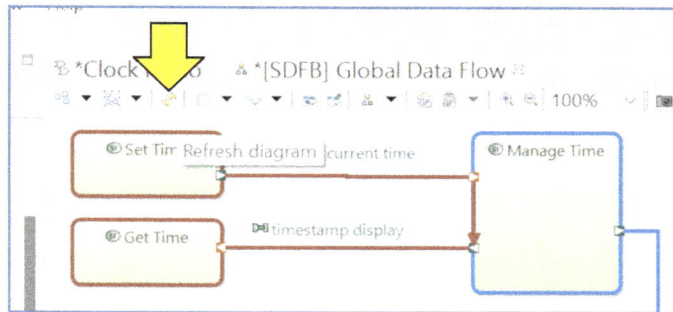

Figure 5-76 – Refresh the diagram

The trick to know is that we just have to ask *Capella* to refresh the diagram by clicking on the corresponding button in the diagram toolbar, or on the F5 key.

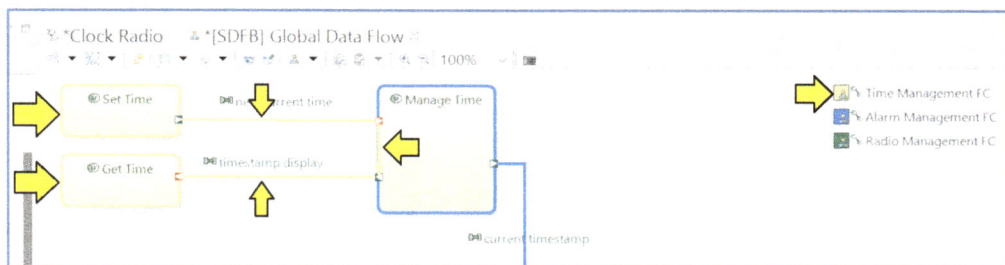

Figure 5-77 – The time management functional chain is now yellow

Now the "Time Management FC" appears in yellow as desired.

Categories

In a larger, more realistic system we will often have a large number of functional exchanges to analyze. This profusion of links can make the diagrams quite cluttered and difficult to read. *Capella* provides **categories** to allow us to group functional exchanges to simplify diagrams and make the system easier to understand.

In order to demonstrate categories, we should first remove the functional chains from the diagram:

1) Select the "Functional Chains" tool in the palette.

2) Click anywhere in the diagram.

3) The transfer dialog will open.

4) The existing functional chains (which are shown in the diagram) will be visible in the right pane of the transfer dialog. You do *not* need to select them.

5) Instead, simply click on the double arrow "<<" add tool to remove all of the functional chains from the diagram.

6) Click "OK".

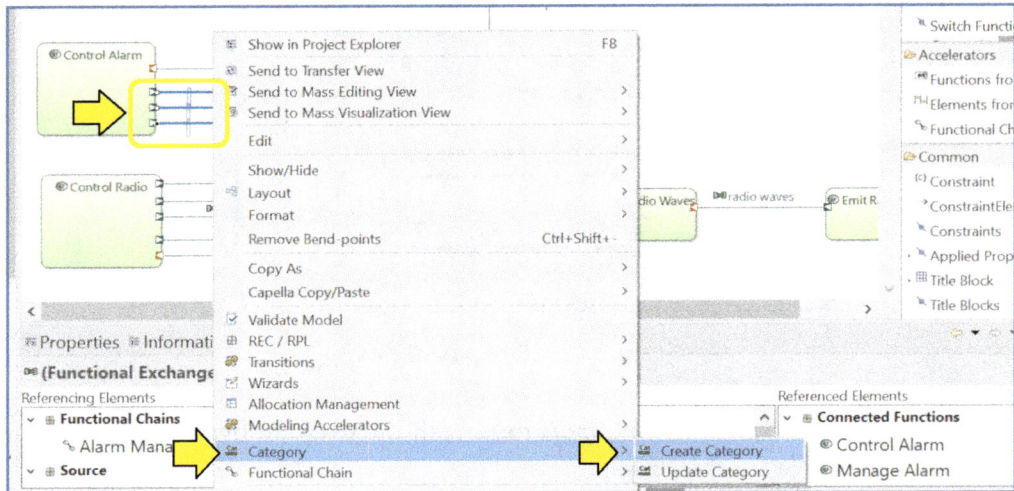

Figure 5-78 – Create a category

Next, select the three functional exchanges that flow from "Control Alarm" to "Manage Alarm". Right-click and select "Category" and then "Create Category".

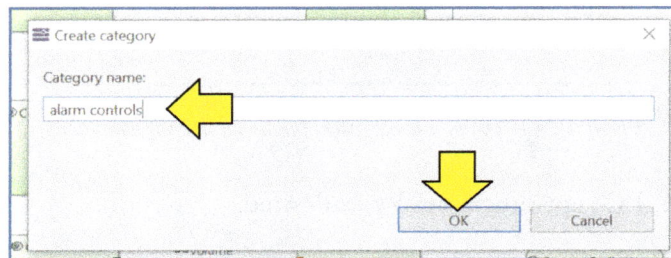

Figure 5-79 – Name the category

Name the category "alarm controls" and click "OK".

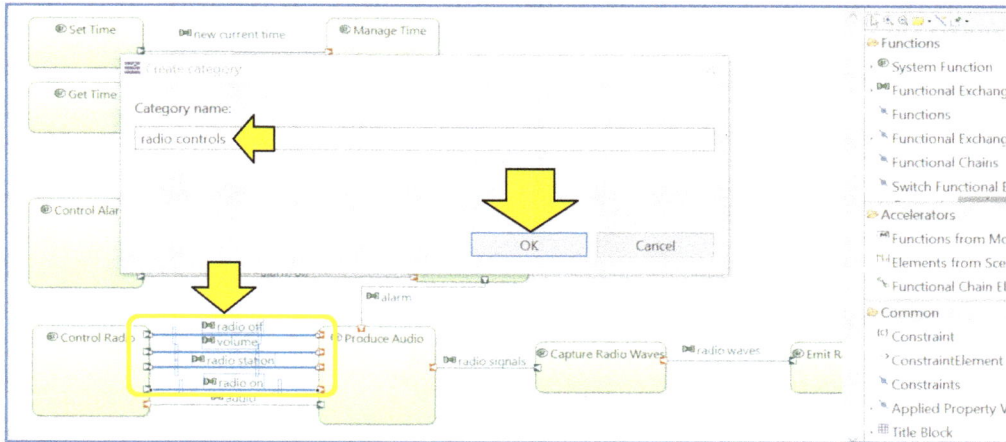

Figure 5-80 – Create a category for radio controls

Do the same for the four functional exchanges that flow from "Control Radio" to "Produce Audio".

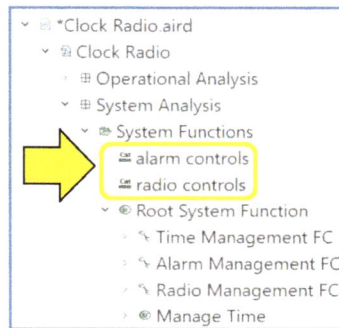

Figure 5-81 – The categories are visible in the project explorer

We can now see the new categories in the project explorer, just below the "System Functions" package, at the same level as "Root System Function".

We can now simplify the diagram by asking *Capella* to display the categories instead of all the individual functional exchanges.

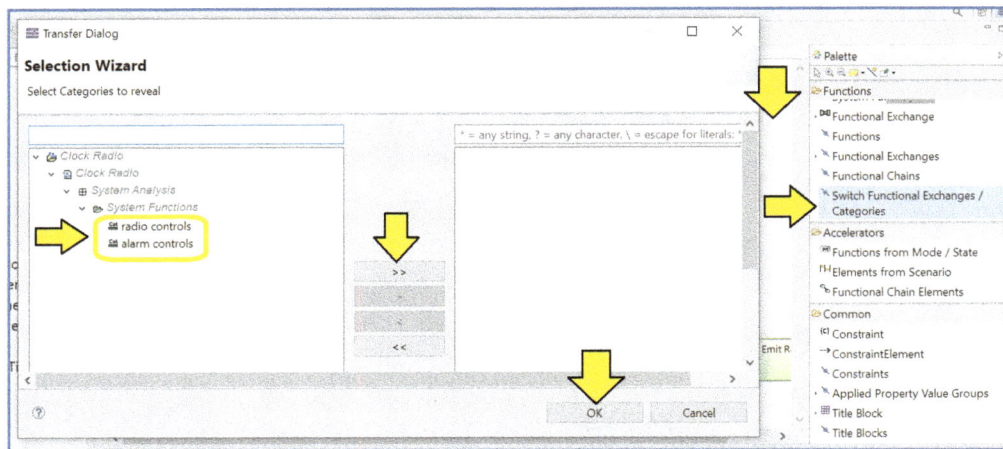

Figure 5-82 – The category selection wizard

1) Select the "Switch Functional Exchanges / Categories" tool in the palette.

2) Click anywhere in the diagram.

3) The selection tool will open.

4) The categories will be visible in the left pane of the transfer dialog. Since we want to display both categories, you do *not* need to select them.

5) Instead, simply click on the double arrow "**>>**" select tool to replace the affected functional chains in the diagram with the categories.

6) Click "OK".

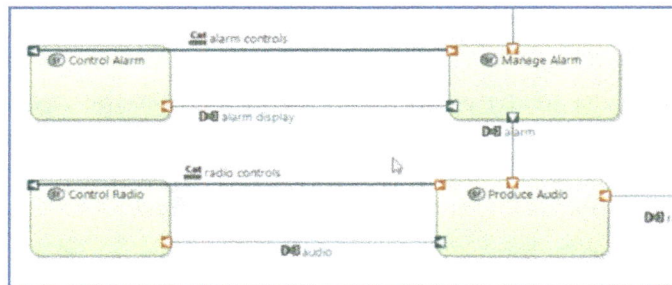

Figure 5-83 – The functional exchanges have been replaced by categories

Now, in the DFB diagram, categories appear instead of the individual functional exchanges. However, the ports are not well positioned.

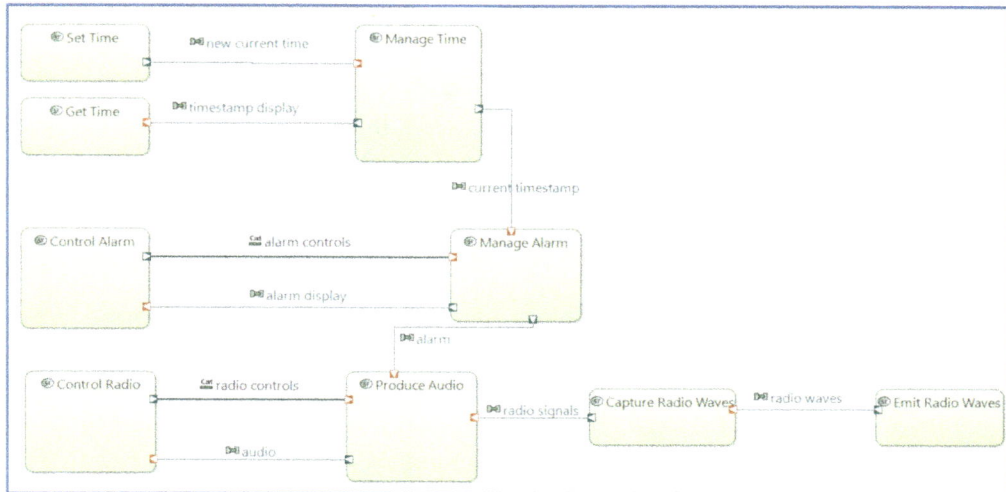

Figure 5-84 – Simplified diagram with categories

We can now move the function ports around a little to obtain a more abstract and clearer data flow diagram.

Further Study

Here is a list of further topics for more in-depth study:

- **Allocation** – We will be covering function allocation in the next chapter *System Architecture Blank Diagrams* on page 315.

- **Data model** – We will be covering how exchange items linked to types can describe the contents of functional exchanges in Chapter 12 *Class Diagram Blank Diagrams (CDB)* on page 315.

- **Functional analysis** – Unlike SysML which inherits object orientation from UML, *Arcadia* is strongly based on functional analysis. See:

    - [Voirin] – Chapter 4, pages 25 to 46

- **Functional chains: Enhancement** – In this video, you will find an advanced explanation of the enhancement of functional chains in Capella, adding control-related information and assembly capabilities. (Slides 15-23 starting at 4:50 in the video).

    - https://url4ap.net/CH-5-DFB-FC

- **Functional chains: Creation and modification** – In this video, you will find a simple explanation of the creation and modification of functional chains in Capella.

    - https://url4ap.net/CH-5-DFB-Chains1

- **Functional chains: Management** – In this video, you will find a more in-depth explanation of the usual way to manage and use functional chains in Capella.

    - https://url4ap.net/CH-5-DFB-Chains2

- **Functional chains: Quantitative evaluation** – In this video, you will find an advanced explanation of a method for quantitative evaluation of functional chains supported by a Capella add-on.

    - https://url4ap.net/CH-5-DFB-Quant

Chapter 6 – System Architecture Blank Diagrams

In this chapter, we will start looking at the **Architecture Blank** diagrams, probably the most important and typical *Arcadia* diagram type. We will start at the **System Analysis** level. Architecture blank diagrams are useful at all *Arcadia* levels. Their main goal at system analysis level is to precisely define

- What the system will do, in terms of functions allocated to it, and
- What the system will not do, in terms of functions allocated to external entities called actors.

Along with as the data flow diagrams presented in the preceding chapter, these diagrams provide a diverse set of mechanisms for managing complexity: display of high-level functions, or high-level components, instead of leaf elements, computed synthetic links between components, etc. Functional chains can also be represented as highlighted paths.

The example used in this chapter is again a simple Clock Radio, exploiting and enhancing the diagrams started in the preceding chapter.

Allocation Relationship

Allocation – The assignment of functions to elements that provide them is called **allocation**. In *Arcadia*, we express this important property by using the **allocation** relationship, with the following golden rules: a function must be allocated to one and only one structural element. But several functions can be allocated to the same structural element.

Let us go back to our clock radio example and have a more precise look at Figure 5-70 on page 114, where we can see both green and light blue system-level functions. The color is automatically computed by *Capella*, depending on whether each function is allocated to the system as a black box (the color remains green), or to an external actor (the color becomes light blue).

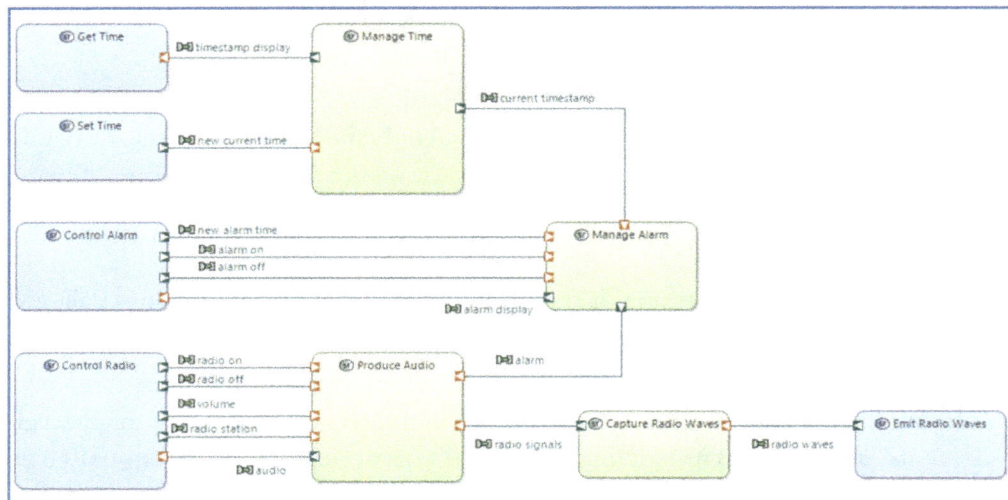

Figure 6-1 – Glimpse of what allocation will look (revisited)

Near the end of Chapter 5, we had provided a preview of this allocation relationship, without actually going through the steps to produce the effect in the model.

In the following sections, we will show you step-by-step how to create these allocation relationships in an *Arcadia* model. The simplest way to set up such allocation relationships is to use an **Architecture Blank** diagram. At the system analysis level, the relevant diagram is called a **System Architecture Blank (SAB)**.

Allocation of Functions to the System

System Analysis Perspective

The System Analysis perspective defines the expectations of the system, that is to say what the system has to perform for users: it builds an external functional analysis, based on the Operational Analysis and input textual requirements, to identify in response functions, services and expected system behaviors, necessary to its users.

The SA is intended to exclusively capture the system needs, excluding any early solution design. The need is consolidated based on the triptych comprising textual requirements, the OA and the SA, which formalizes most requirements to which the solution design will have to respond.

[Voirin] – Chapter 6

We have already defined the concept of actor in Chapter 4, but we did not give the *Arcadia* definition of "system". Let's clarify both fundamental concepts, especially because they are visible in the Architecture Blank diagrams.

System – The **system** is an ordered set of elements functioning as a whole, responding to customer and user demand and need, and subject to engineering supported by *Arcadia*.

System Actor – A **system actor** is an entity that is external to the system (human or not), interacting with the system, especially via the system's interfaces.

The "System" is a model element that was created by *Capella* with the model, and it appears under the "Structure" folder in the System Analysis level. Notice that is not possible to use "Delete from Model" to remove the System component element, and that it is not possible to create another one!

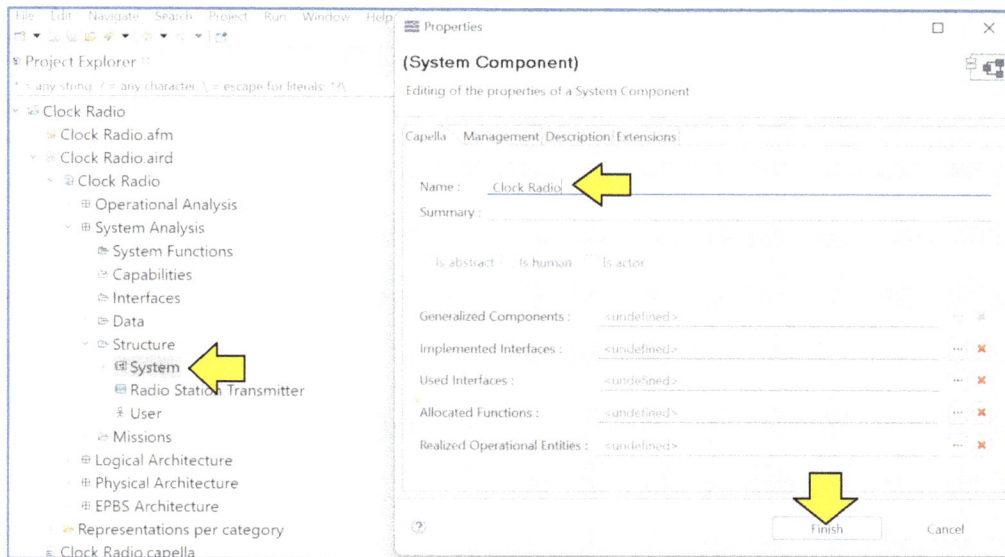

Figure 6-2 – Rename the system component

Rename the system component to "Clock Radio".

We can then create a "System Architecture Blank (SAB)" diagram from the activity explorer, by selecting the topic "Allocate System Functions to System and Actors".

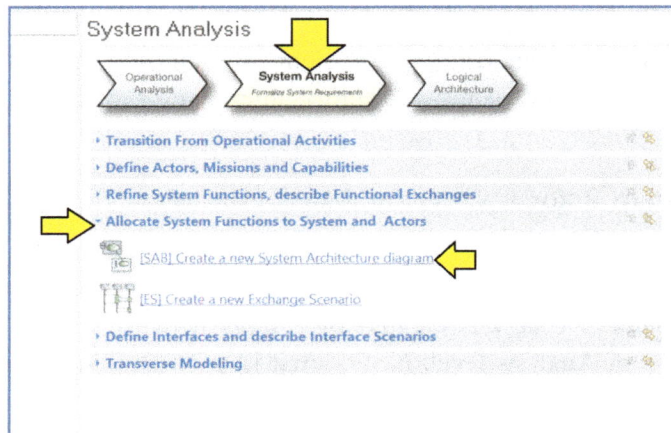

Figure 6-3 – Create a system architecture blank

Click on "[SAB] Create a new System Architecture diagram". A window will appear prompting us to set the name of the diagram. By default, *Capella* sets the diagram to the name of the folder in the project explorer where the diagram will be located. In this case, the diagram will be located in the "Structure" folder, along with the System model element.

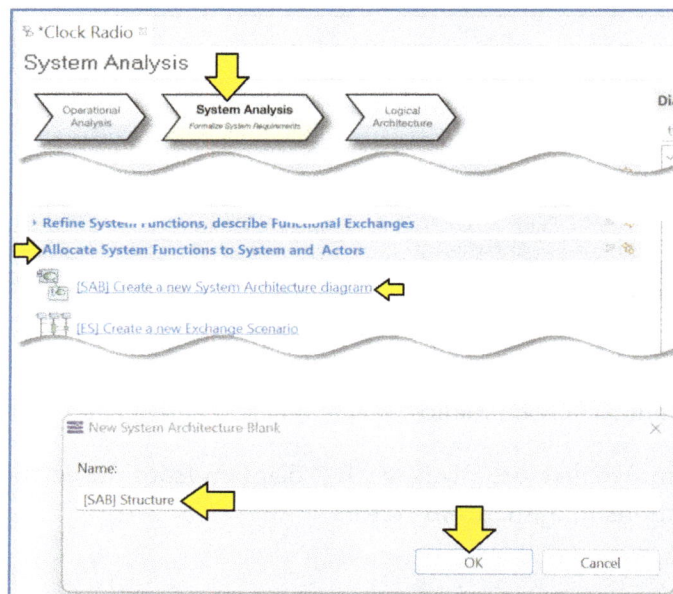

Figure 6-4 – Keep the default name

Keep the default name for the moment. We strongly recommend that you keep the acronym before the name, as it is meaningful for experienced *Capella* users. We also suggest that you change the name of a new diagram as soon as there are several diagrams of the same type with the same name.

As we explained in Chapter 2, "Blank" diagrams are usually empty just after creation. There is an exception for System Architecture Blank diagrams! In this specific type of diagram, if you do not display the (unique)

System model element, containing allocated functions and linked to external actors, you completely miss the point of creating such a diagram.

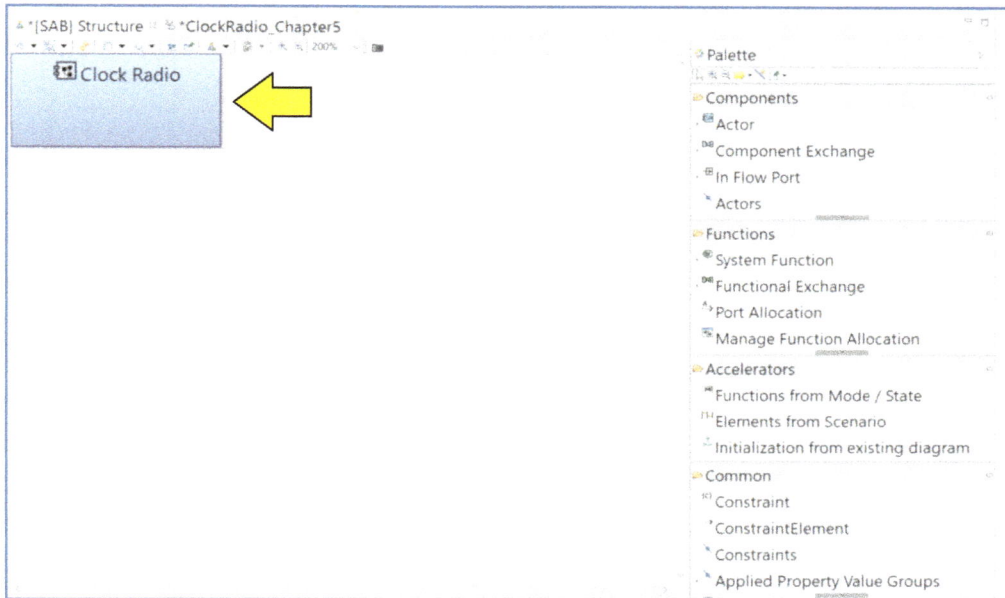

Figure 6-5 – The System model element is at the top left

Capella inserts the system component automatically for you, in the top left corner of the diagram. As usual, a specific palette appears on the right.

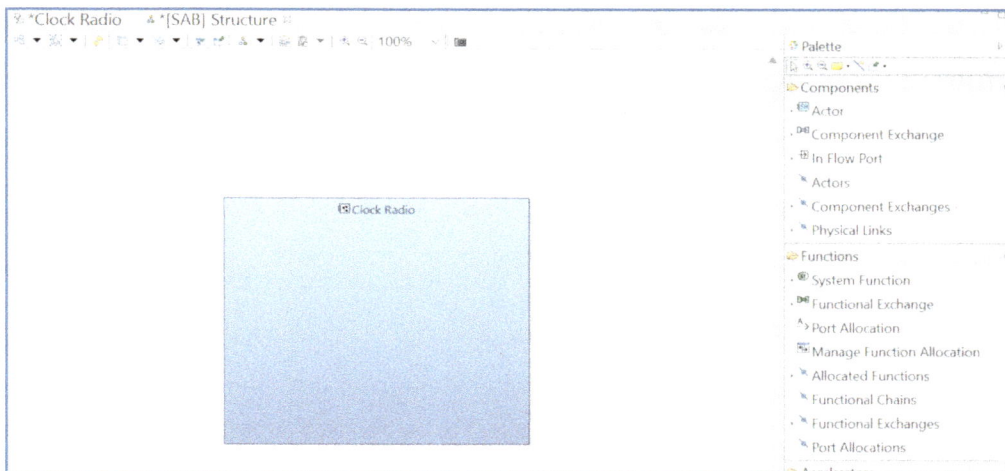

Figure 6-6 – Move the clock radio to the center

Move the clock radio box to the center of the diagram and resize it so that we can allocate the four functions: Manage Time, Manage Alarm, Produce Audio and Capture Radio Waves in it.

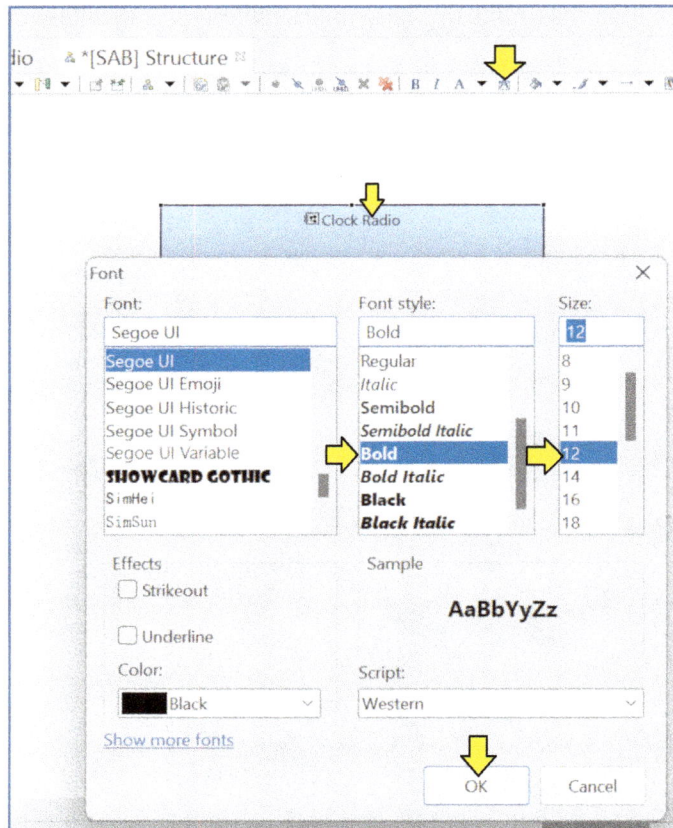

Figure 6-7 – We can change the font of the element

By clicking in the "Clock Radio" text field in the diagram and selecting the "font" tool in the top toolbar, we can bring up a panel to set the font size, style, and color. In this case, we will make the text a little larger and make it bold as well.

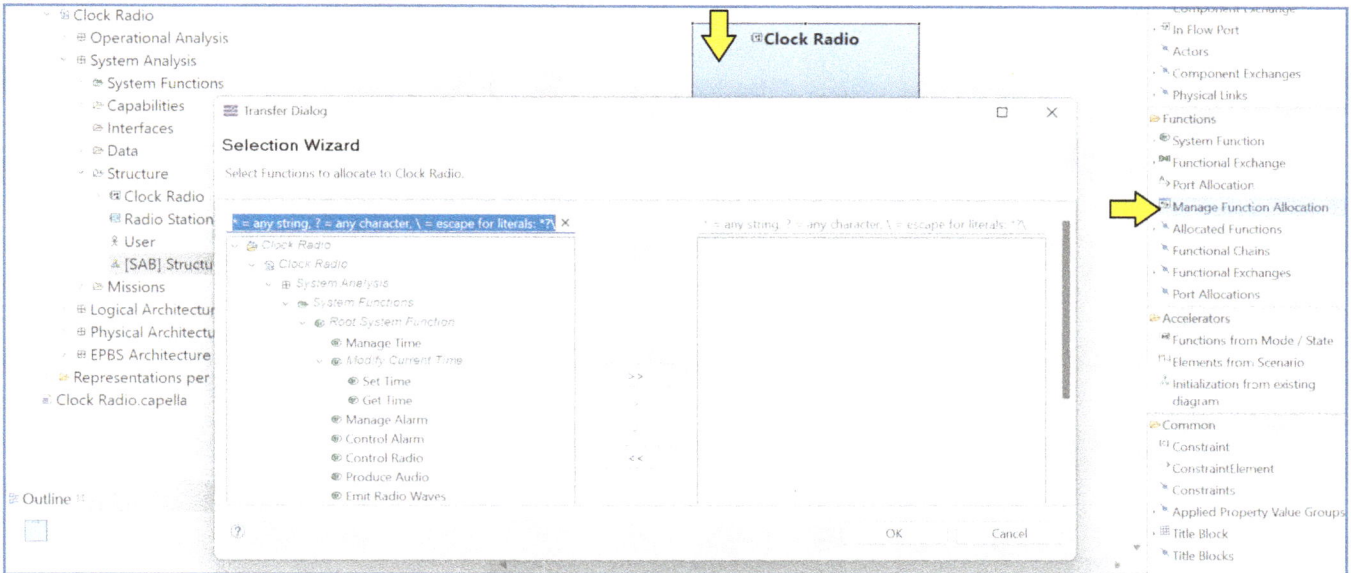

Figure 6-8 – Use Manage Function Allocation

As the functions already exist [19], we just need to allocate them to the system. We can allocate the functions by selecting the "Manage Function Allocation" tool in the palette and clicking inside the "Clock Radio" system component.

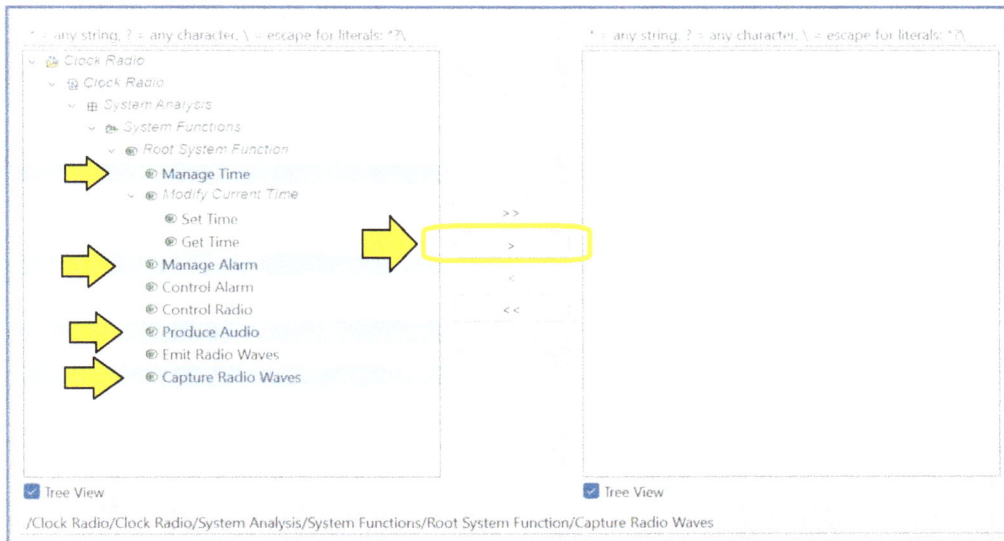

Figure 6-9 – Select only the desired functions

We do not want to allocate all of the functions to the system component. Instead select only the following functions:

---

[19] We defined the functions in Chapter 5.

- Manage Time

- Manage Alarm

- Produce Audio

- Capture Radio Waves

Click on the single arrow ">" add tool to allocate just these four functions to the system component.

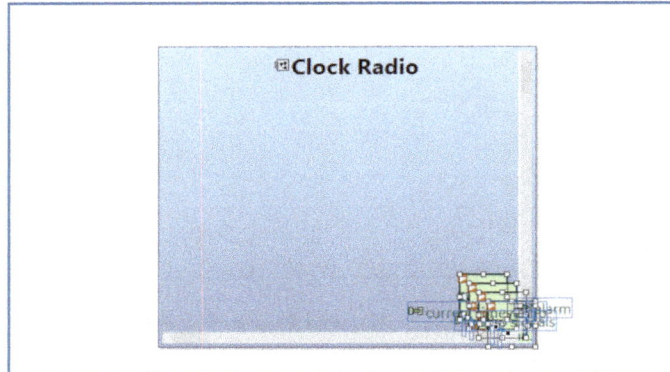

Figure 6-10 – Four functions are now shown inside the system element

As soon as we click "OK" to exit, *Capella* inserts all four functions inside the system box.

The default layout has the functions jumbled on top of each other.

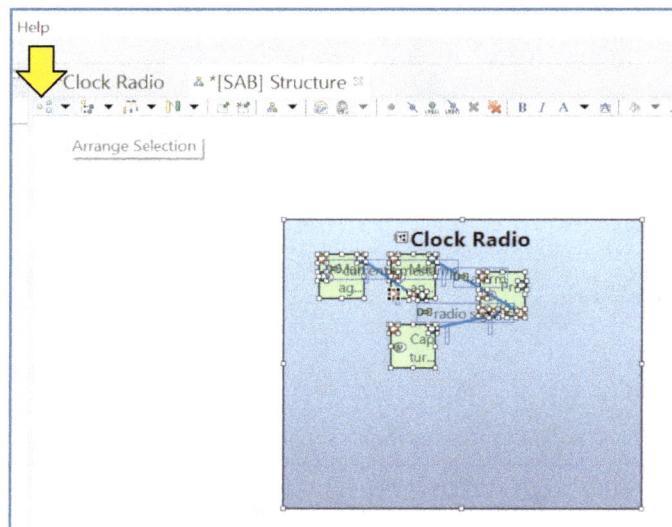

Figure 6-11 – Use arrange selection

First, use the "Arrange Selection" tool at the top left of the diagram tool bar.

The automatic layout improves the appearance, but still leaves a bit of manual work to do. We would need to resize all of the functions and arrange the ports to minimize crossings.

Fortunately, there is a much more efficient way to fix the layout problem. Since we have already worked on the layout of functions and functional exchanges in the Data Flow diagrams, we can just go back to the "Global Data Flow" diagram and copy the format to this diagram.

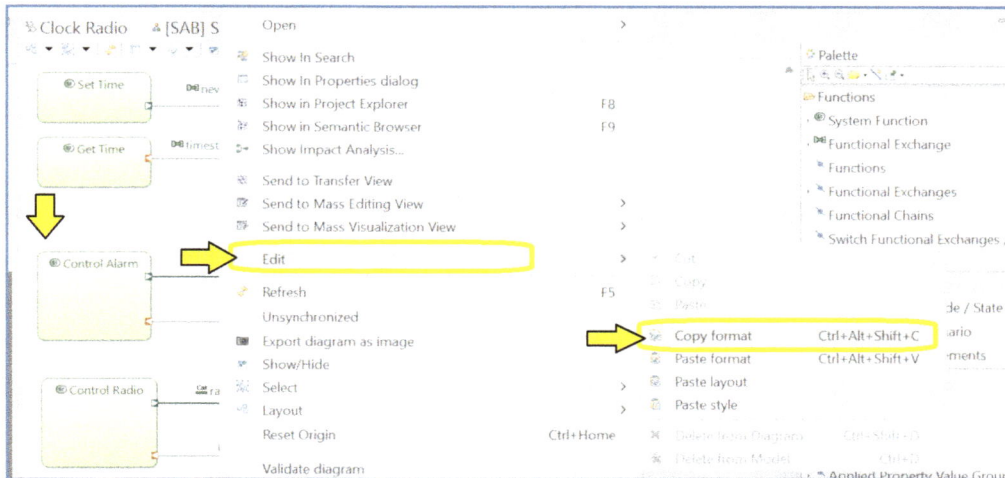

Figure 6-12 – Copy format from the global data flow diagram

Return to the global data flow diagram. Right-click anywhere in the diagram and select "Edit" followed by "Copy format".

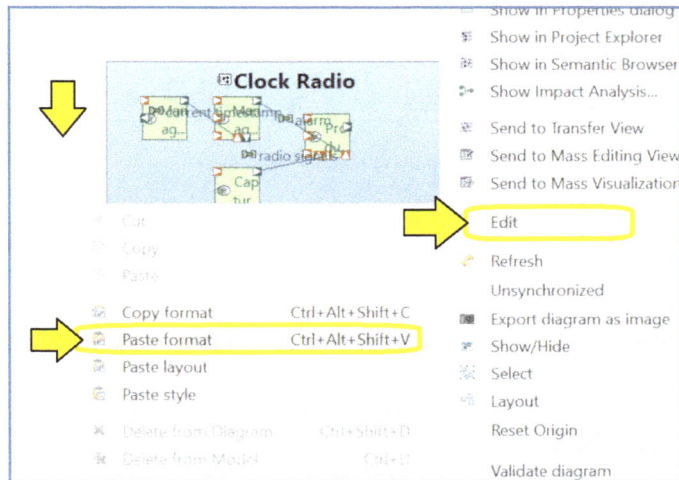

Figure 6-13 – Use the paste format button

Returning to the system architecture blank diagram, right-click anywhere in the diagram and select "Edit" followed by "Paste format". There is also a "Paste format" tool in the diagram top tool bar.

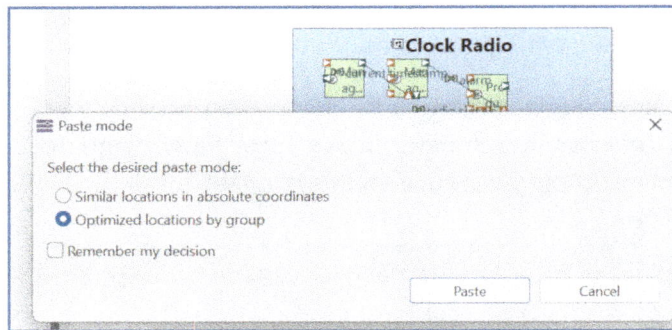

Figure 6-14 – Paste Mode Panel

Next, a "Paste mode" panel will give you two options for pasting the format with slightly different nuances about the formatting approach. You can experiment to see which option you like better. You can use Ctrl-Z to undo a paste and try the other formatting approach.

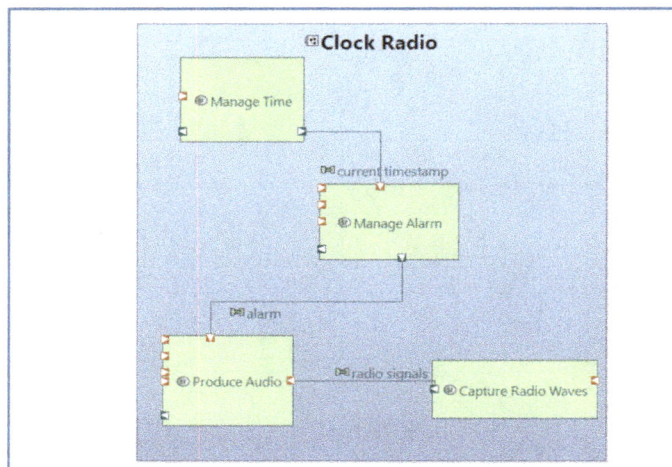

Figure 6-15 – Cleaned up diagram

You will need to move the functions around and perhaps resize the functions and system component shapes a bit to perfect the layout.

Notice that the functions automatically appear with their owned ports, as previously mentioned.

Of course, if the functions had not been created yet, we could have just used the "System Function" command of the palette, which both creates new functions and allocates them to the selected structural element.

We can also add the functional chains to the diagram any time we want using the "Functional Chains" tool in the palette as shown in Figure 5-71 on page 114.

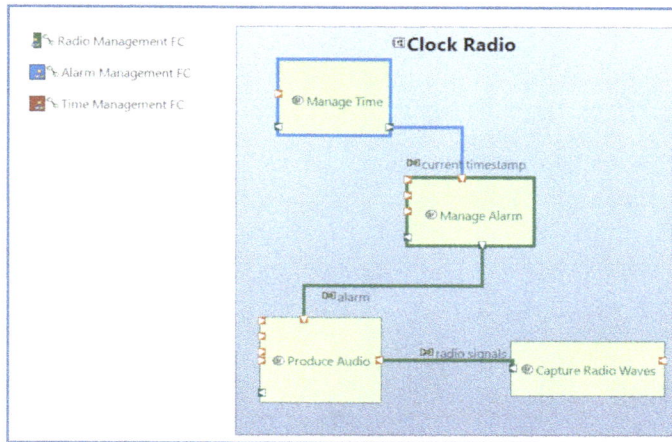

Figure 6-16 – Diagram with functional chains added

However, as a practical matter, it is easier to work on diagrams without the functional chains displayed. We recommend that you do all the major editing of your diagrams with functional chains hidden and add them as the last step in formatting the diagram.

Actors and Function Allocation

We already created two actors in Chapter 4, Section *Actors* on page 63 :

- a human actor called "User",

- a non-human actor called "Radio Station Transmitter".

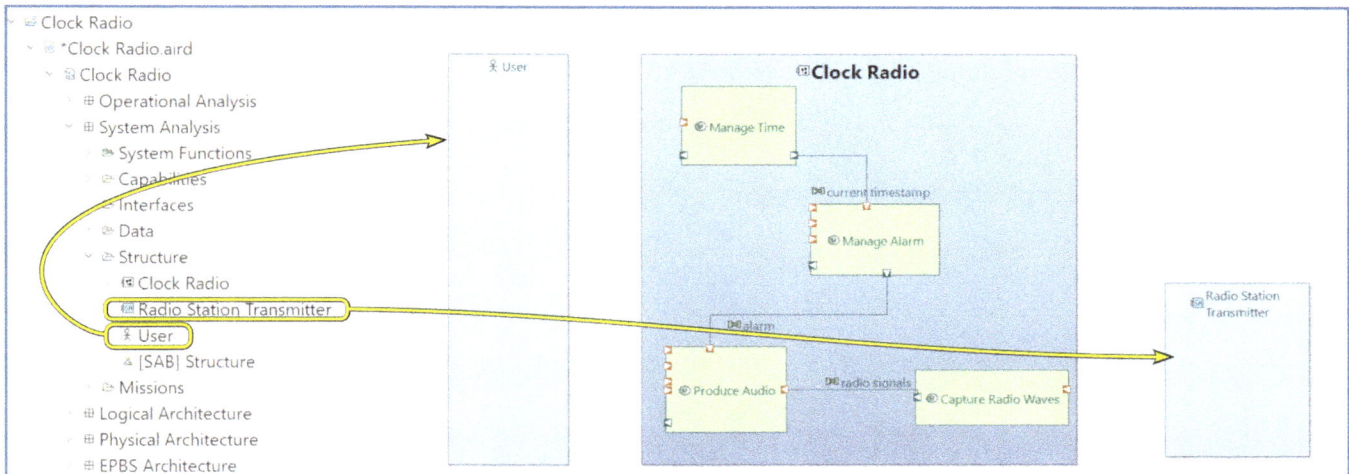

Figure 6-17 – Add two previously defined actors to the diagram

Drag the two previously defined actors from the project explorer to the diagram and resize as shown.

Go ahead and change the font for the actor names to 12-point, italic using the technique shown in Figure 6-7 on page 128.

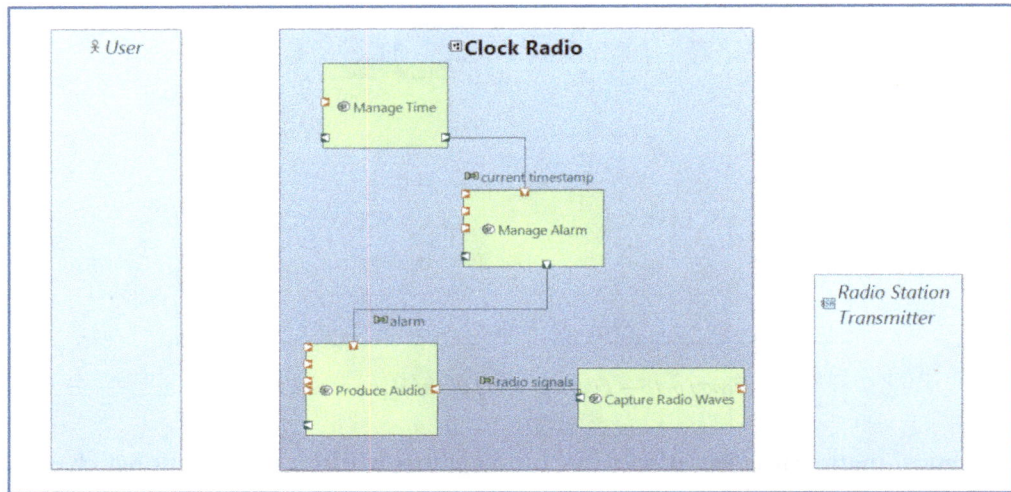

Figure 6-18 – Icons and fonts adjusted

After adjusting the actor icons, the font size for the actor names, and the size of the actor rectangles, the diagram and model are ready for us to continue.

Next, we want to allocate the functions "Get Time", "Set Time", "Control Alarm", and "Control Radio" to the user actor. The function "Emit Radio Waves" has to be allocated to the radio station transmitter actor.

There are two ways to know whether we still have functions which are not allocated. The simplest is to try again to allocate functions with the "Manage Function Allocation" command of the palette. If the left part of the selection wizard is not empty, it means that some functions are not yet allocated.

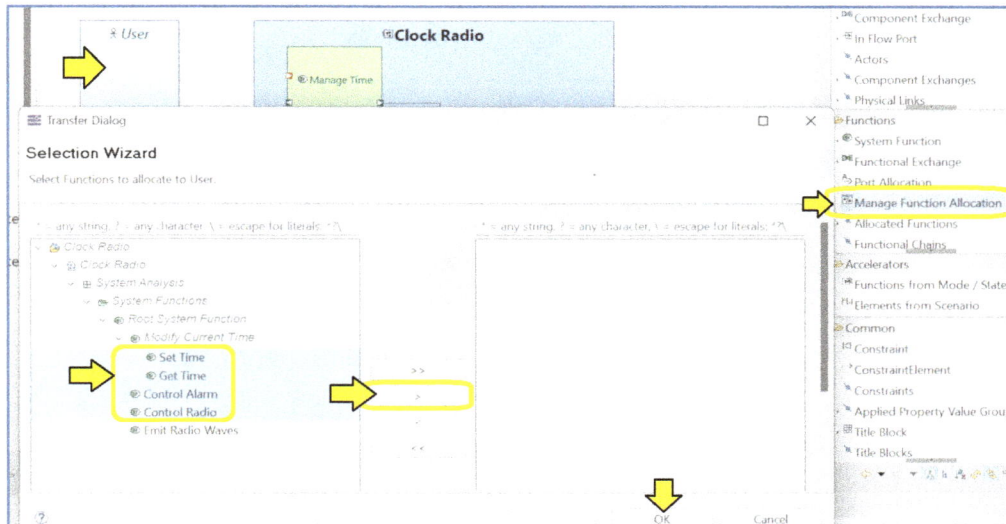

Figure 6-19 – Select from functions available to allocate

Select the "Manage Function Allocation" tool in the palette. Click in the user actor box. The transfer dialog box will appear with the functions available to allocate.

If you look closely, you will notice that the decomposed function "Modify Current Time" cannot be selected. Only the child functions "Get Time" and "Set Time" can be selected. Here *Capella* is enforcing a subtle *Arcadia* rule: only leaf functions can be allocated.

Go ahead and select the four functions that should be allocated to the user actor. Use the single arrow ">" to allocate the functions. Click "OK".

> **Arcadia on Functional Allocation**
>
> We have already said that a function must be allocated to one and only one structural element, but that several functions can be allocated to the same structural element.
>
> Another very important rule states that *only leaf functions can be allocated.*
>
> We have also seen in Chapter 5 that only leaf functions can own ports directly. These two rules means that non-leaf functions are just grouping constructs, interesting for instance for documentation purposes, but are not full functions in *Arcadia*.

Leaf Function – A **leaf function** is a function that is not decomposed. All levels of parent functions are said to be non-leaf.

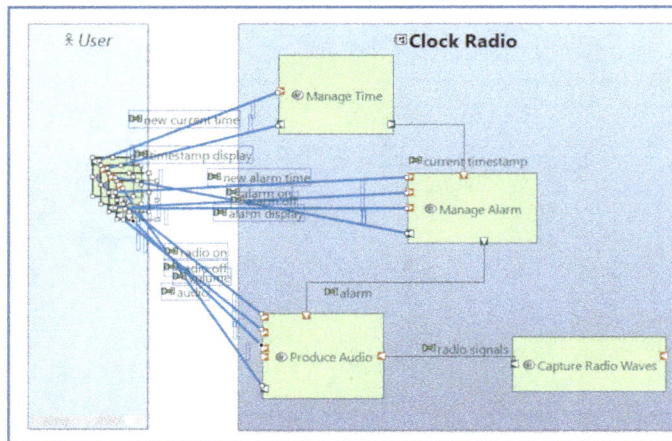

Figure 6-20 – The functions appear in the light blue box

The functions appear inside the user actor's light blue box, along with all the relevant functional exchanges.

We can polish the layout by resizing and moving the functions in the user actor's light blue box. We would also need to manually drag the ports into position.

There is another, quicker, solution! We can first copy the format of the "Global Data Flow" SDFB where we already optimized the size and location of functions and exchanges.

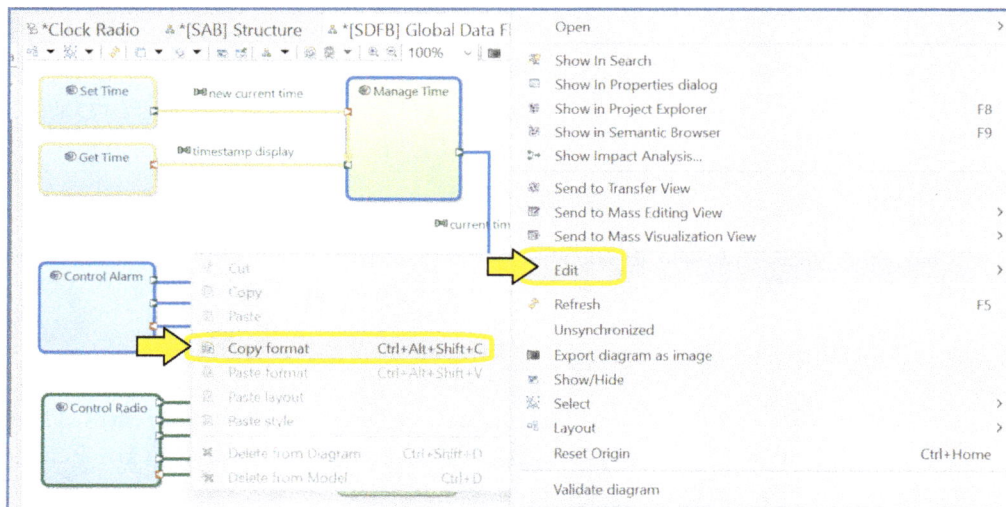

Figure 6-21 – Copy the format of the SDFB

Return to the "Global Data Flow" SDFB. You may need to perform a few formatting steps to get it back to where we had it in Figure 5-77 on page 117.

- Turn off the display of categories.
- Add the functional chains to the diagram.

- Set the color of the time management functional chain to yellow.

Once again, right-click anywhere in the diagram and select "Edit" followed by "Copy format".

Then we can paste this format in the SAB, either by right-clicking in the background of the diagram, or by using a specific command in the diagram toolbar.

You will very likely need to make the drag the user and clock radio rectangles to make them large enough to display all of the allocated functions and functional exchanges.

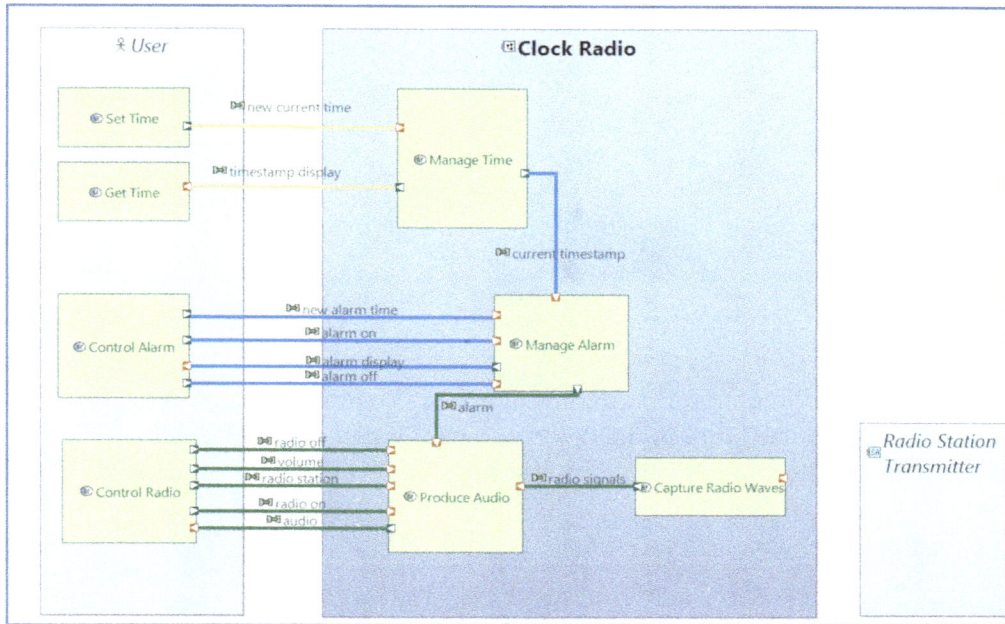

Figure 6-22 – Diagram after pasting and resizing rectangles

Repeat the procedure shown in Figure 6-19 on page 135 to allocate the remaining function "Emit Radio Waves" to the "Radio Station Transmitter" actor.

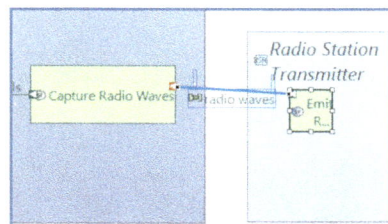

Figure 6-23 – The function appears inside the radio transmitter

The function appears inside the radio station transmitter's light blue box, along with the relevant functional exchange. The format needs some improvement.

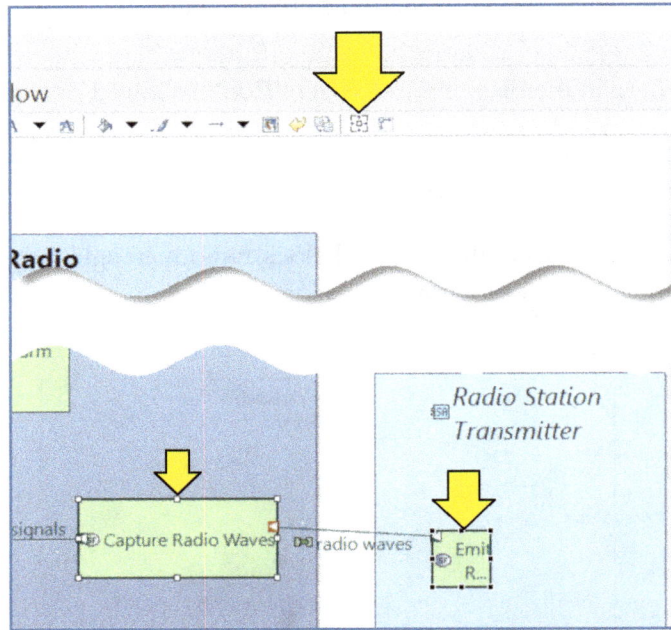

Figure 6-24 – Make boxes the same size

Select the "Emit Radio Waves" function first, followed by the "Capture Radio Waves" function. Click on the "Make height and width same size" tool in the upper toolbar. Notice that the order of selection is important – we want the smaller box to become the same size as the larger box, not the other way around.

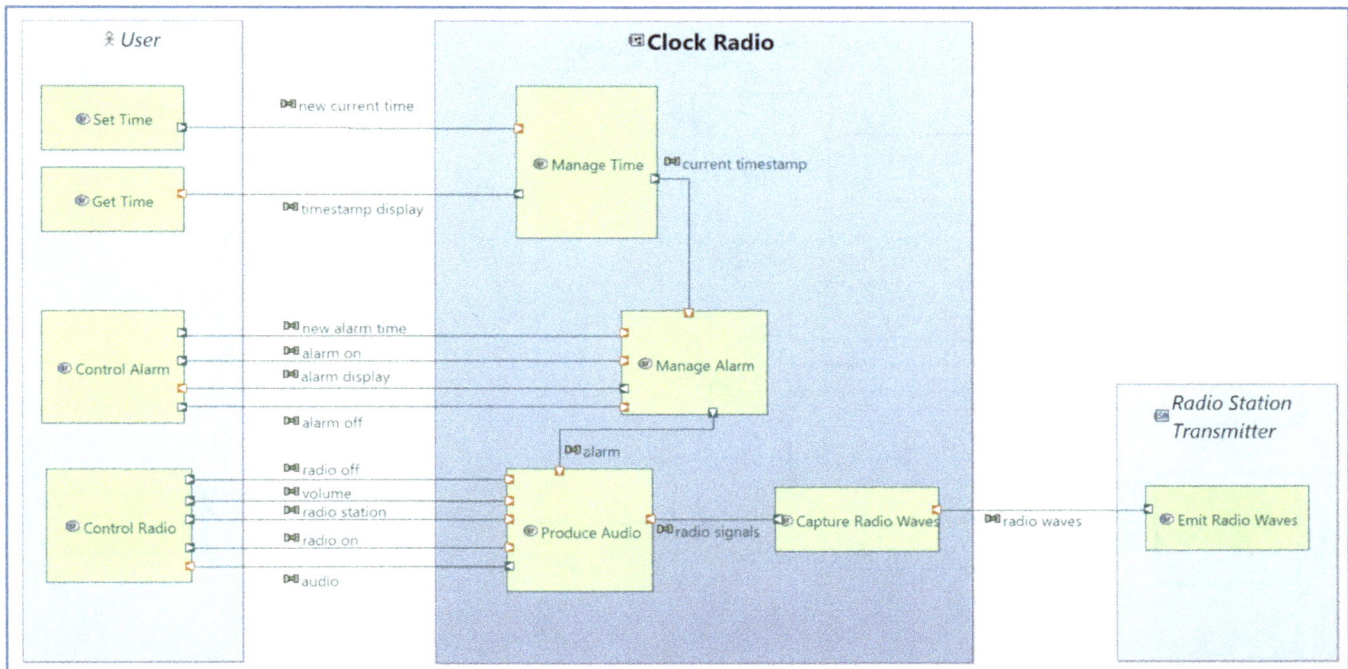

Figure 6-25 – Final diagram

With a little more adjustment of the spacing, we have the finalized diagram. Notice that we have applied a consistent layout rule for labels on the functional exchanges:

- Output to input labels are above the line.
- Input to output labels are below the line.

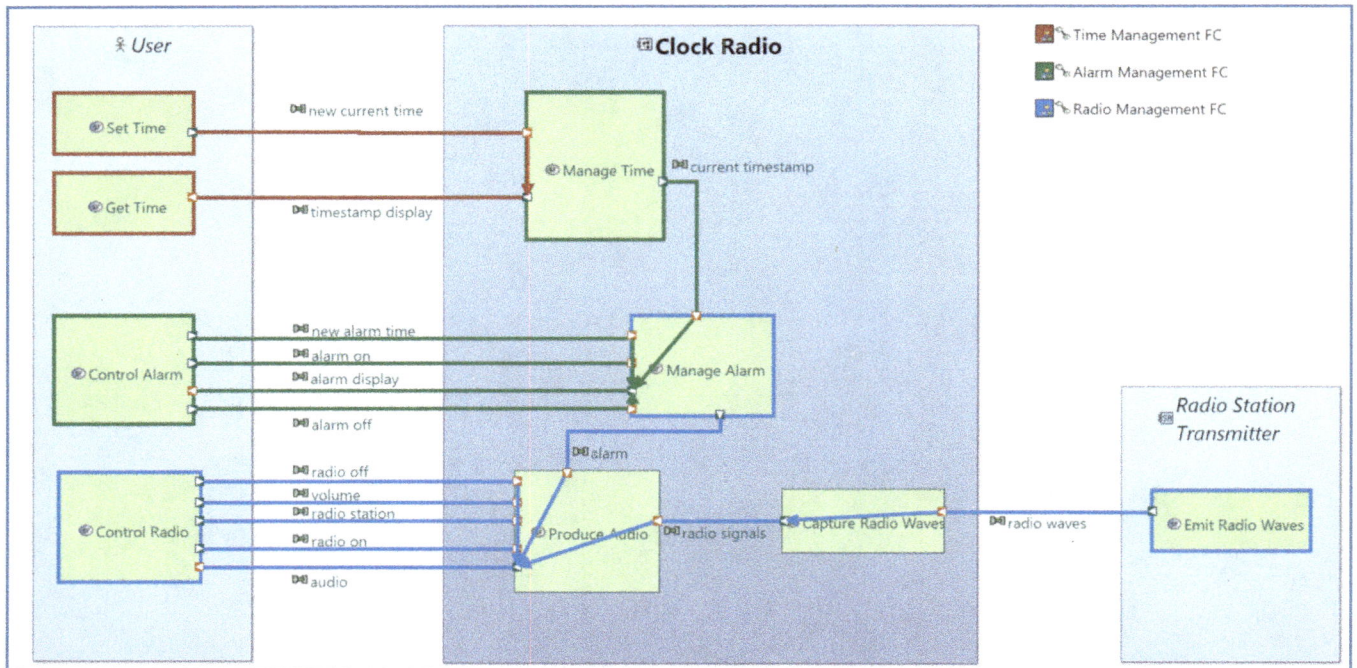

Figure 6-26 – Diagram with functional chains added

Again, we can also add the functional chains to the diagram any time we want using the "Functional Chains" tool in the palette as shown in Figure 5-71 on page 114.

Component Exchanges

So far, we have been working with functional exchanges. These are abstract flows between functions which themselves are abstract. We have also discussed functional chains and categories, but these are likewise abstract groupings composed of abstract functional exchanges. What about the real system? We have taken the first step by defining components and external actors and allocation the (abstract) functions to them. Next, we have to start tying the functional exchanges to some more specific means of moving from point A to point B. *Arcadia* provides the **component exchange** for this purpose.

Component Exchange – A **component exchange** is a structural connector between structural elements.

- at System Analysis level: between the system and is actors

- at Logical Architecture level: between the logical components and the actors or between the logical components themselves

- at Physical Architecture level: between the physical components and the actors or between the physical components themselves

A **component exchange** links a single component port of the source structural element to a single component port of the target structural element. A **component port** can be oriented: *in*, *out*, or *inout*. That is, a **component exchange** can be bidirectional.

Note: we are about to introduce component exchanges and show how to link functional exchanges to component exchanges. These diagrams are going to get rather *intricate!* Functional chains have the effect of making diagrams even more intricate and really are not important for this discussion. Before proceeding, remove the functional chains from the "[SAB] Structure" diagram. Remember, we can easily bring them back any time we want them.

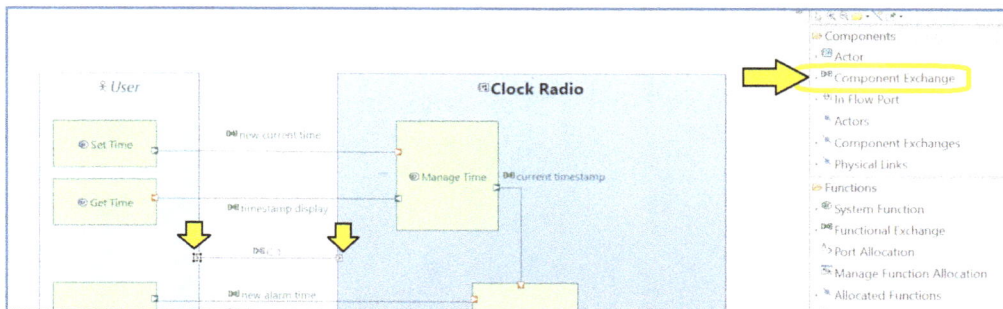

Figure 6-27 – Create a component exchange

In the palette, select the "Component Exchange" tool. Next, click inside a source structural element, for instance the user actor, move the mouse to the target structural element, for instance the clock radio, and click the mouse button once more *Capella* creates one component port in the source structural element (with direction set to output by default), one component exchange named "C1", and one component port in the target structural element (with direction set to input by default).

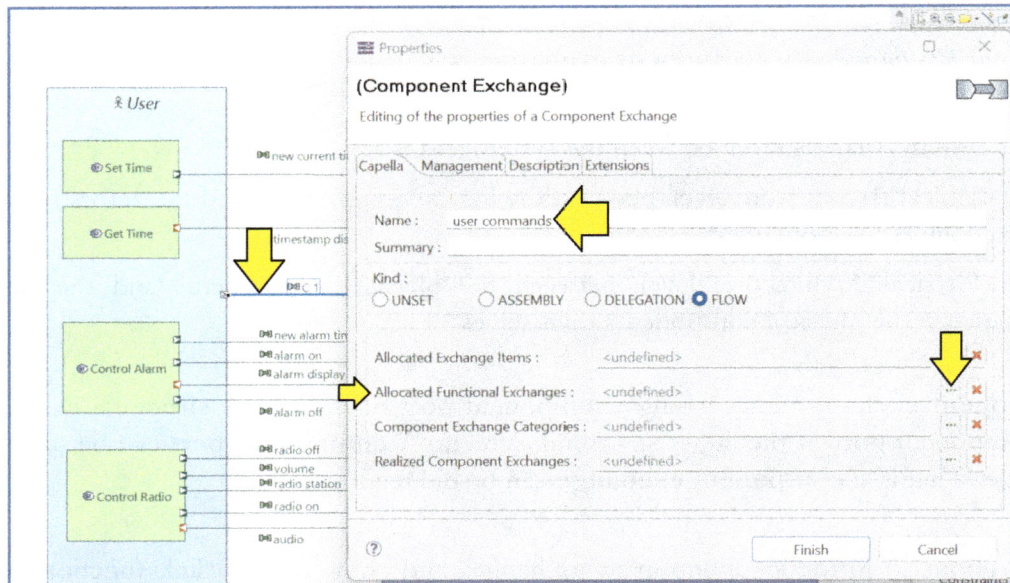

Figure 6-28 – Rename the component exchange and click on the three dots

First, we will want to give the component exchange a meaningful name.

1) Double-click on the "C 1" component exchange connector.

2) Rename the component exchange connector "user commands".

3) Click on the three dots icon to the right of "Allocated Functional Exchanges:".

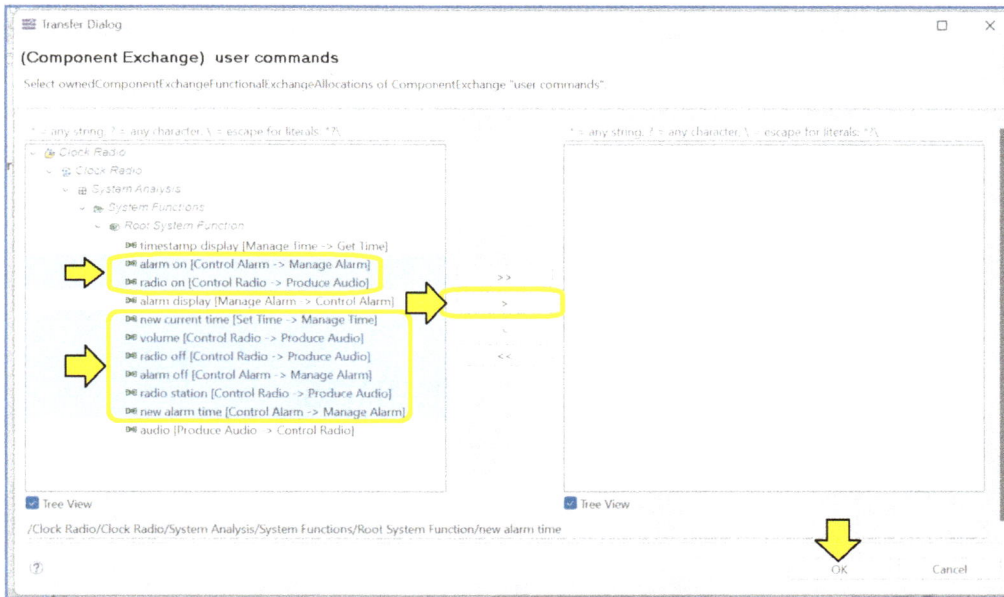

Figure 6-29 – Select all of the commands from the user to the clock radio

Select all of the commands from the user to the clock radio. Use the single arrow "**>**" to allocate the functional exchanges to the component exchange. Click "OK". Then, click "Finish".

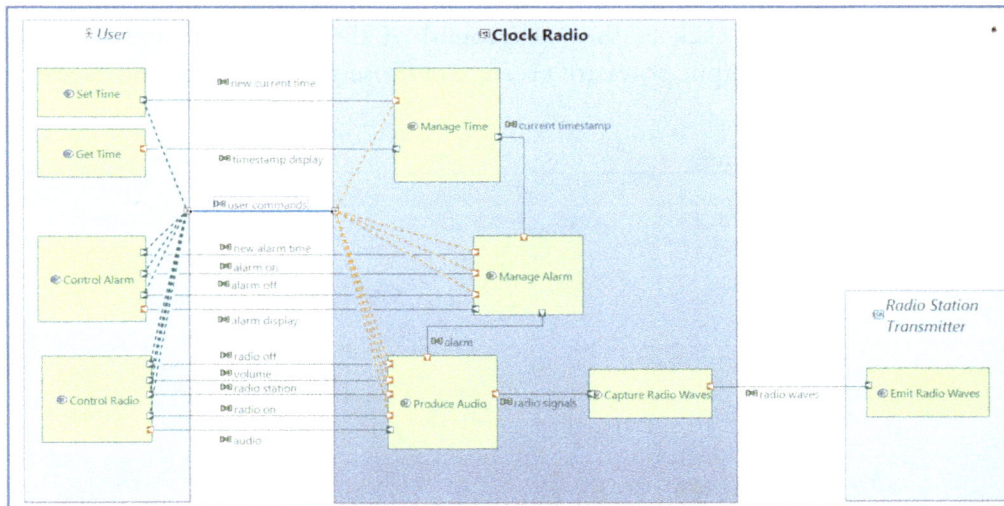

Figure 6-30 – The tool creates dashed lines

Capella creates dashed lines to represent the allocations of function ports to component ports.

Notice the interesting nuance here: we asked *Capella* to allocate functional exchanges to the component exchange and the tool has indeed done so. However, in the process *Capella* has also automatically allocated the related function ports to the component exchange ports and these relationships are what the tool is showing on the diagram. You can verify all of the above by selecting the different elements in the diagram and viewing the relationship information in the semantic browser.

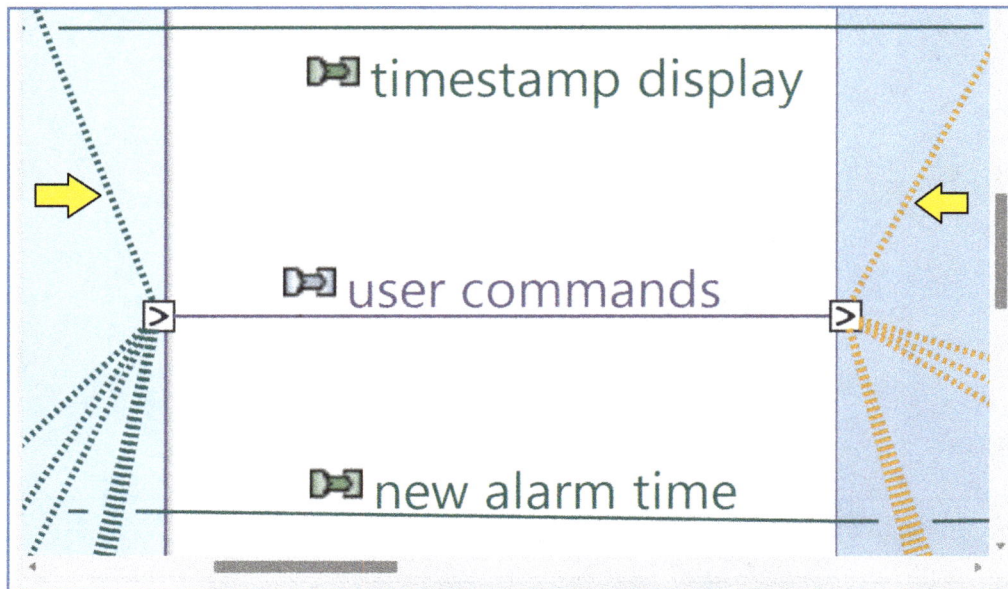

Figure 6-31 – The dashed lines change color

After *Capella* creates the dashed port-to-port allocation lines, they are in the "selected" mode and appear in blue. However, as soon as you click in the background of the diagram to deselect them, the colors change. The allocations from the output ports are green, and those from the input ports are orange, like the function port colors.

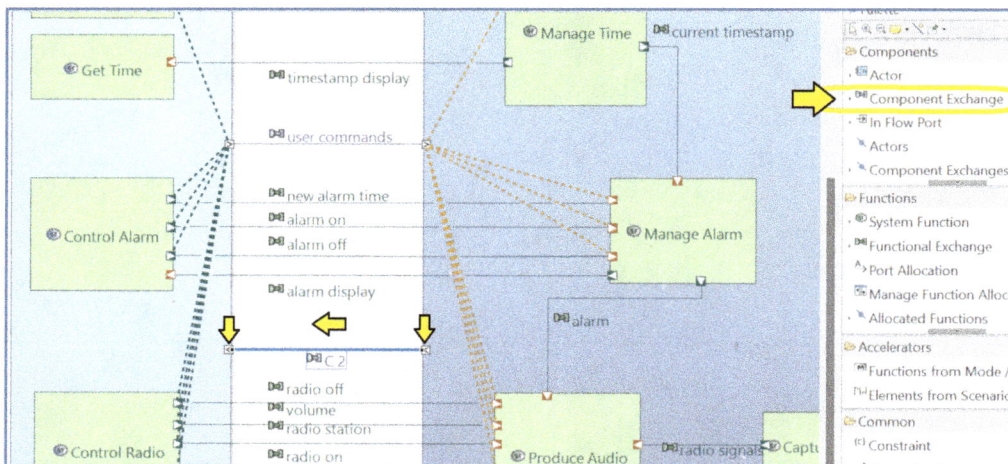

Figure 6-32 – Create a second component exchange

Next, create a component exchange that flows in the opposite direction. After selecting the "Component Exchange" tool in the palette, click first in the clock radio system component and then in the user actor.

Name the second component exchange "clock radio outputs" and assign the remaining three available function exchanges to it using the technique shown in Figure 6-28 on page 142.

Arcadia on Component Exchanges

We have already said that a function must be allocated to one-and-only-one structural element, but that several functions can be allocated to the same structural element.

Similarly, *a functional exchange must be allocated to one-and-only-one component exchange*, but several functional exchanges can be allocated to the same component exchange. It is important to understand that different component exchanges can allow the architect to separate different categories of service (performance, security, frequency, etc.).

In our case, we could have grouped all functional exchanges between the user and the clock radio into only one bidirectional component exchange, but we chose to differentiate inputs from outputs in particular to improve readability of the diagram.

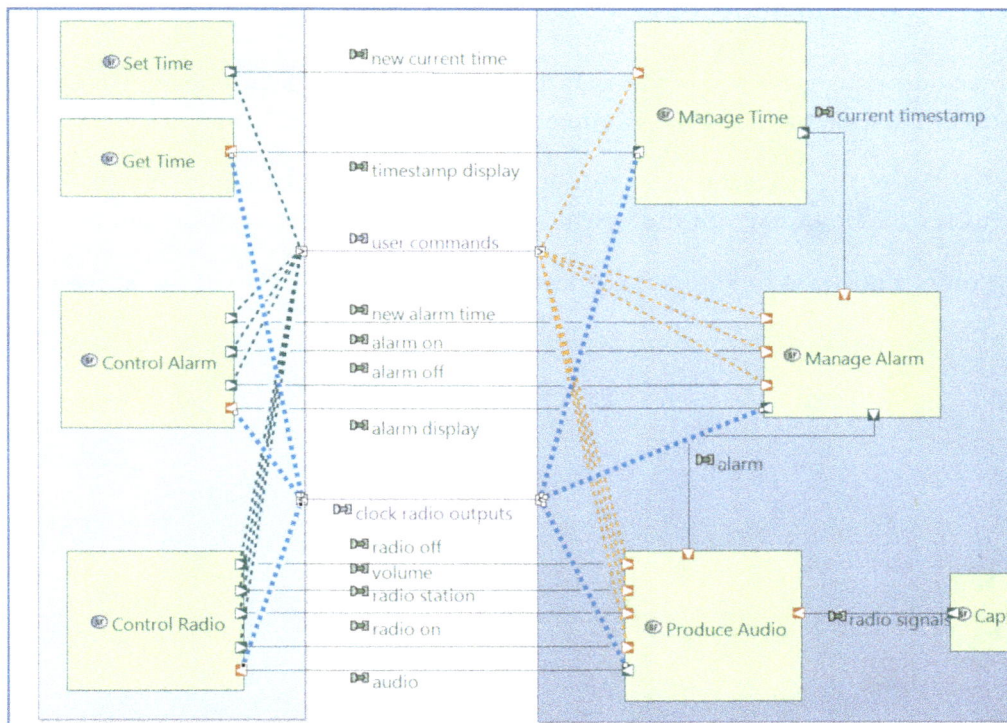

Figure 6-33 – Functional exchanges grouped by inputs and outputs

The resulting diagram is shown in Figure 6-28. Notice that the new port-to-port allocation relationships are blue because they are still selected just after *Capella* creates them.

The last step is to create a component exchange from the radio station transmitter actor to the clock radio system component.

We could create this final component exchange by using the palette again. However, it is also possible (and quicker) to create it using a "Modeling Accelerator".

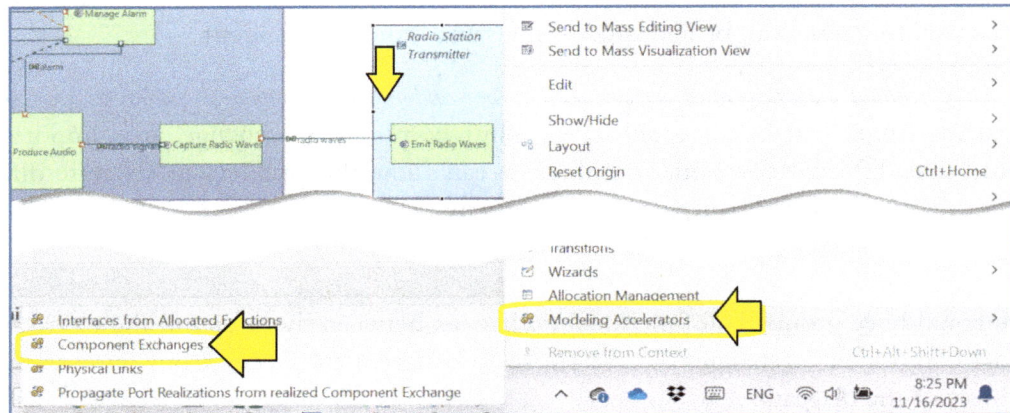

Figure 6-34 – Use the modeling accelerator

Select the radio station transmitter actor and right-click on it. There is a "Modeling Accelerator" command with subcommands. We have to select the "Component Exchanges" subcommand.

The advantage of using the accelerator is that it creates one component exchange for any functional exchange that crosses the boundaries of the structural element. Even better, it also handles the allocation.

Capella always positions ports on the top left corner of a box, so the result is shown on the next figure.

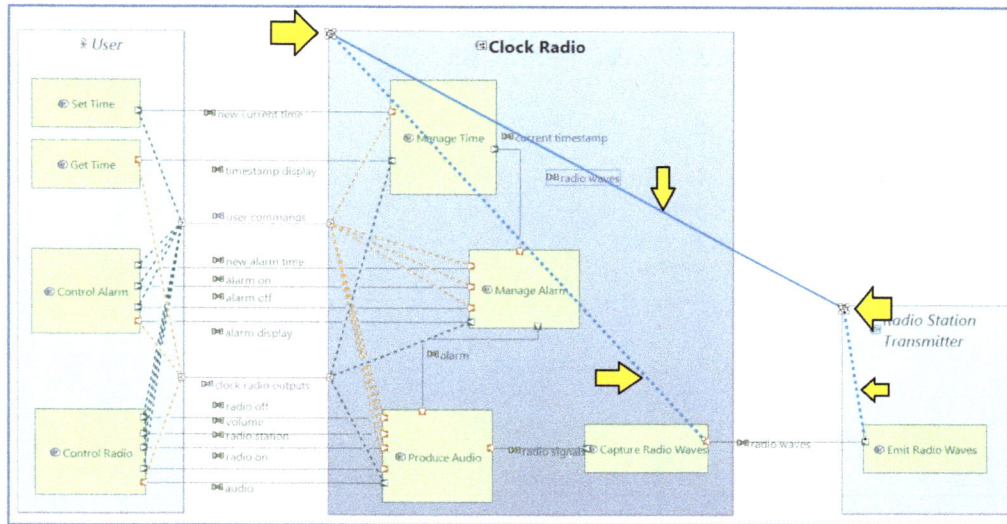

Figure 6-35 – Component exchange created automatically

Capella has automatically created the component exchange and allocated the available functional exchange to it. Unfortunately, *Capella* makes no attempt to figure out where the best place to put the ends of the component exchange would be, instead connecting the top left corner of the radio station transmitter actor to the top left corner of the clock radio system component.

Move the ports around to put the component exchange in a more convenient location, slightly above the "radio waves" functional exchange. Since there was only one available functional exchange, *Capella* named the component exchange to match the functional exchange. This default name is a good guess. However, it is better modeling practice to keep the names unique to differentiate the elements of different types. Rename the component exchange to "radio transmitter interface".

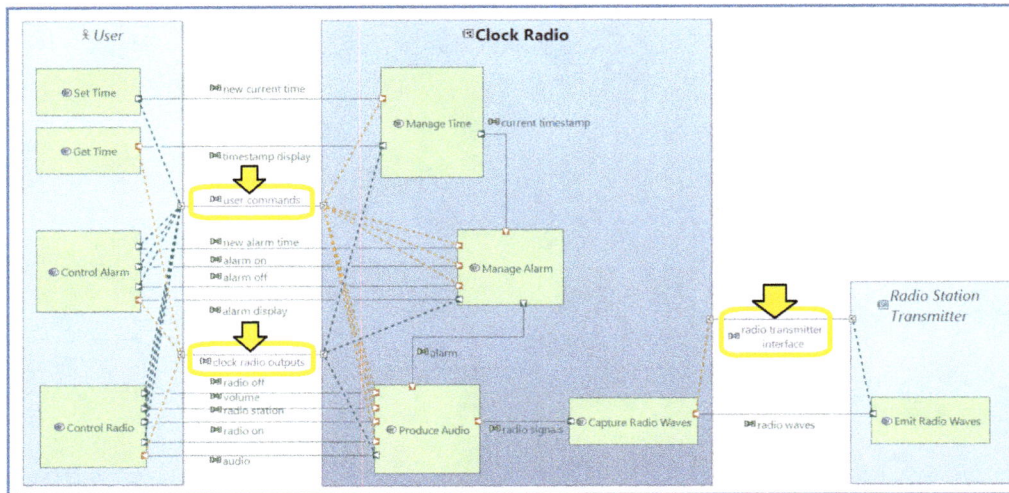

Figure 6-36 – Cleaned up final result

Now we have all the functional exchanges correctly allocated to component exchanges. Note, from a correct *Arcadia* modeling practice point-of-view, this step is not optional. To have a semantically correct model, you must define component exchanges and allocate all of the functional exchanges to the appropriate component exchanges.

Cloning Diagrams to Provide Different Views

Views and Model-Based Systems Engineering

Understanding your audience is a key part of model-based systems engineering, regardless of the methodology and language used. Fundamentally, before doing any modeling activity, you should stop, take a deep breath, and answer three questions for yourself:

1) What question is this model going to answer?

2) Who is the audience for the answer?

3) What business benefit do we expect the model to produce for that audience?

In some cases, it may be worthwhile to make a specific model to answer one single question for one single stakeholder. More often, you will have multiple questions for multiple stakeholders. However, the fact that there are multiple questions to be answered does not relieve the modeler of the need to think clearly about what the specific questions are, and which type of audience is going to need the answer. In general, for a model that answers multiple questions for multiple stakeholders, you want to produce a **view** for each question of interest to each stakeholder. Such a view should have the information needed to answer the specific stakeholder question with as little extraneous clutter as possible.

In this section, we are going to learn how *Capella* supports the creation of diagrams to answer specific questions for specific stakeholders. The basic strategy is:

1) **Superset Diagram** – Make a "Master" diagram that has all of the elements needed to answer several questions or types of question.

2) **Clone Diagrams** – Make as many clones of the diagram as needed.

3) **Filters** – Use *Capella* filters to remove all of the information from each clone of the diagram that is not needed to answer that specific clone's intended question.

Return to the activity explorer. Select the "System Analysis" activity. The diagram viewer will appear on the right. If you expand all groups, we can see all the diagrams of the current *Arcadia* level.

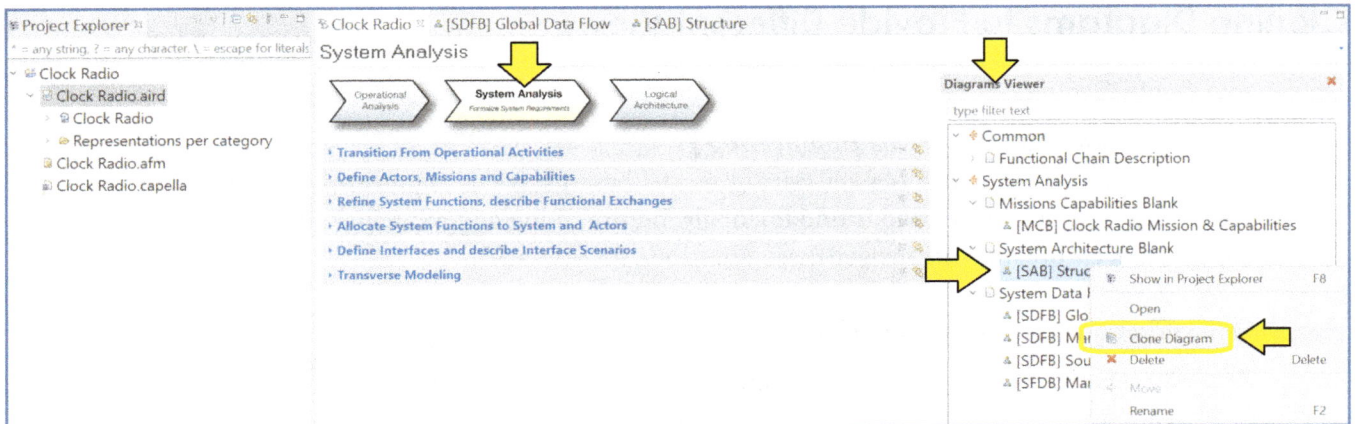

Figure 6-37 – Select clone diagram command

Select the [SAB] structure diagram. Right-click and select "Clone Diagram".

Capella creates a new diagram, called "Clone of [SAB] Structure".

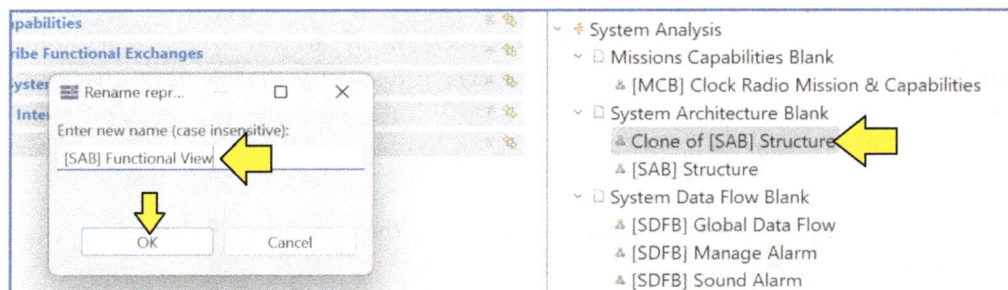

Figure 6-38 – Rename the cloned diagram

Rename the cloned diagram "[SAB] Functional View".

A clone is really a pure copy of a diagram, with the same model elements represented and the graphical layout kept identical. We can then modify each diagram independently, like all other diagrams, but still with the consistency ensured by the underlying model managed by *Capella*.

Rename the original diagram into "[SAB] External View". If we double-click on the [SAB] functional view, it looks exactly like the [SAB] external view. Now we are in position to hide specific model elements in the functional view without disturbing the external view. We can use the diagram filters to do such selective hiding of elements.

Figure 6-39 – Select filter

On the top command bar of the diagram, click on the down arrowhead beside the "Filters" icon. A long list of available filters appears. Note that these filters depend on the type of the diagram and also the *Arcadia* level. From this list, select "Hide Component Exchanges".

The result is immediately visible: all component exchanges have disappeared. We still see the component ports and the port allocations, which are not very useful at this point.

Figure 6-40 – Closer look at the icon

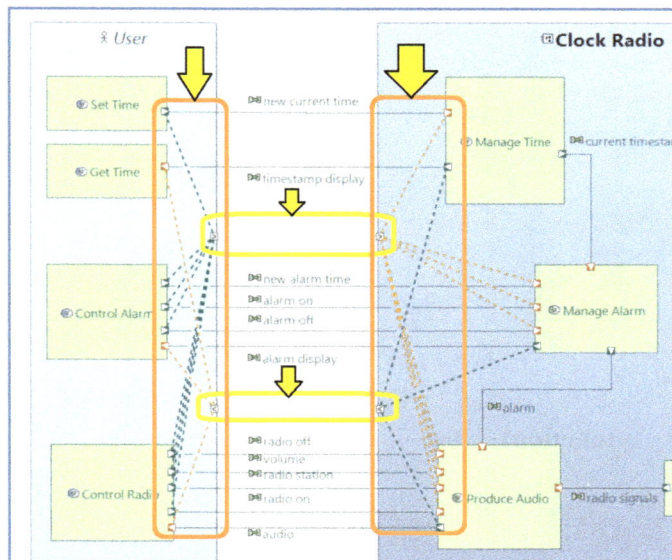

Figure 6-41 – Component exchanges have disappeared

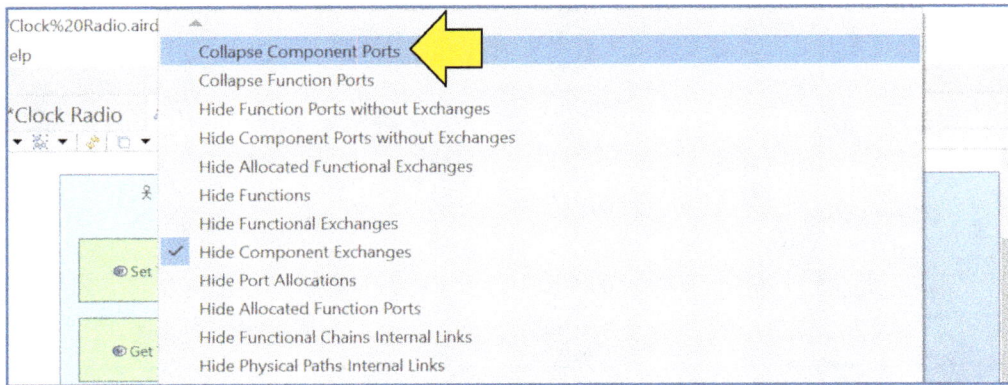

Figure 6-42 – Select collapse component ports

Apply a second filter: "Collapse Component Ports". Note that *Capella* only allows you to activate one filter each time. This restriction is by design, as some filters may be contradictory, and others could be redundant.

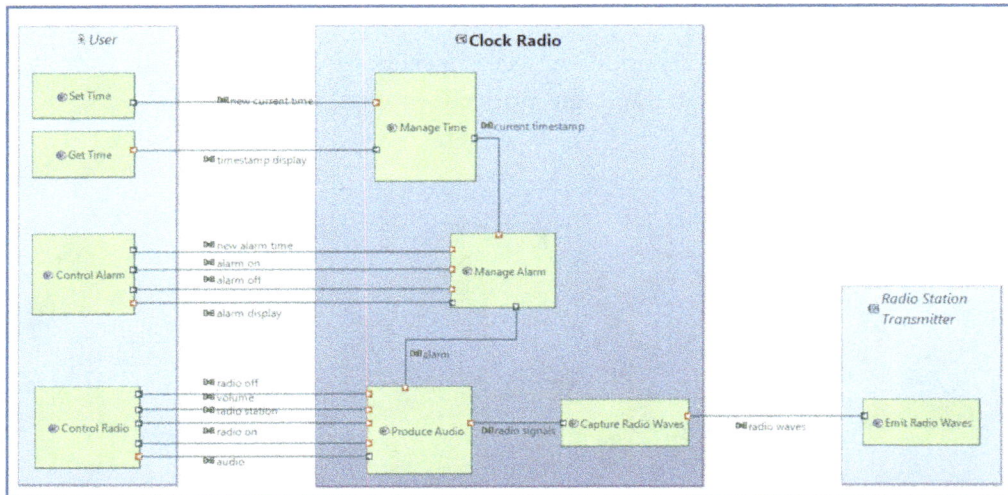

Figure 6-43 – Component ports collapsed

The diagram is back to a simple functional view.

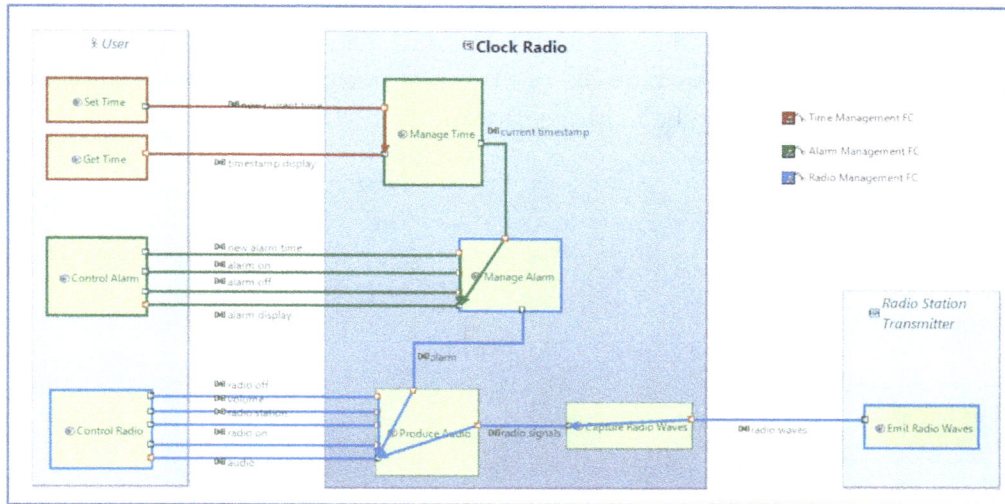

Figure 6-44 – Functional chains inserted

Add the functional chains to the diagram using the "Functional Chains" tool in the palette as shown in Figure 5-71 on page 114.

Capella displays the functional chains with standard colors (blue, red, green, etc.). We just need to align the square icons of the functional chains to get a nice functional view of the System Analysis level. Remember, we can easily change the colors of the functional chains if needed, as previously shown in *Functional Chain Colors* on page 115.

Let's turn our attention back to the external view.

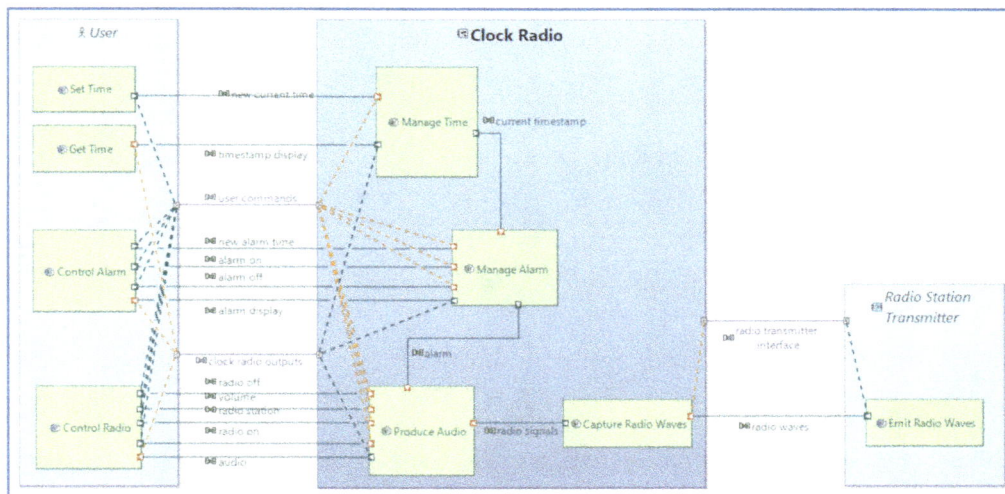

Figure 6-45 – Return to the external view

Reopen the diagram "[SAB] External View". At the moment, it is the original diagram with both functional and component exchanges.

In this diagram, we would like to focus on the external interface of the system, without showing at all functions and functional exchanges. Adjusting the focus of a diagram in this manner is very easy with *Capella*: all we have to do is to apply a specific filter to hide the functions. If we hide the functions, function ports and functional exchanges will also be hidden automatically by *Capella*.

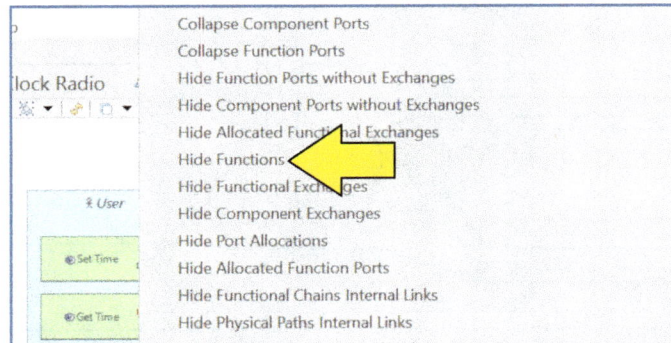

Figure 6-46 – Select hide functions

Select "Hide functions".

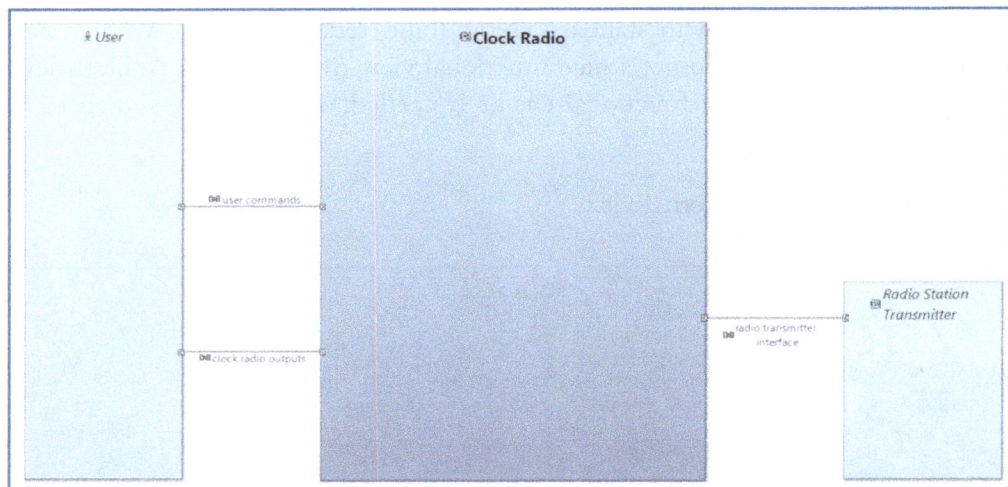

Figure 6-47 – The result is drastic!

The complexity of the diagram has been reduced drastically. Only the component exchanges are still displayed.

To better see the external interfaces of the system, we will want to make the Clock radio rectangle narrower and also ask *Capella* to display the list of names of the functional exchanges, instead of the name of each component exchange. *Capella* has a specific filter to achieve this: "Show Allocated Functional Exchanges on Component Exchanges".

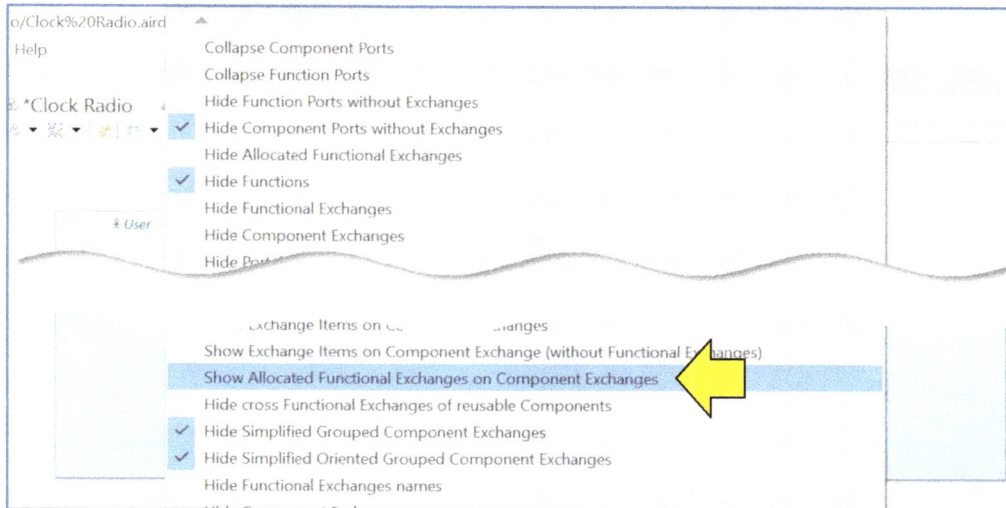

Figure 6-48 – Select filter to show allocated functional exchanges

Select the filter to show allocated functional exchanges.

Hmmm... Nothing happens.

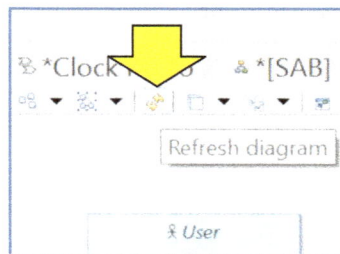

Figure 6-49 – Refresh the diagram

We have to use the "Refresh" tool to prompt *Capella* to compute the required elements for the diagram. The filters which are described as "Show…" often need manual refresh to update the diagram.

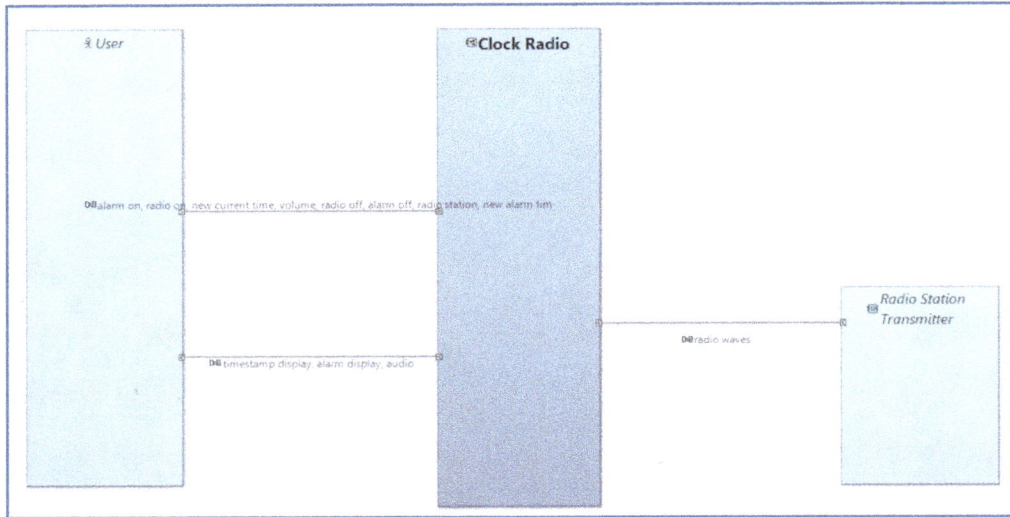

Figure 6-50 – Allocated functional exchanges are shown

The names of the component exchanges have been replaced by lists of the names of allocated functional exchanges.

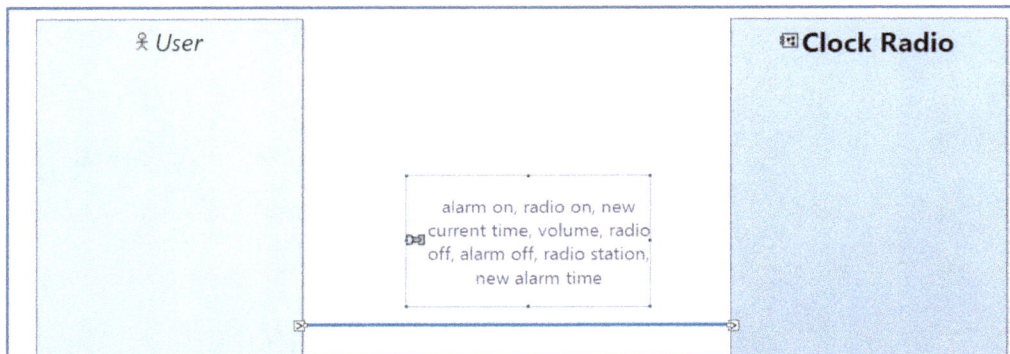

Figure 6-51 – Resize the label

To improve the layout, we can resize the rectangles of the string names to force *Capella* to display them on several lines.

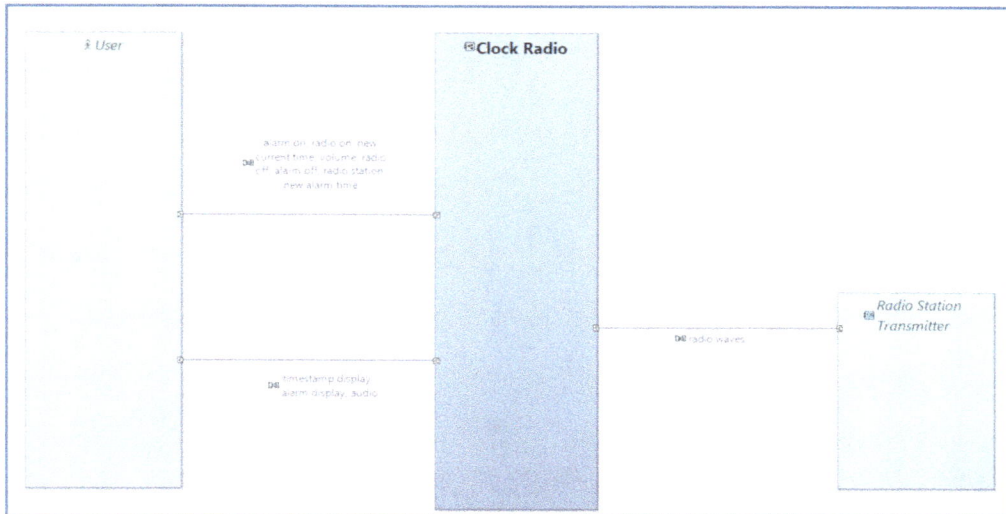

Figure 6-52 – External view completed

The completed external view is shown above. If you compare this diagram to the functional view shown in Figure 6-44 on page 153, you will see that the two views are complimentary.

We can make a synthesis of the two views in a third diagram by adding only functions to the external view, without function ports and functional exchanges.

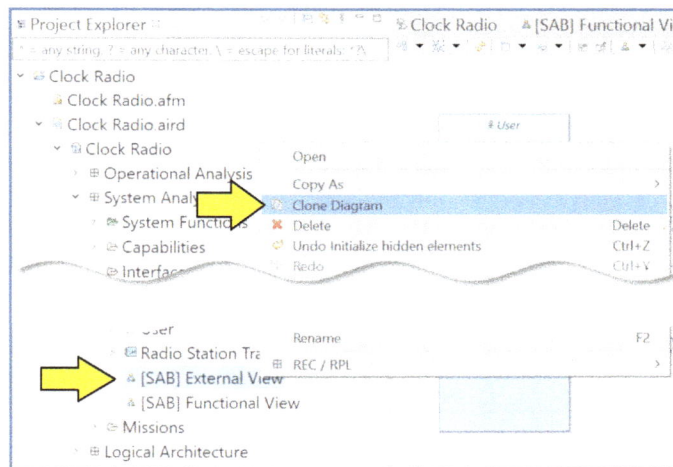

Figure 6-53 – Clone function is also available in project explorer

The clone function is also available in the project explorer. Clone the "[SAB] External View" diagram and rename it "[SAB] Global View". Open the new diagram. Use the filter tool to remove the "Hide Functions" filter.

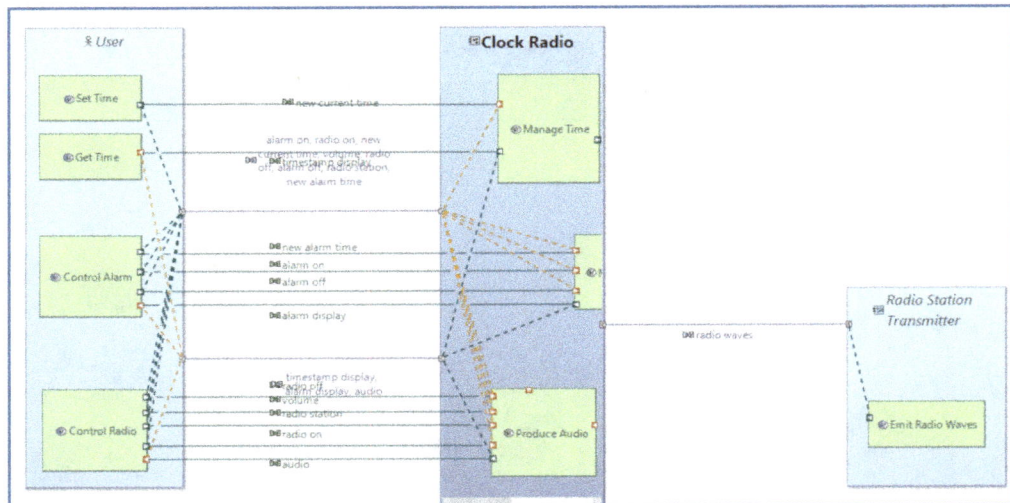

Figure 6-54 – Functions are back, but so are the functional exchanges

The result is a little messy. The functions are back, but so are the functional exchanges. We need to apply two more filters to clean this diagram up. First, apply the "Hide Functional Exchanges" filter. Second, apply the "Collapse Function Ports" filter.

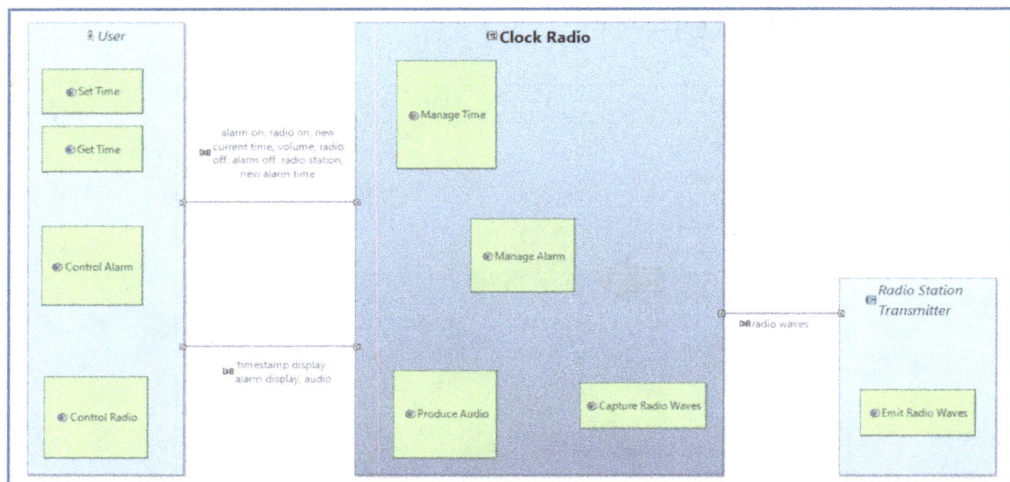

Figure 6-55 – Global view format improved

We may need to resize the boxes a bit.

We previously learned how to change the font for an element in Figure 6-7 on page 128. What if we want to increase the font size for all of elements of a single type? For example, what if we increase the font size for all of the function names?

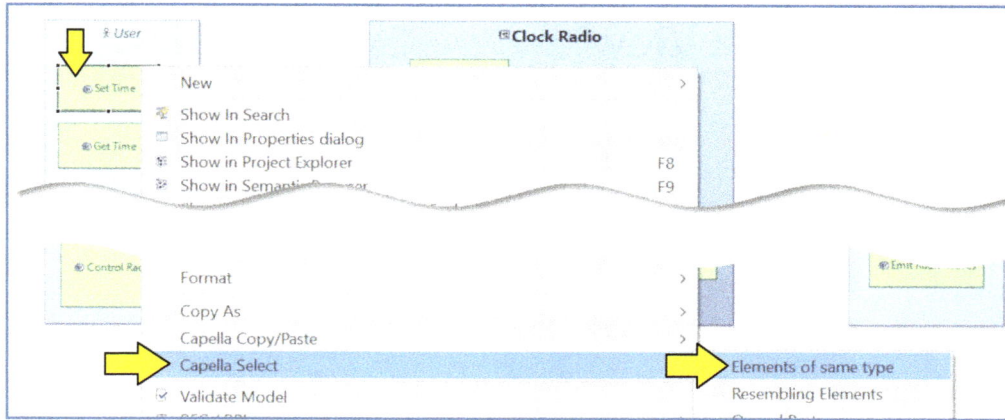

Figure 6-56 – Select similar elements

Select of one the green functions. Right-click and select "Capella Select" followed by "Elements of same type". All functions are now selected, and you can increase the size of the font of all of them to 10 points using the tool in the top toolbar as shown in Figure 6-7 on page 128.

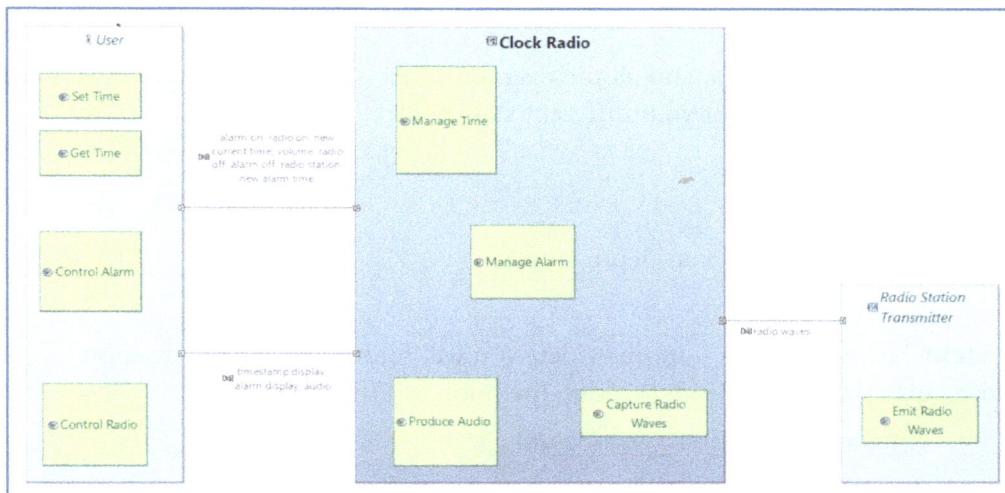

Figure 6-57 – Final global view

Done. All the function names are larger.

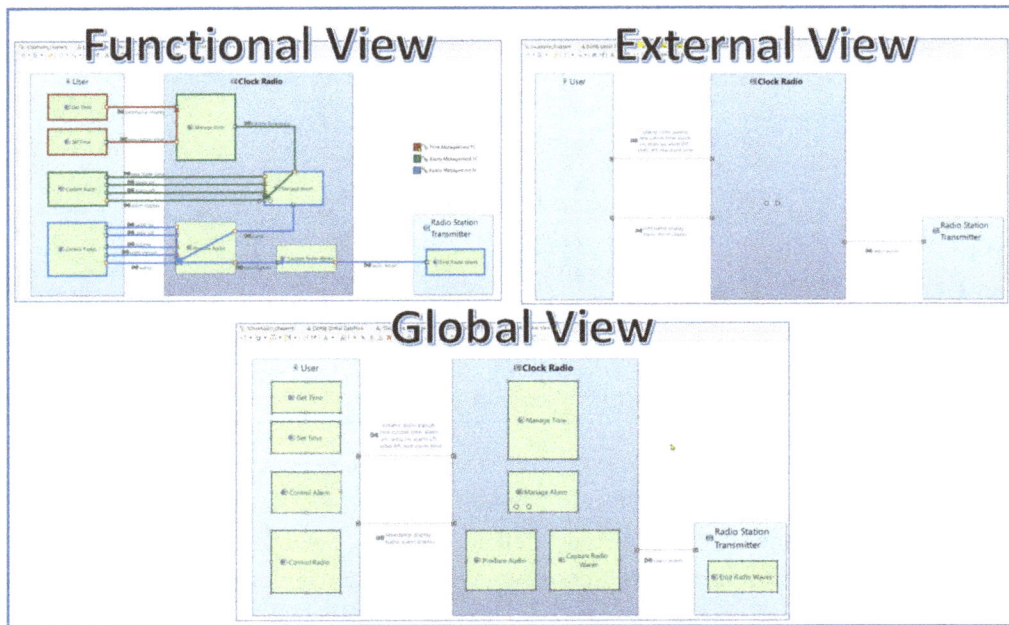

Figure 6-58 – Summary of the three views

With the "Clone Diagram" command, and all the diagram filters supported by *Capella*, it is easy to create as many filtered diagrams as you need for your different stakeholders.

Further Study

Here is a list of further topics for more in-depth study:

- **System Need Analysis** – In this video, you will find a commented example of "System Need Analysis" model, from [Voirin], the book of J.L Voirin.

  - https://url4ap.net/CH-6-SAB-Needs

- **Example of SAB Diagram** – In the second part of this short video, you will find a real example of a SAB diagram as well as a comparison with the OAB of the same model.

  - https://url4ap.net/CH-6-SAB-System

Chapter 7 – Logical Architecture Blank Diagrams

In this chapter, we will continue looking at the architecture blank diagrams, this time at the logical architecture level. Architecture blank diagrams are useful at all *Arcadia* levels. Their main goal at logical architecture level is to refine what the system should do, in terms of functions allocated to it, by defining local components within the system and allocating functions to those components. The architecture blank diagram for the logical architecture level is called the **Logical Architecture Blank (LAB)**.

The logical architecture blank (LAB) is very similar to the system architecture blank diagram (SAB). The key difference between the LAB and the SAB is that the SAB shows the system as a "black box" – we can allocate functions to the system, but we can't see the internal structure of the system.

In the LAB, we move to a "white box" view of the system. We can define components within the system and allocate logical architecture level functions directly to those components.

It is also worth noting that at the logical architecture level we usually have not made final choices about specific vendor components. In a sense, the logical architecture level is usually the **trade study** level at which we are still evaluating multiple options for the final implementation.

Logical Architecture Perspective

> The *logical architecture perspective formalizes the first design choices of the architecture of the solution: first, through an internal functional analysis describing the behavior chosen for the system, then through the identification of the principal components implementing these solution functions, integrating therein the non-functional constraints that have been chosen to address at this level.*
>
> The *functional analysis performed at this stage should not be regarded as a simple refinement of the system analysis, but as the result of the system design in terms of its behavior in response to this need.*

[Voirin] – Chapter 7

As in the preceding chapter, these diagrams provide a diverse set of mechanisms for managing complexity: display of high-level functions, or high-level components, instead of leaf elements, computed synthetic links between components, and so on. Functional chains can still be represented as highlighted paths.

The example used in this chapter is again a simple Clock Radio for domestic usage, exploiting and enhancing the diagrams started in the preceding chapter. For our simple case study, we will just create five logical components, enabling us to separate the direct user interface and the clock, alarm and radio management. Creating logical components inside the system box will also lead us to break down some of the functions that were described globally at the system level. Remember the strong *Arcadia* rule: a function can be allocated only once!

Transition of System Functions

The first step is to ask *Capella* to copy the system-level functions and actors to the logical architecture level. This feature is called "transition" in Capella.

Note that we usually do **not** expect the logical architecture level to end up being a simple copy of the system analysis level! Functions will need further elaboration at the logical level as we start thinking about how to break the system into components. Nevertheless, it is convenient to start with a clone of the functions at the system level and modify them as needed.

We will start by transitioning all system functions. *Capella* systematically creates one logical architecture level function for each system analysis level function. It also clones the function ports, categories, and functional exchanges. *Capella* also adds a "realization" relationship between each new logical architecture level model element and the corresponding element at the system analysis level.

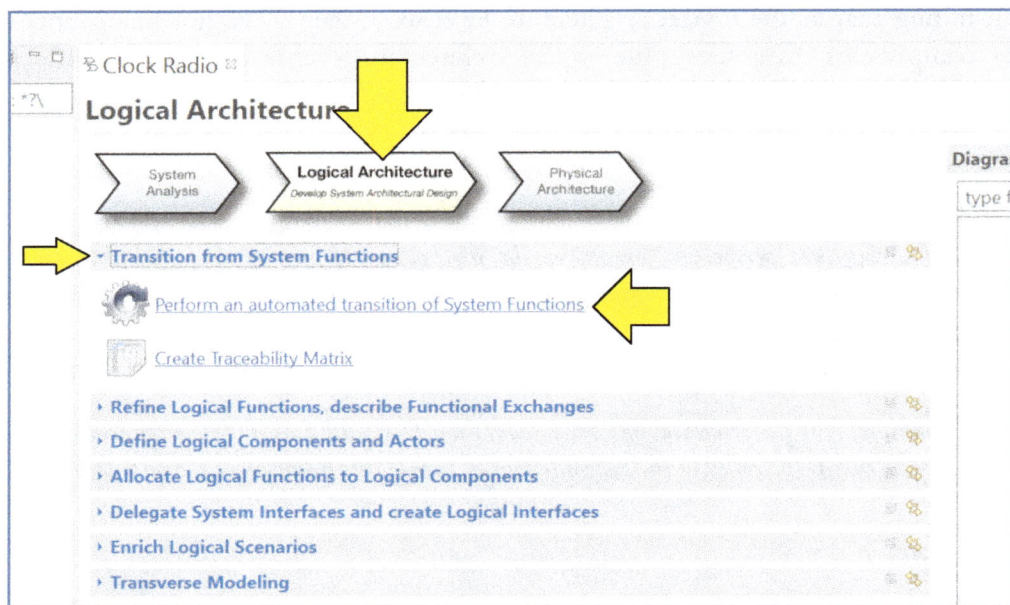

Figure 7-1 – Automated transition of system functions

To perform this transition, simply go back to the activity explorer, and open the "Transition from System Functions" group. Then, select "Perform an automated transition of System Functions".

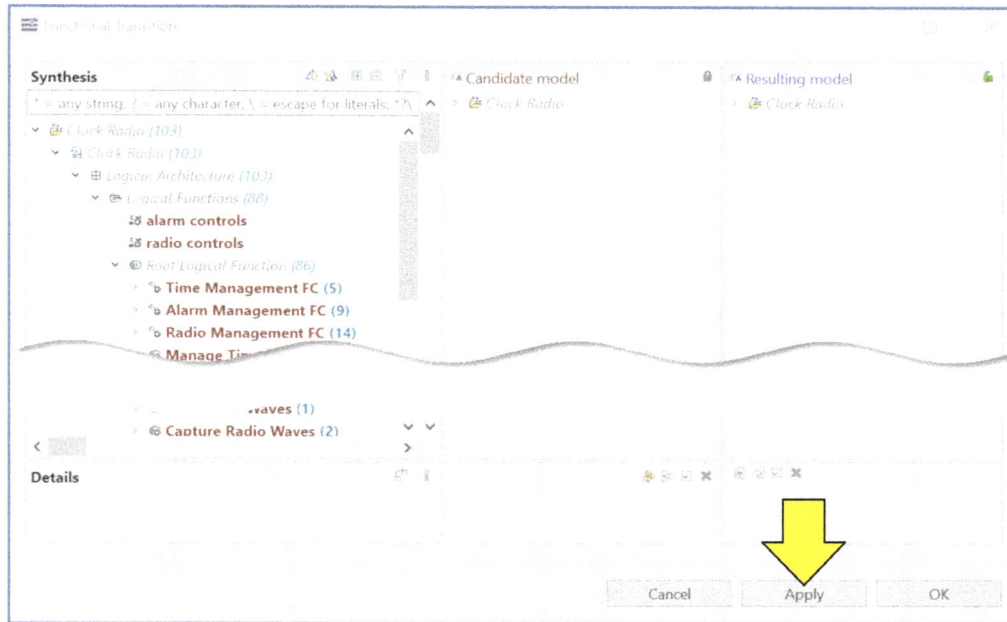

Figure 7-2 – Functional transition window

Capella opens a big and complex window called "Functional Transition". *Capella* automatically identifies the candidate functions to transition. Of course, you can select a subset of functions to transition. For our purposes, however, we will want to transition all of the available functions. Click on "Apply". [20]

---

[20] Be careful to click "Apply" rather than "OK". If you click on "OK", nothing will happen. (It is not clear what the intended function of the "OK" button is; it might be there to separate novice modelers from expert modelers).

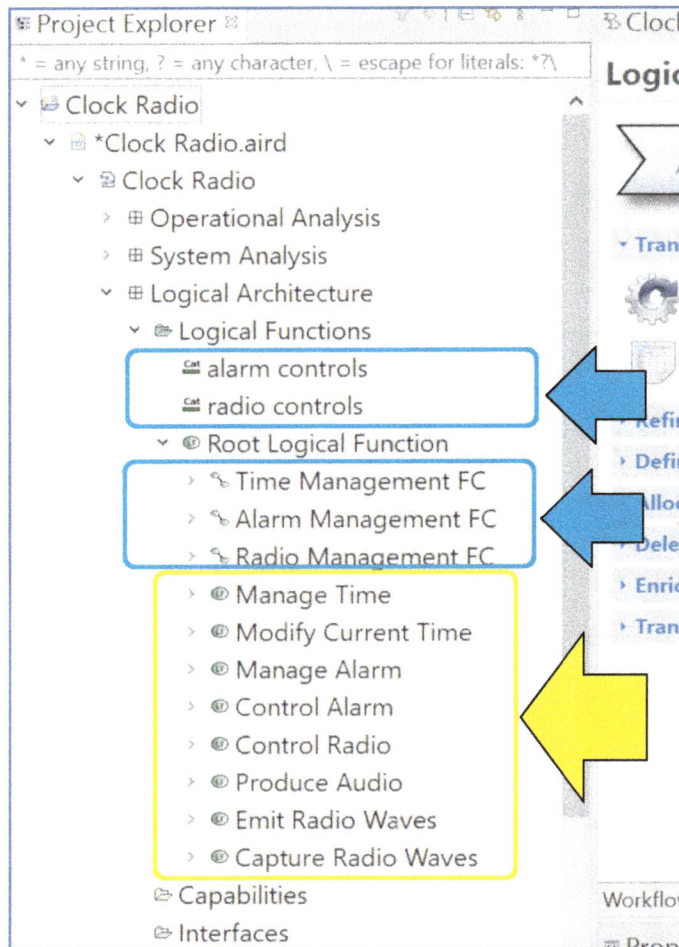

Figure 7-3 – Result of the functional transition in the project explorer

As soon as you click "Apply", the transition window will disappear. Expanding the "Logical Architecture" section in the project explorer, under "Logical Functions" we can see the cloned functions. Notice that in addition to the functions, *Capella* has also cloned our previous categories and functional chains.

Transition of System Actors

The second transition is the transition of actors.

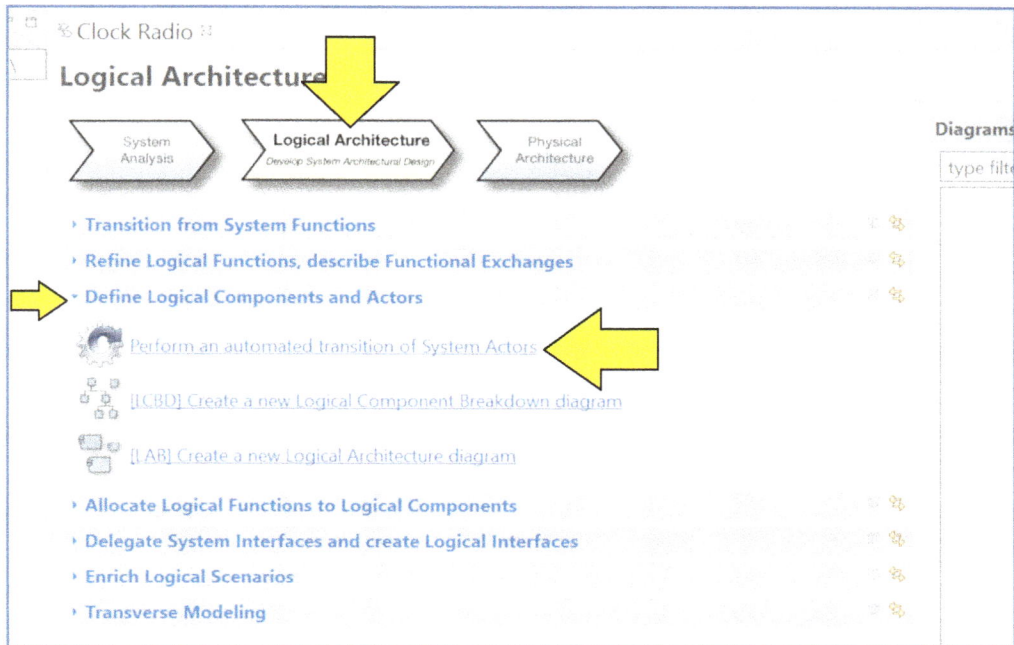

Figure 7-4 – Automated transition of system actors

To perform this transition, simply go back to the activity explorer, and open the "Define Logical Components and Actors" group. Then, select "Perform an automated transition of System Actors". Here also, *Capella* systematically creates one logical actor for each system actor. It also clones the component ports and component exchanges. Finally, *Capella* adds a "realization" link between each logical architecture level model element and the corresponding system analysis level element. Notice that *Capella* defaults to keeping the same function allocations inside the logical actors.

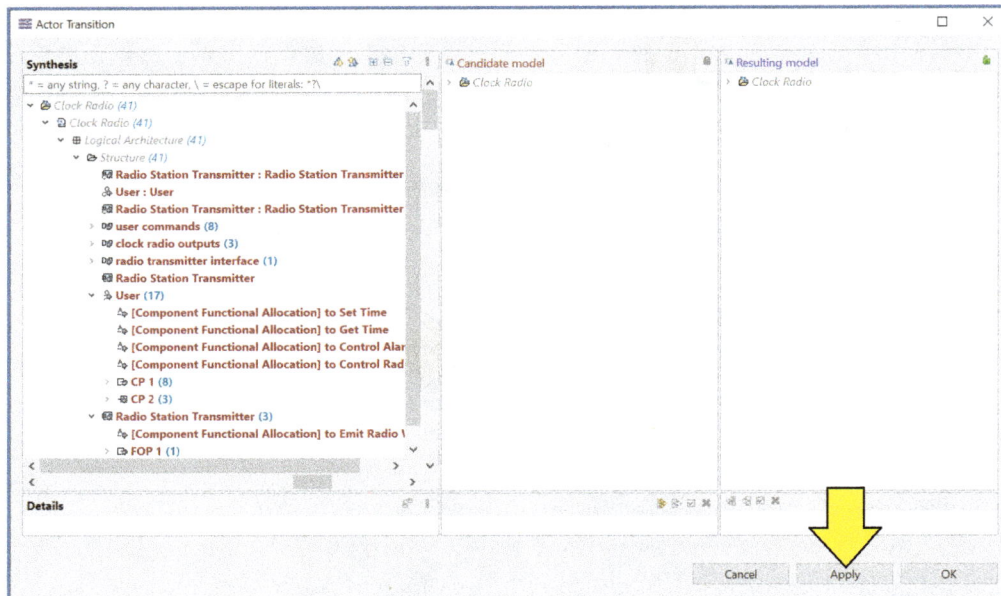

Figure 7-5 – Actor transition window

Again, *Capella* opens another big and complex window called "Actor Transition". *Capella* automatically identifies the candidate system actors and relationships to transition. Of course, you can select a subset of the offered elements. For our purposes, however, we will want to transition all of the available elements. Click on "Apply".

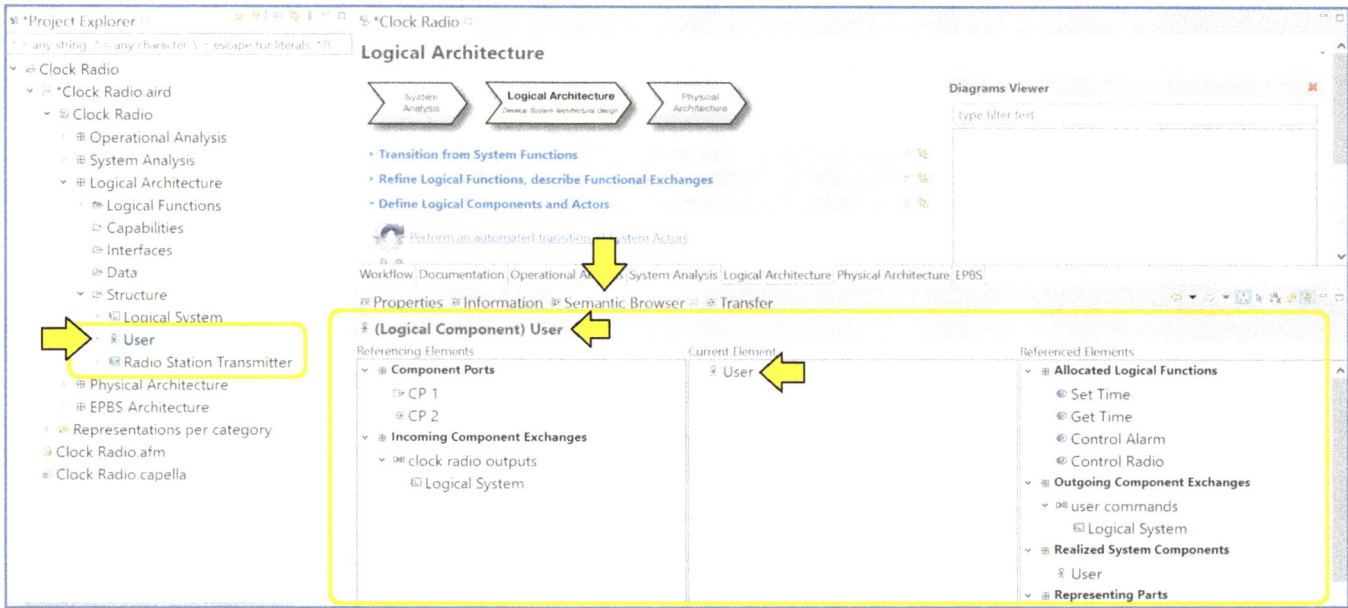

Figure 7-6 – Two logical actors are now visible

The two logical actors are now visible in the project explorer. You can see their allocated functions in the semantic browser.

Creation of the LAB Diagram and the Logical Components

We can now proceed with the creation of the logical architecture blank (LAB) diagram.

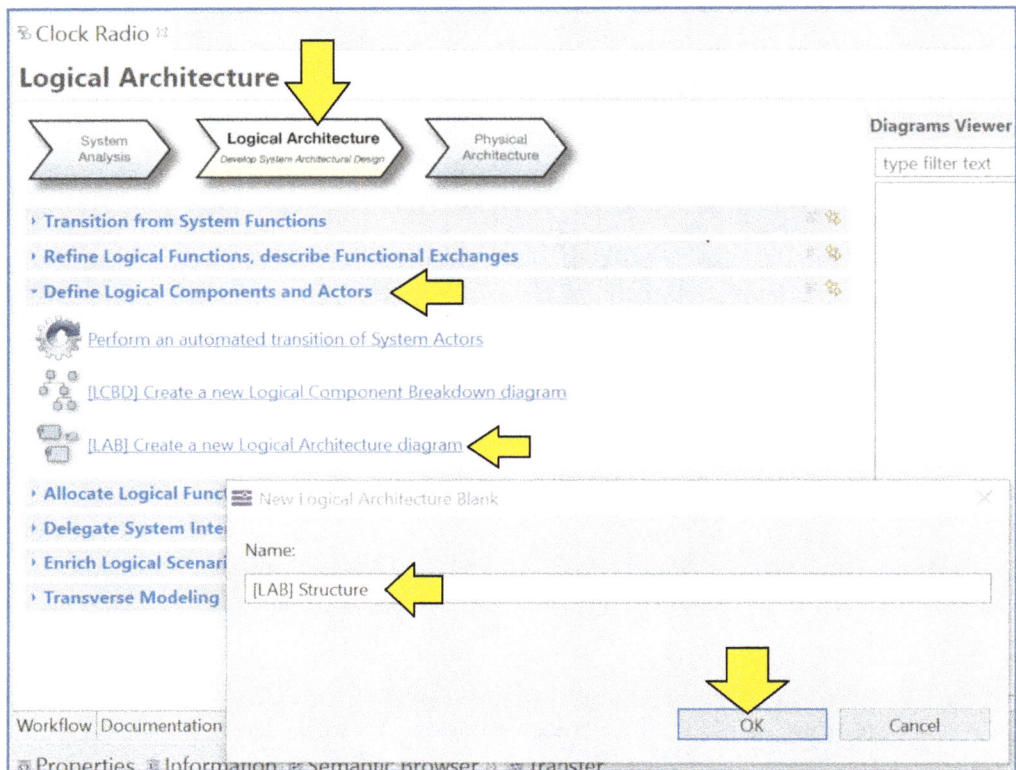

Figure 7-7 – Create new logical architecture blank diagram (LAB)

In the same "Define Logical Components and Actors" group of the activity explorer, just click on "[LAB] Create a new Logical Architecture diagram". Accept the default diagram name of "[LAB] Structure" for the moment and click "OK".

Capella creates an empty (blank) diagram, but we will use a smart accelerator to initialize it cloning one of the similar diagrams from the system analysis level (SAB).

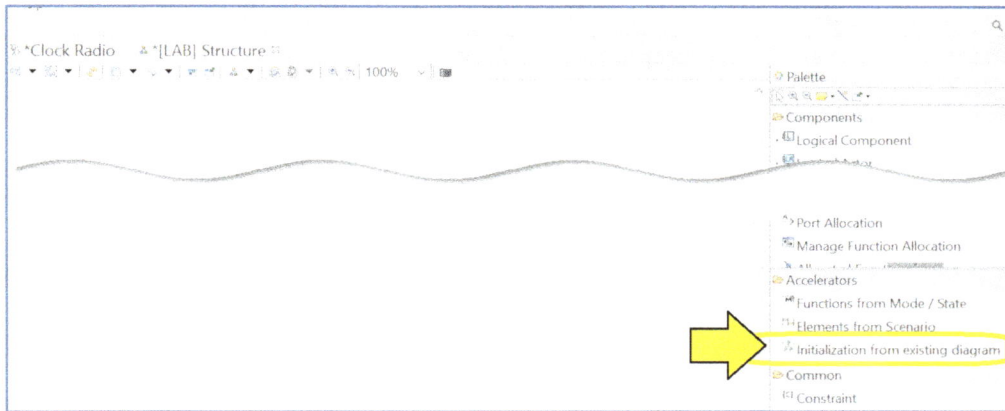

Figure 7-8 – Accelerators panel

Go to the palette and scroll as needed to the "Accelerators" group. Click on "Initialization from existing diagram" and then in the background of the diagram.

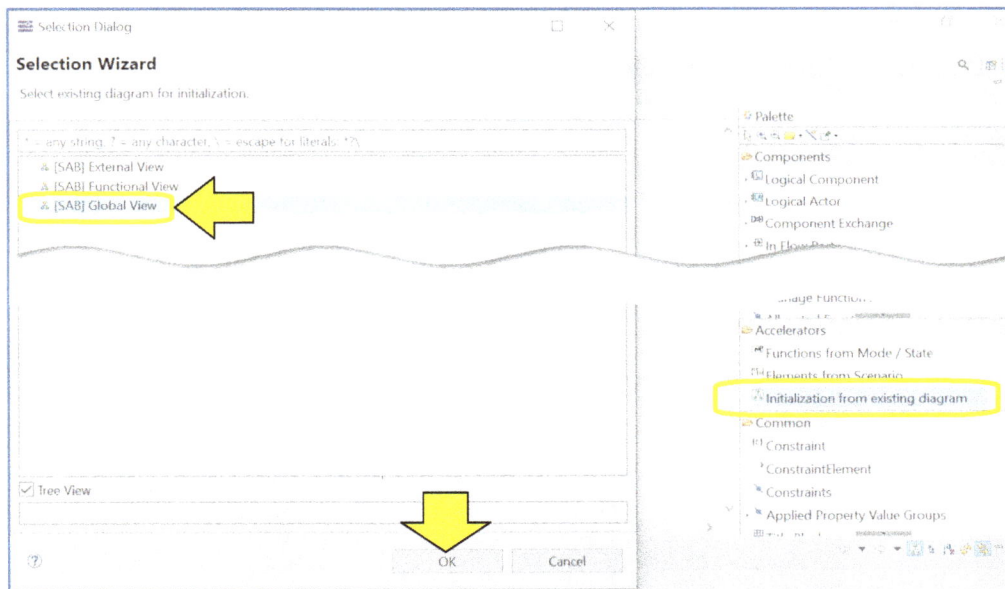

Figure 7-9 – Select an existing diagram

The selection dialog will open. Select one of the existing SAB diagrams, for instance the "Global View". Click "OK".

Capella will build the diagram using logical elements laid out in the same format as the system-level elements in source SAB.

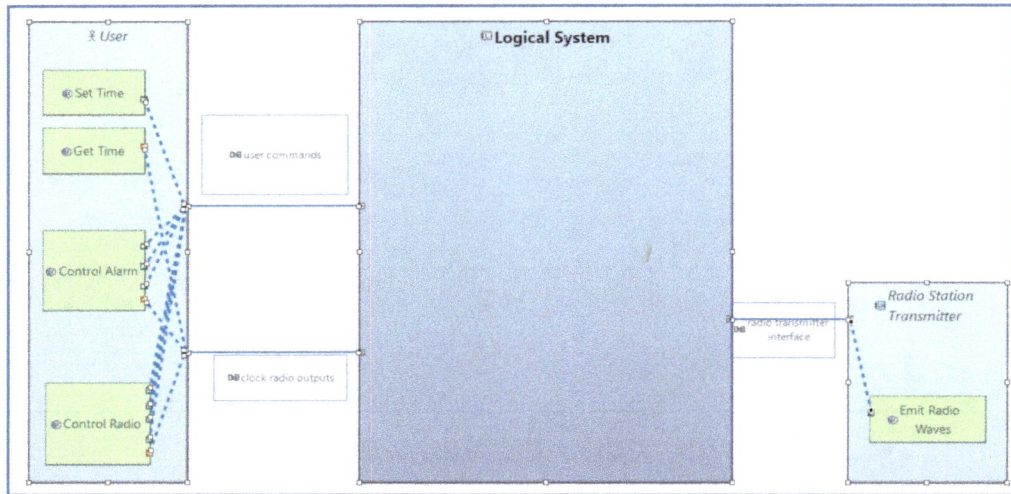

Figure 7-10 – Automatically constructed diagram

The result appears immediately, showing the logical system and the two logical actors, as well as the allocated functions, function ports and even component exchanges and component ports (with dashed lines indicating allocating function ports).

Notice that the functions which were allocated to the system as a black box at the system level are not allocated to the logical system at the logical architecture level. This tool design is deliberate. At the system analysis level, the system is a black box – no internal structure. Anything allocated to the system at the system analysis level is simply allocated to the entire system. At the logical architecture level, our intent is to make the system a white box. We will be defining the logical components inside the system, and we will allocate the logical functions to the more detailed components.

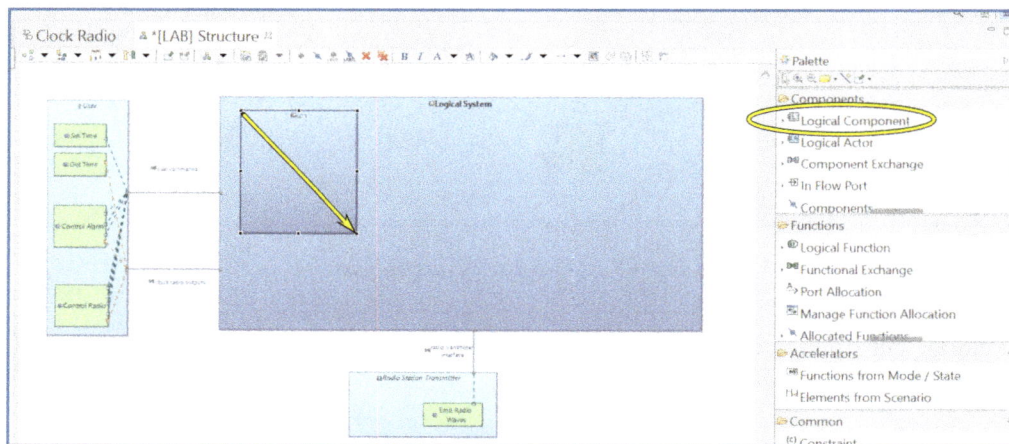

Figure 7-11 – Create first logical component

We are now ready to create logical components inside the logical system. However, first we need to do some diagram housekeeping to make space.

- Zoom out a little bit.

- Move the "Radio Station Transmitter" around to a position below the "Logical System" component to make space. Resize it and move the ports around as needed.

- Make the "Logical System" blue box quite a bit wider.

Having made space for some new logical components inside the "Logical System", select the "Logical Component" tool from "Components" group of the palette and drag to draw a new component inside the "Logical System".

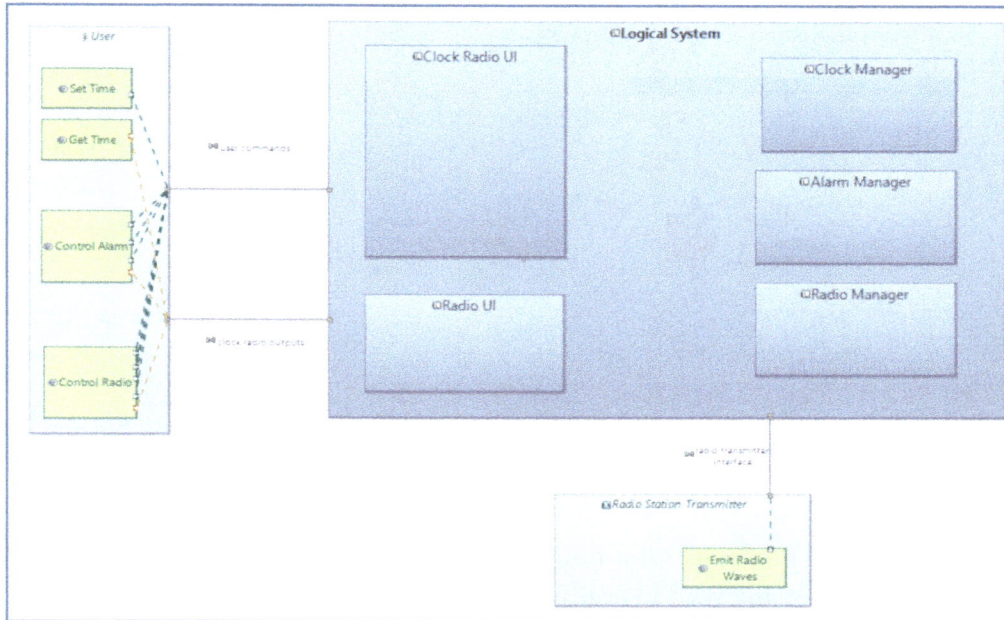

Figure 7-12 – Five components created

- Rename the first logical component to "Clock Radio UI".

- Create four more logical components:

  - Radio UI

  - Clock Manager

  - Alarm Manager

  - Radio Manager

- Increase the font size to 12 for all five component names.

- Adjust the sizes of the logical components as needed.

Allocation of the Logical Functions

Next, we need to allocate the relevant function(s) to each logical component, by using the "Manage Function Allocation" tool in the palette.

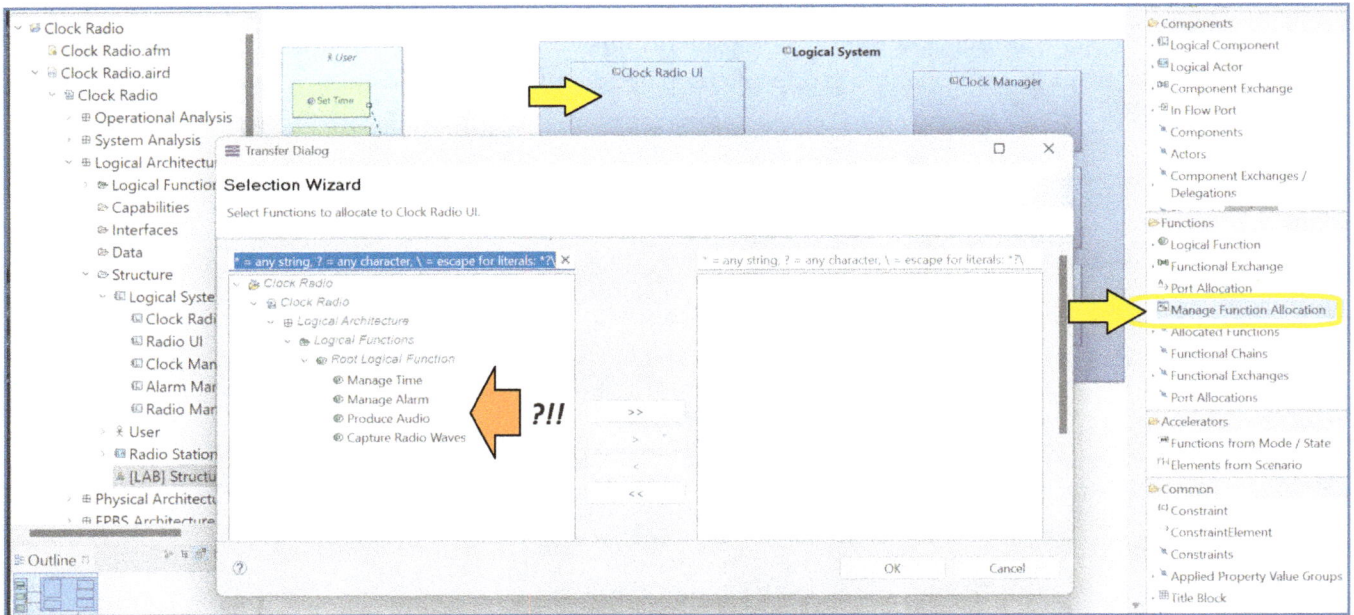

Figure 7-13 – Manage function allocation

We could select the "Manage Function Allocation" tool in the palette and click on the "Clock Radio UI" component to allocate some of the root logical functions to it. However, we would quickly see that these functions are at too high a level of abstraction. They need to be refined into some smaller pieces so that the pieces can be allocated to the more granular logical components.

For instance, the "Manage Time" function needs to be decomposed: a part of it will be allocated to the "Clock Radio UI", another part to the "Clock Manager". It is quite frequent to refine a system-level function into several subfunctions at logical architecture level to precise the detailed behavior responsibilities of each logical component.

To break down a function into subfunctions and work on the dataflow, we will need to create a Logical Data Flow Blank diagram (LDFB) for the function. In Chapter 5, we learned how to create a System Data Flow Blank (SDFB) diagram at the System Analysis level. The LDFB is the same type of diagram created at the logical architecture level.

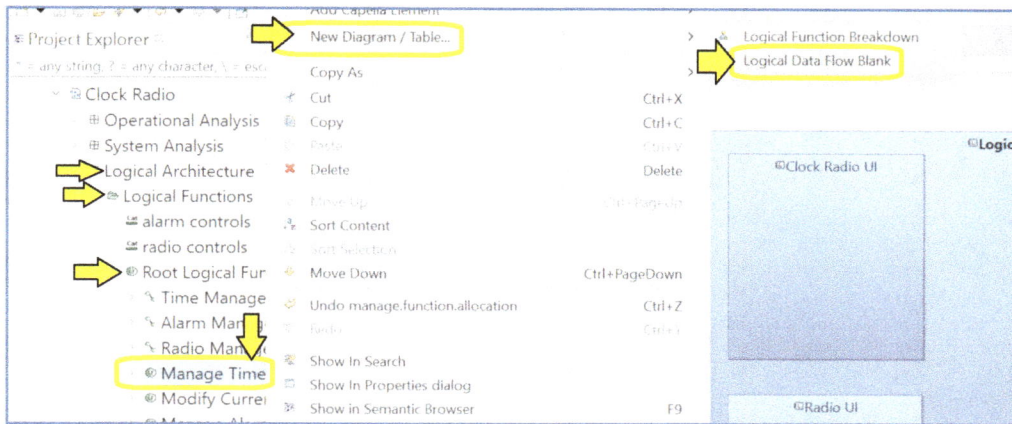

Figure 7-14 – Create logical data flow blank

1) In the project explorer, expand "Logical Architecture", then "Logical Functions", and finally "Root Logical Functions".

2) Right-click on "Manage Time".

3) Select "New Diagram/Table..." and then "Logical Data Flow Blank".

4) Accept the default name of "[LDFB] Manage Time".

Next, we need to perform a few steps to identify the contextual element for the diagram (which will be the "Manage Time" function) and add that element and its related elements to the diagram.

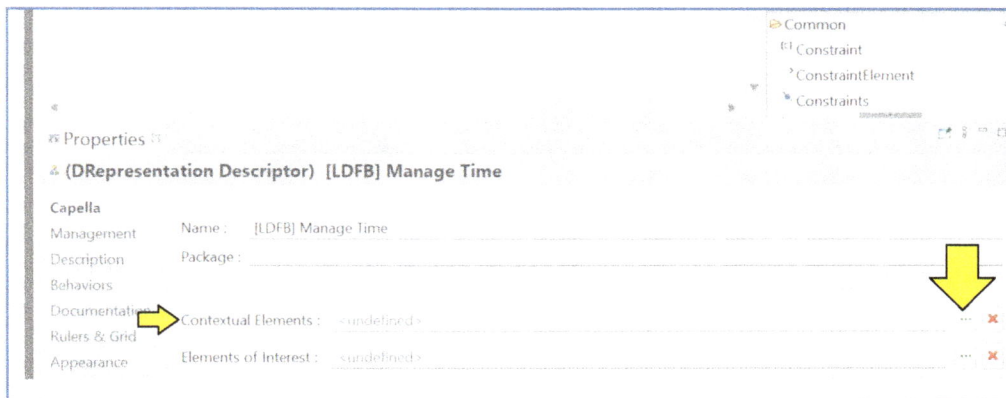

Figure 7-15 – Click on three dots to open transfer dialog

In the properties panel for the new diagram, click on the three dots to the right of the "Contextual Elements:" field to open the transfer dialog.

What if the Properties Panel is Missing?

What if you get to this point and find that the properties panel is missing? You can force it to display again by selecting "Windows" from the top menu, then selecting "Show View" and then "Properties".

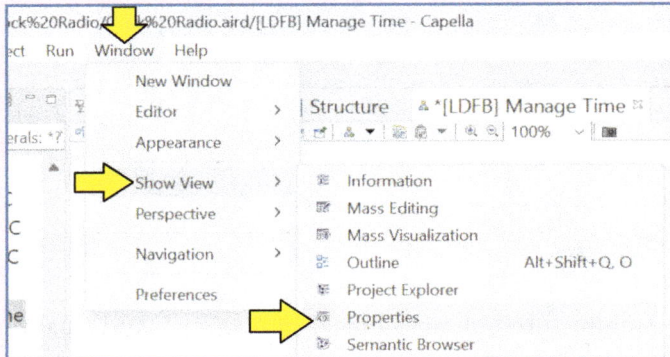

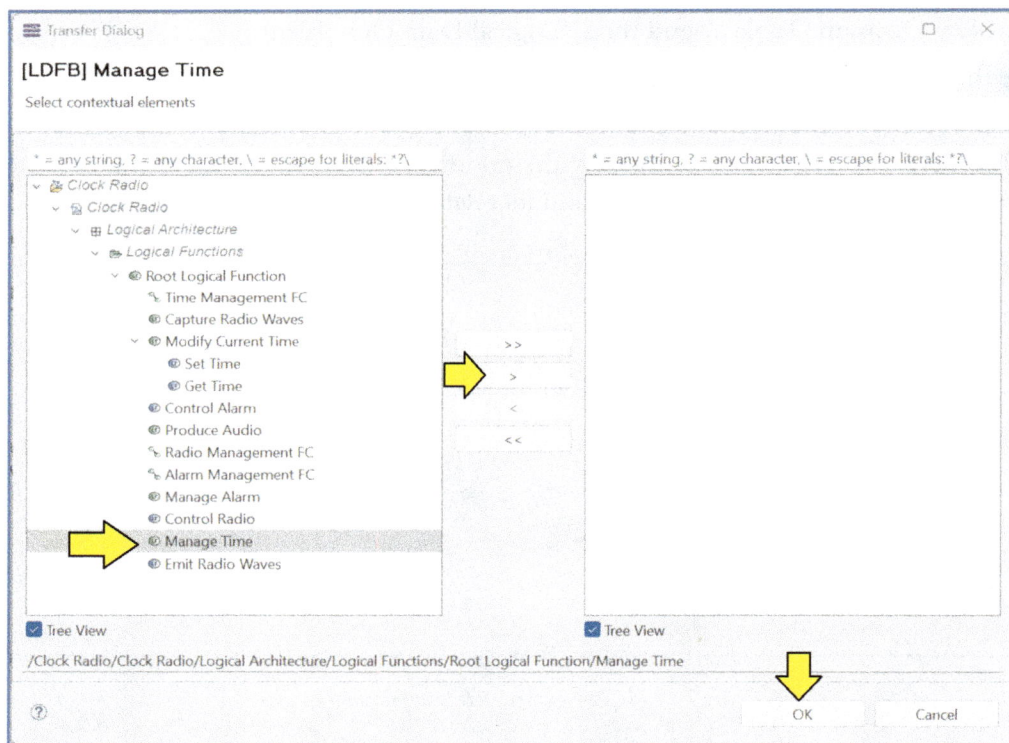

Figure 7-16 – Add the function to the diagram

In the transfer dialog, select the "Manage Time" function, click on the single arrow (>) to move it into the diagram, and click "OK".

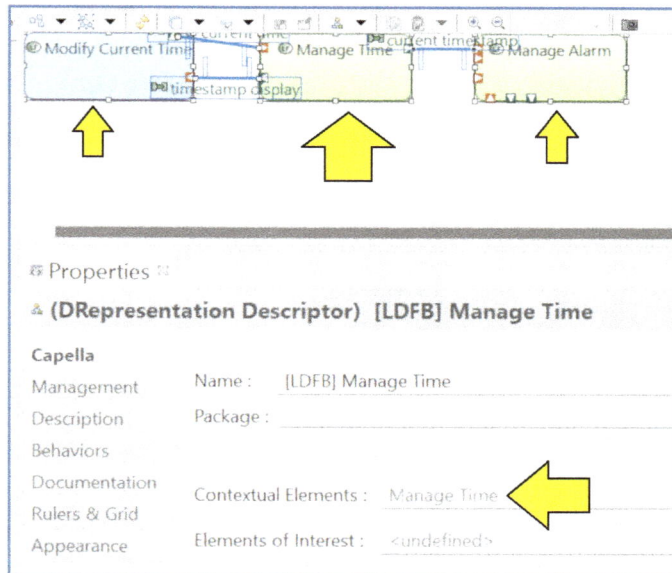

Figure 7-17 – The functions have been added to the diagram

Capella has added "Manage Time" to the diagram as well as two other elements that are connected by functional exchanges.

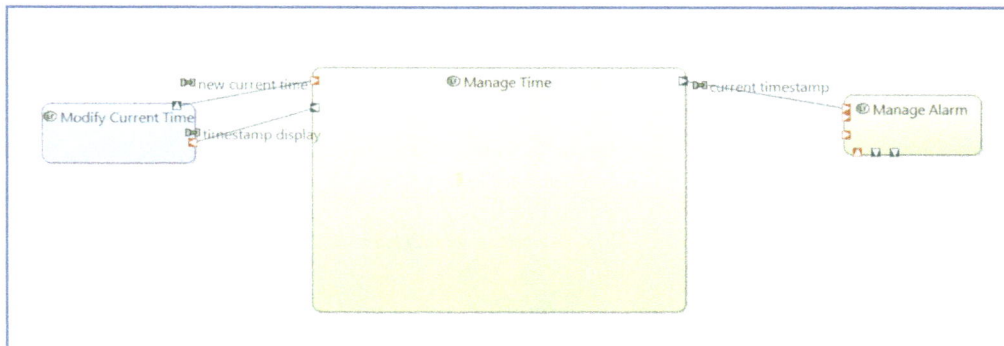

Figure 7-18 – Resize and rearrange

We will want to rearrange the diagram to make it more readable. Since we are going to decompose the "Manage Time" function, we will want to make that larger in order to make space to draw child functions inside of it.

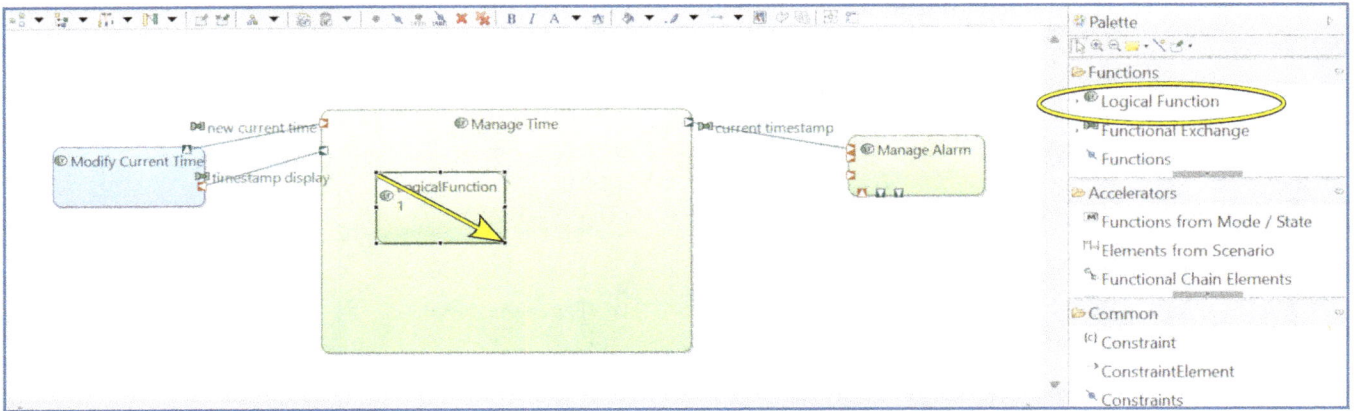

Figure 7-19 – Create a new logical function

In the palette, select the "Logical Function" tool and drag to create a new subfunction inside "Manage Time". Name this subfunction "Display Time".

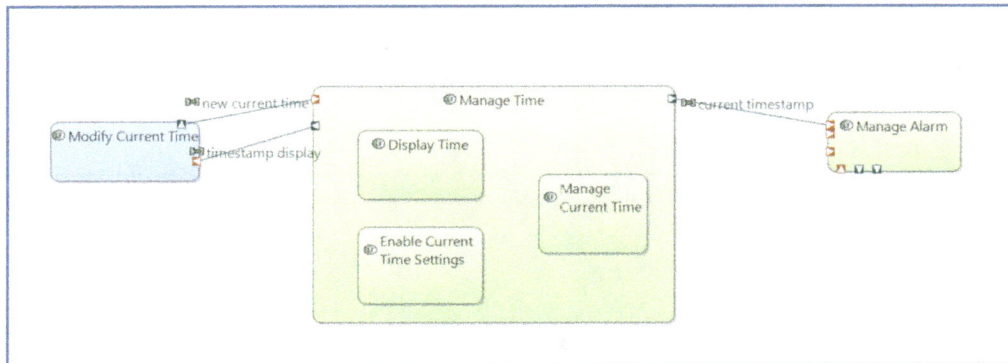

Figure 7-20 – Create two more subfunctions

Add two additional subfunctions "Enable Current Time Settings" and "Manage Current Time".

We have now broken down "Manage Time" into three subfunctions:

- "Display Time" and "Enable Current Time Settings" will be allocated to "Clock Radio UI".
- "Manage Current Time" will be allocated to "Clock Manager".

Remember that *only leaf functions can own ports*. We need to move the function ports from "Manage Time" to its subfunctions.

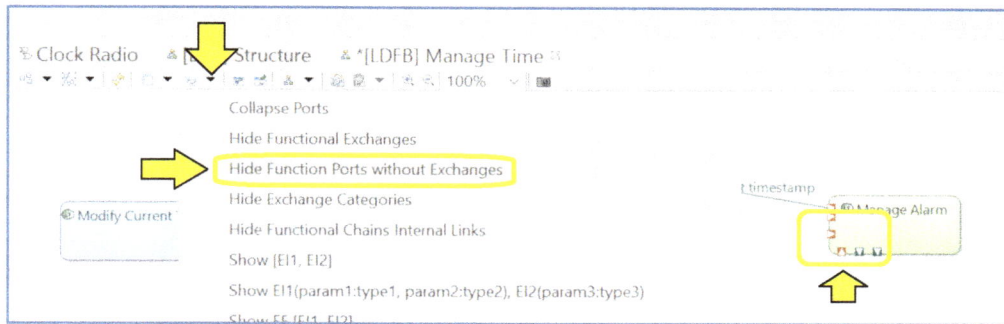

Figure 7-21 – Filter the unused ports

Filter the unused ports on "Manage Alarm" using the filter: "Hide Function Ports without Exchanges" as shown previously in on page 90.

After applying the filter, "Manage Alarm" shows only one port. The other ports (connected to functions not displayed in the diagram) have been hidden by the filter.

We are now ready to move ports from "Manage Time" to its lead subfunctions.

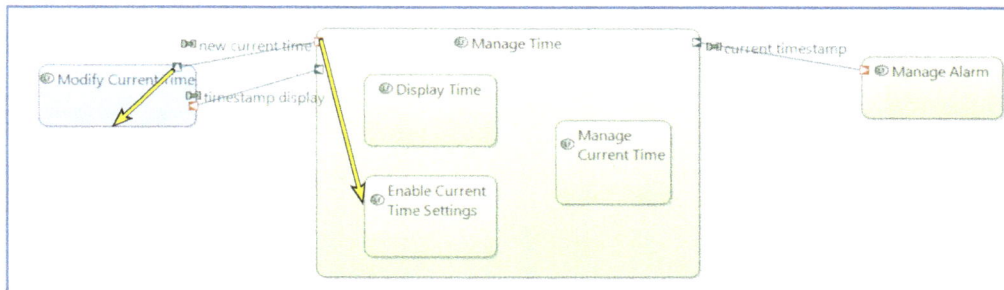

Figure 7-22 – Move ports

The "new current time" functional exchange should be an input of the "Enable Current Time Settings" function. Move the ports as shown. You may also want to drag the connector to make it bend into a right-angle to make the alignment neater.

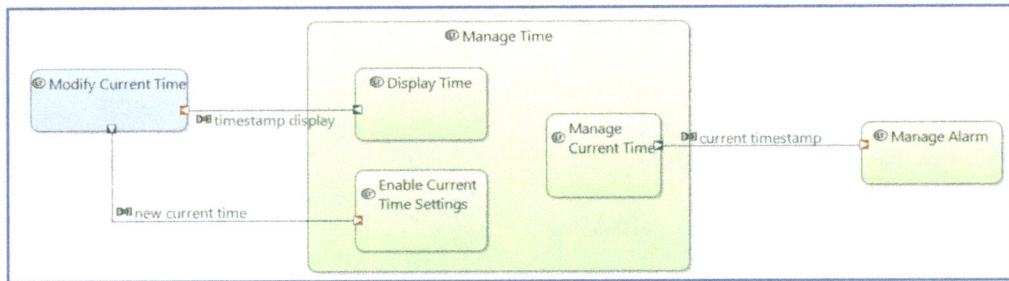

Figure 7-23 – Move the other two ports

Move the ports for the other two functional exchanges to their relevant leaf subfunctions. "timestamp display" is produced by "Display Time", and "current timestamp" by "Manage Current Time".

We need now to connect the subfunctions together. For instance, the "new current time" functional exchange should be relayed to the "Manage Current Time" subfunction by the "Enable Current Time Settings" subfunction.

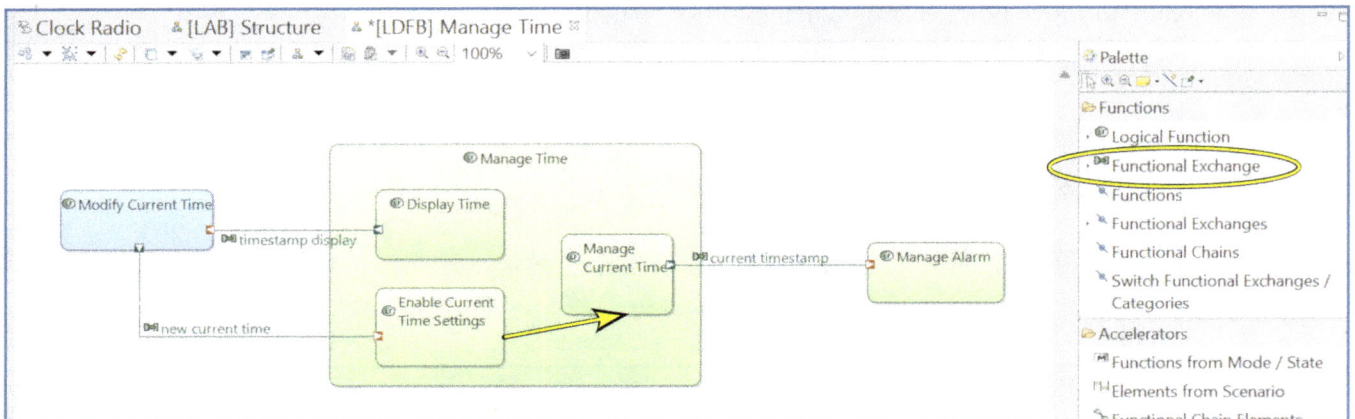

Figure 7-24 – Create a new functional exchange between the subfunctions

Create a new functional exchange with the palette flowing from "Enable Current Time Settings" to "Manage Current Time" and name it "new current time" again. [21]

---

[21] Notice that it is acceptable (best practice in fact) to have more than one functional exchange with the same name.

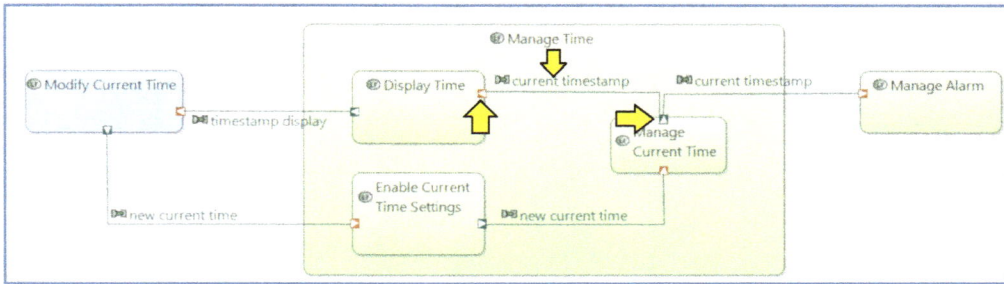

Figure 7-25 – Use existing output port

In the way, "current timestamp" should be transmitted also to "Display Time". As it is the same information as the one already going to "Manage Alarm", it is a good practice to start the new functional exchange from the existing output port, instead of creating a pair of new function ports.

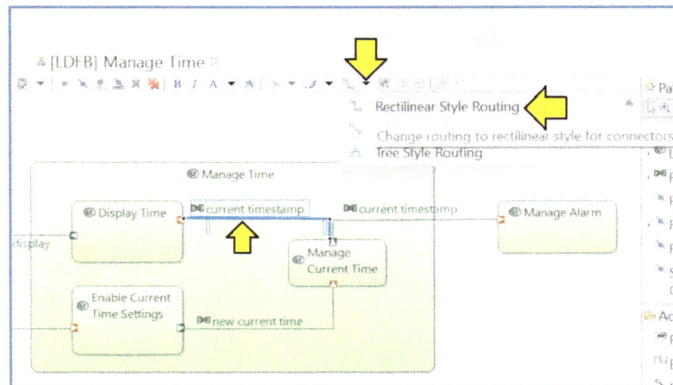

Figure 7-26 – Use rectilinear line style

At this point, you may have already spent several minutes struggling to get *Capella* to line up the two copies of the "current timestamp" functional exchange. Fortunately, there is an easier way to line these up. Make sure the new functional exchange is still selected and select the "Rectilinear Line Style" command from the diagram top palette to make it nicer.

Figure 7-27 – Closer look at the icon

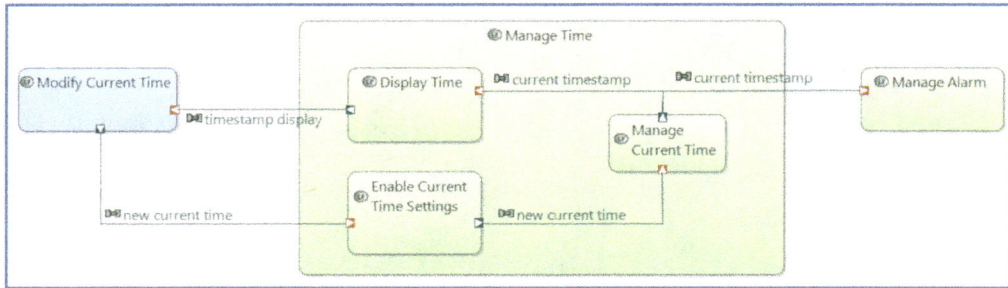

Figure 7-28 – Finished LDFB diagram

We are now ready to go back to the LAB in order to allocate the three leaf functions to the relevant logical component. Remember that the functional exchanges will automatically be displayed as soon as the source and target functions are inserted in a blank diagram.

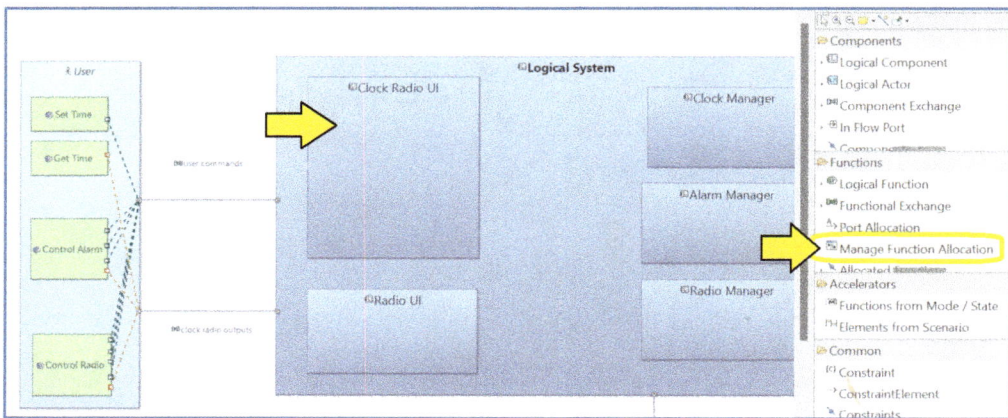

Figure 7-29 – Manage functional allocation of the clock radio user interface

1) Return to the "[LAB] Structure" diagram.

2) In the palette select the "Manage Function Allocation" tool.

3) Click inside the "Clock Radio UI" component.

4) The transfer dialog will appear.

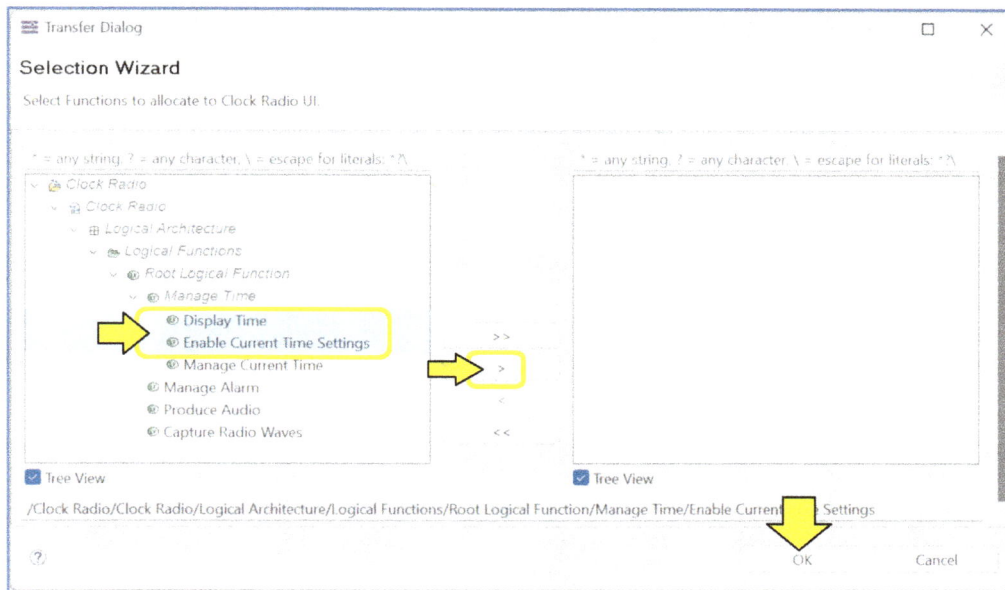

Figure 7-30 – Transfer dialog

Select the "Display Time" and "Enable Current Time Settings" functions, click on the single arrow (>) to allocate them to the selected component, and click "OK". Notice that the decomposed function "Manage Time" can no longer be selected *(only leaf functions can be allocated)*.

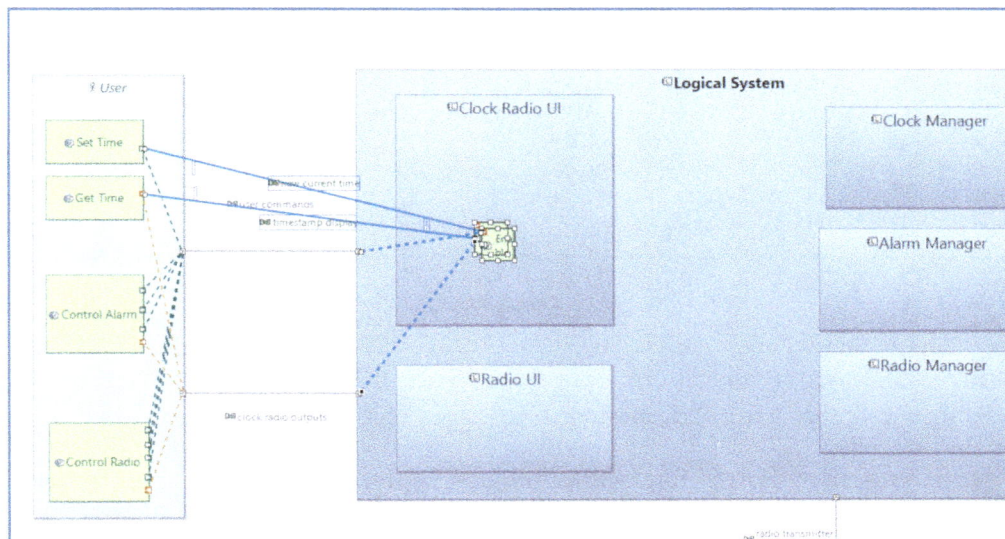

Figure 7-31 – Allocated functions shown inside the logical components

Capella displays the allocated functions inside the logical component as square boxes. Of course, the layout is not very elegant.

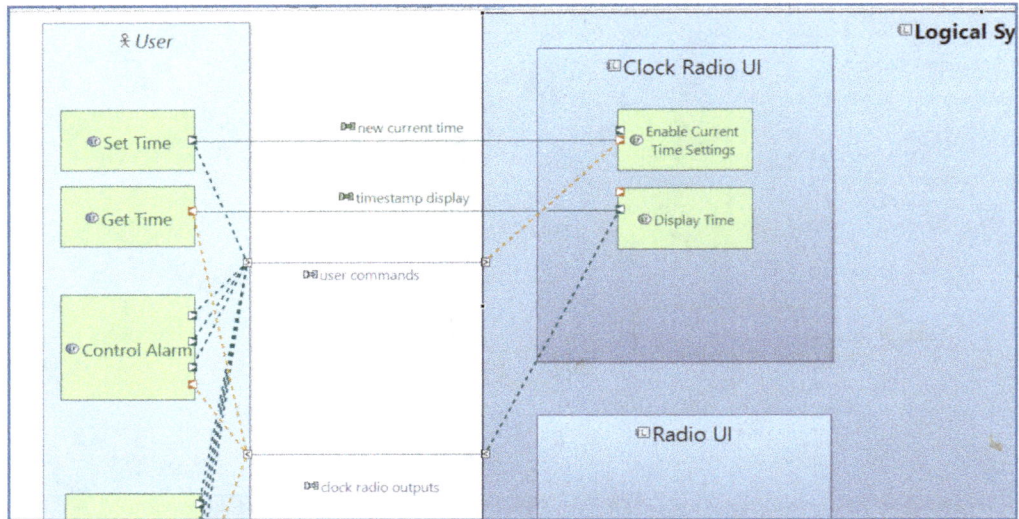

Figure 7-32 – Improved layout

Resize the elements and drag ports around as needed to obtain a nice diagram.

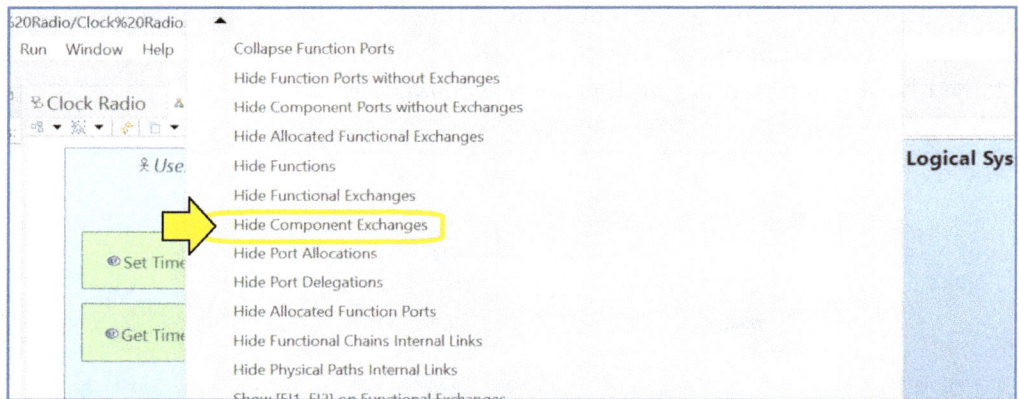

Figure 7-33 – Hide component exchanges

Before we continue the function allocations, let's the diagram a little bit. We don't really need to see the component exchanges. Use the "Hide Component Exchanges" filter to hide them.

Use the "Collapse Component Ports" filter (the first one in the list) to hide the ports used by the (now hidden) component exchanges. Remember only one filter can be applied at a time. This restriction is a deliberate design by *Capella* developers, as different filters could be redundant or even contradictory.

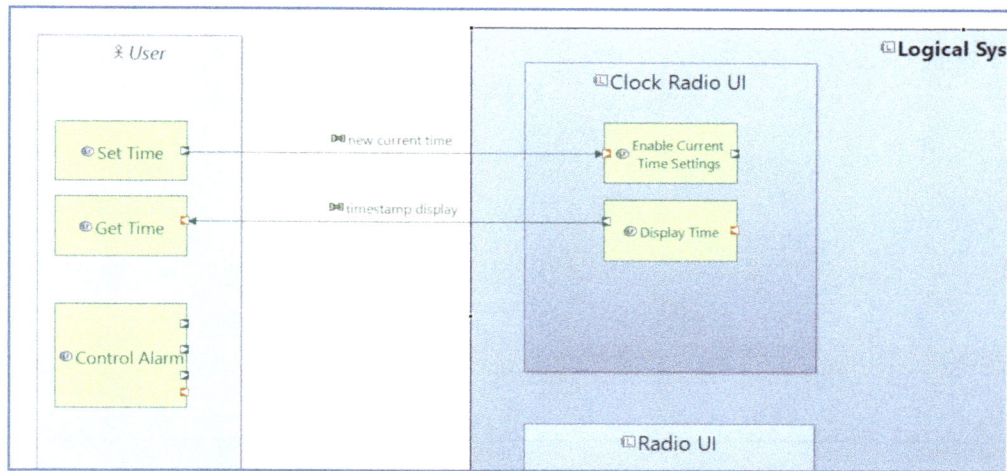

Figure 7-34 – Simplified logical architecture blank (LAB)

This now simplified LAB focuses on functions and functional exchanges.

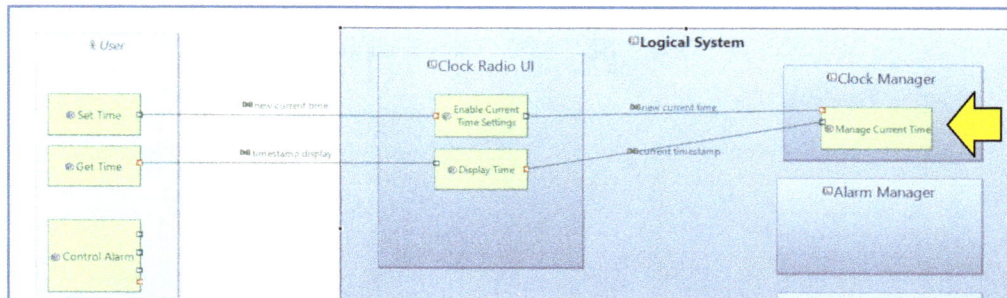

Figure 7-35 – Allocate the function manage current time

Allocate "Manage Current Time" to "Clock Manager".

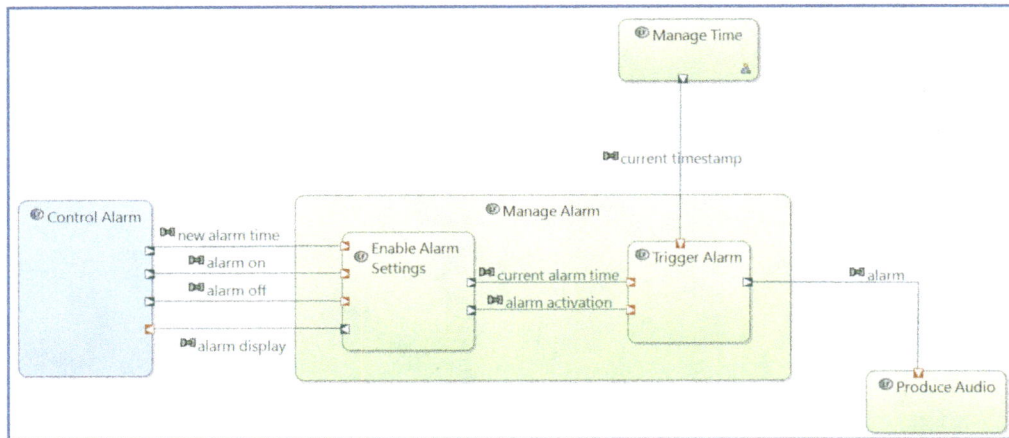

Figure 7-36 – Create another logical data flow blank diagram (LDFB)

We will also need to decompose the "Manage Alarm" function. Create an LDFB for "Manage Alarm" and add two subfunctions "Enable Alarm Settings" and "Trigger Alarm".

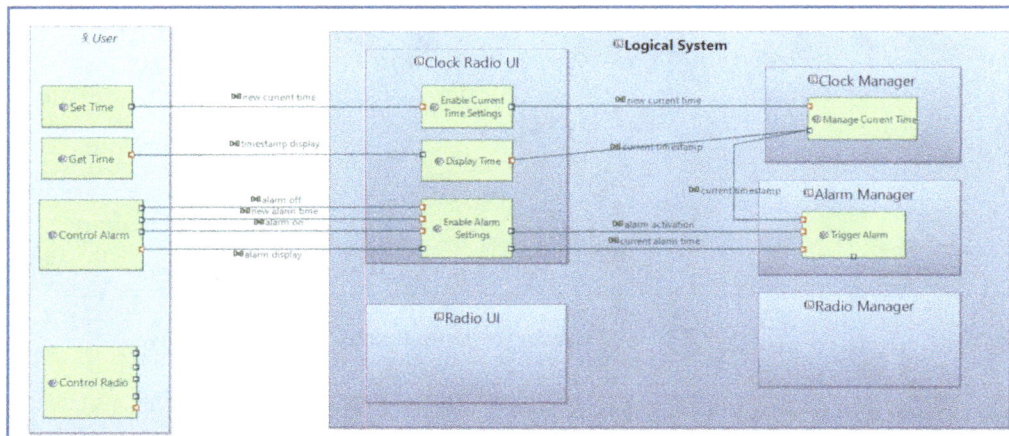

Figure 7-37 – Return to the LAB and allocate the two new functions

Next, return to the LAB in order to allocate the two new functions. Allocate "Enable Alarm Settings" to "Clock radio UI" and "Trigger Alarm" to "Alarm Manager".

At this point, only two functions remain to be allocated: "Produce Audio" and "Capture Radio Waves". "Capture Radio Waves" can be directly allocated to "Radio Manager". "Produce Audio" needs to be decomposed into subfunctions split between the "Radio UI" and "Radio Manager" logical components.

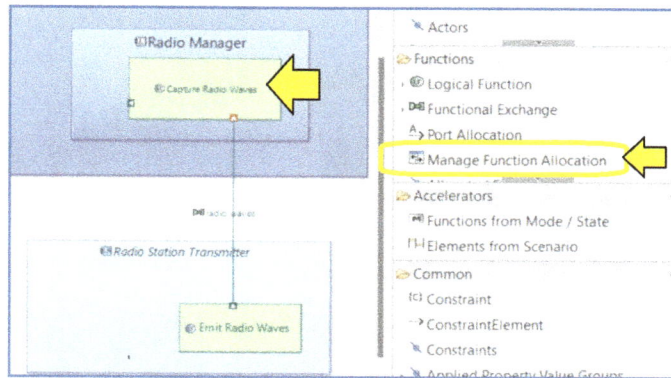

Figure 7-38 – Allocate capture radio waves

Once again, use the "Manage Function Allocation" tool in the palette to allocate "Capture Radio Waves" to the "Radio Manager" component.

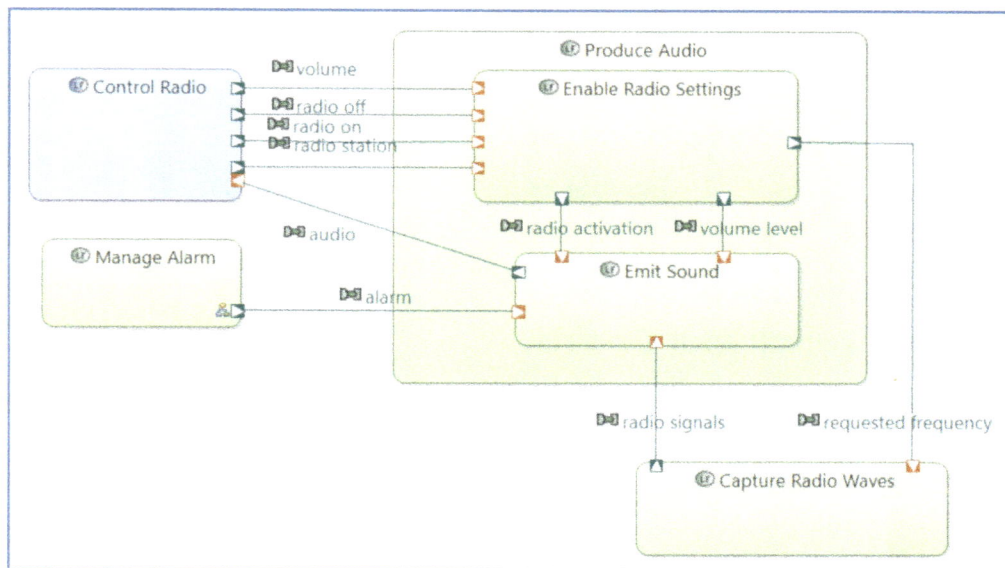

Figure 7-39 – Create a third logical data flow blank diagram (LDFB)

Create a contextual Logical Data Flow Blank diagram (LDFB) for "Produce Audio".

1)   Add subfunctions to "Produce Audio" for "Enable Radio Settings" and "Emit Sound".

2)   Add new functional exchanges for "radio activation", "volume level", and "requested frequency".

3)   Move the function ports around as shown in Figure 7-39.

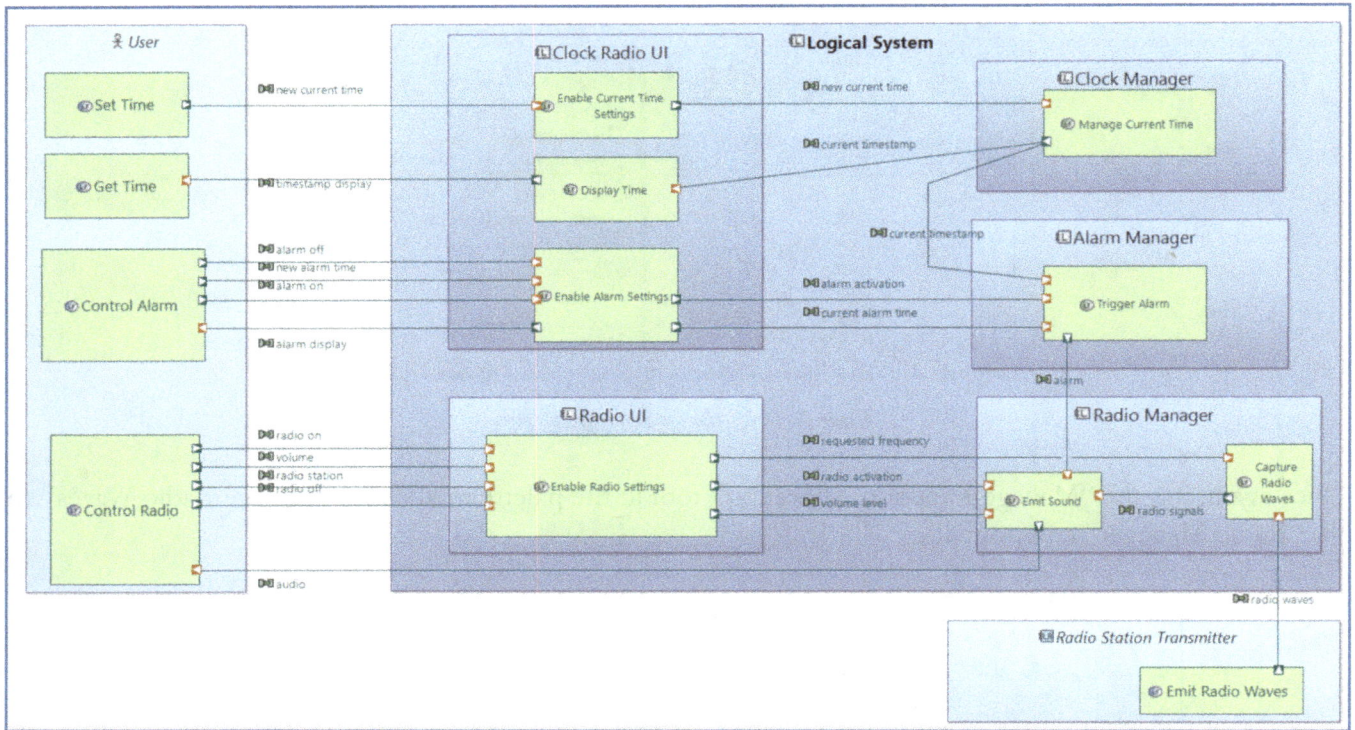

Figure 7-40 – Completed logical architecture blank diagram (LAB)

We can then complete the function allocation in the LAB to obtain the finalized diagram.

Refinement of the Functional Chains

When we transitioned the model elements from the system analysis level to the logical architecture level, the functional chains "came along for the ride". Not surprisingly, however, after all of the rearrangements we have made in the logical function decomposition and path of the logical architecture level functional exchanges (which the logical architecture level functional chains are built on) these new transitioned logical architecture level functional chains are no longer correct. Some fixups will be required.

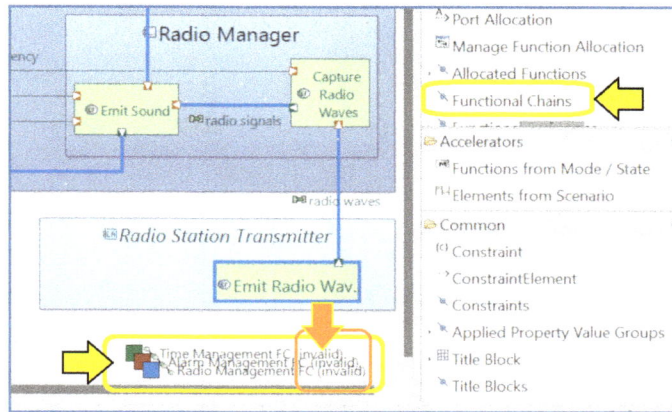

Figure 7-41 – Inserted functional chains show as invalid

If we use the "Functional Chains" tool in the palette to add all of the functional chains to the diagram, we will see that *Capella* unsurprisingly tells us that all three are invalid.

We will need to fix the functional chains, one-by-one. Select each functional chain, right-click on it and select "New – Functional Chain Description", to create a diagram which we can use to add or remove involvements.

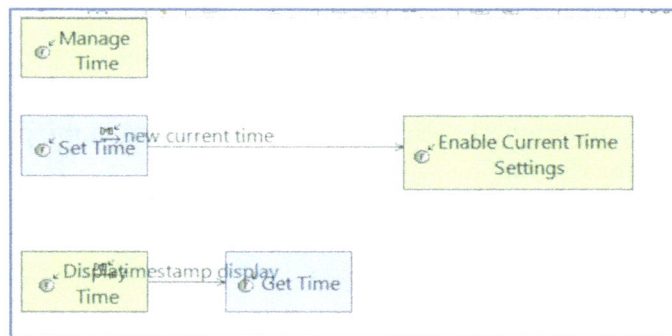

Figure 7-42 – Fix the time management functional chain

The time management functional chain presents initially as three separate pieces. The problem is that the "Manage Time" function at the System Analysis level has been decomposed into pieces at the Logical Architecture level. We need to replace this function with the child function "Manage Current Time" and reconnect the chain.

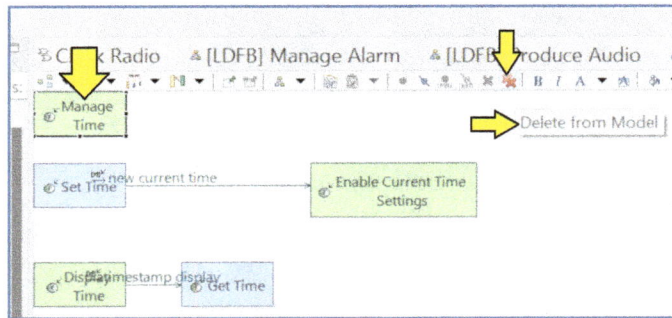

Figure 7-43 – Delete the Manage Time function

Select the "Manage Time" and click on the "Delete from Model" tool in the toolbar – remember, what we are deleting from the model here is the reference to the "Manage Time" function within the functional chain, **not** the "Manage Time" function itself.

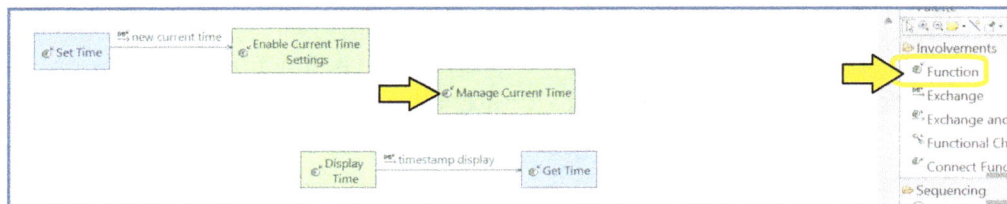

Figure 7-44 – Add the Manage Current Time function

Rearrange the diagram a little to make space. Select the "Function" tool in the palette and add the "Manage Current Time" subfunction to the diagram.

Figure 7-45 – Use the exchange tool

Select the "Exchange" tool in the palette and drag a relationship from the "Enable Current Time Settings" subfunction to the "Manage Current Time" subfunction. A transfer panel will appear. Select the "new current time" functional exchange and add it to the diagram. Add the "current timestamp" functional exchange between "Manage Current Time" and "Display Time".

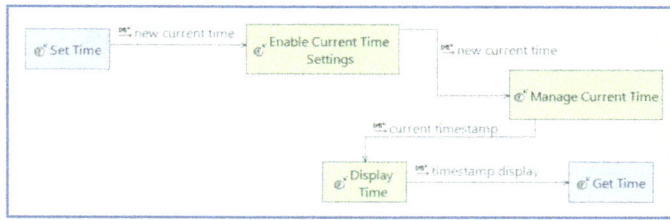

Figure 7-46 – Fixed time management functional chain

We have now fixed the time management functional chain in the LFCD diagram. If you return to "[LAB] Structure" diagram, you can confirm that the warning "invalid" next to the time management functional chain has disappeared and the path of the functional chain is properly highlighted in the diagram.

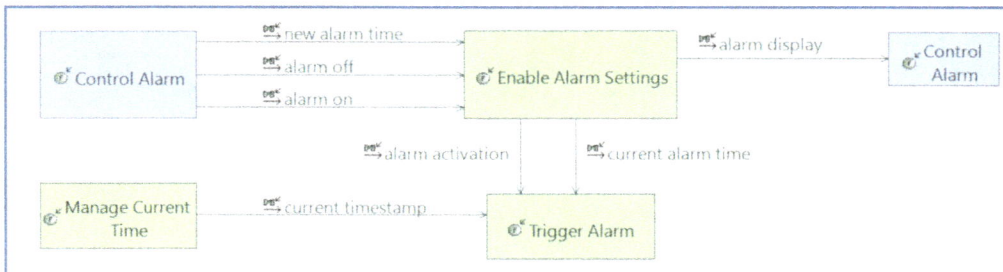

Figure 7-47 – Fix the alarm management functional chain

Next, let's fix the alarm management functional chain. Use the same techniques to adjust the functional chain as shown in Figure 7-47.

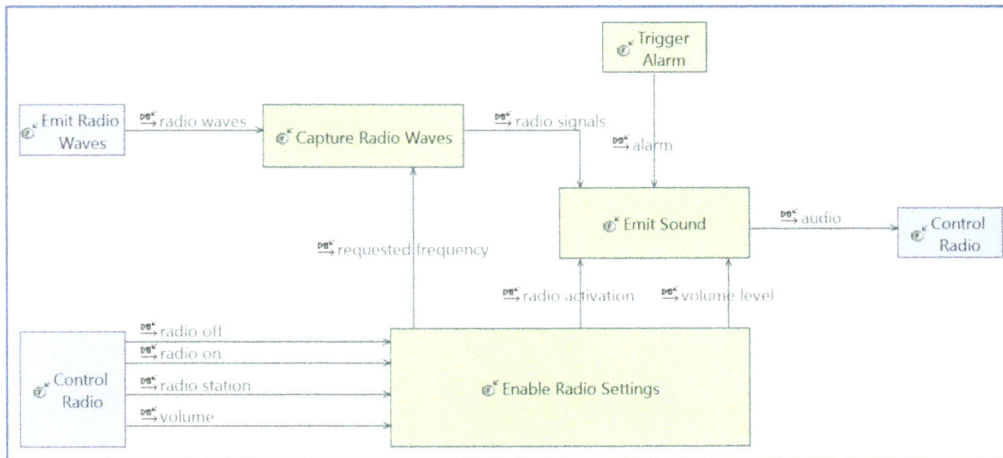

Figure 7-48 – Fix the radio control functional chain

Finally, let's fix the radio control functional chain. Use the same techniques to adjust the functional chain as shown in Figure 7-48.

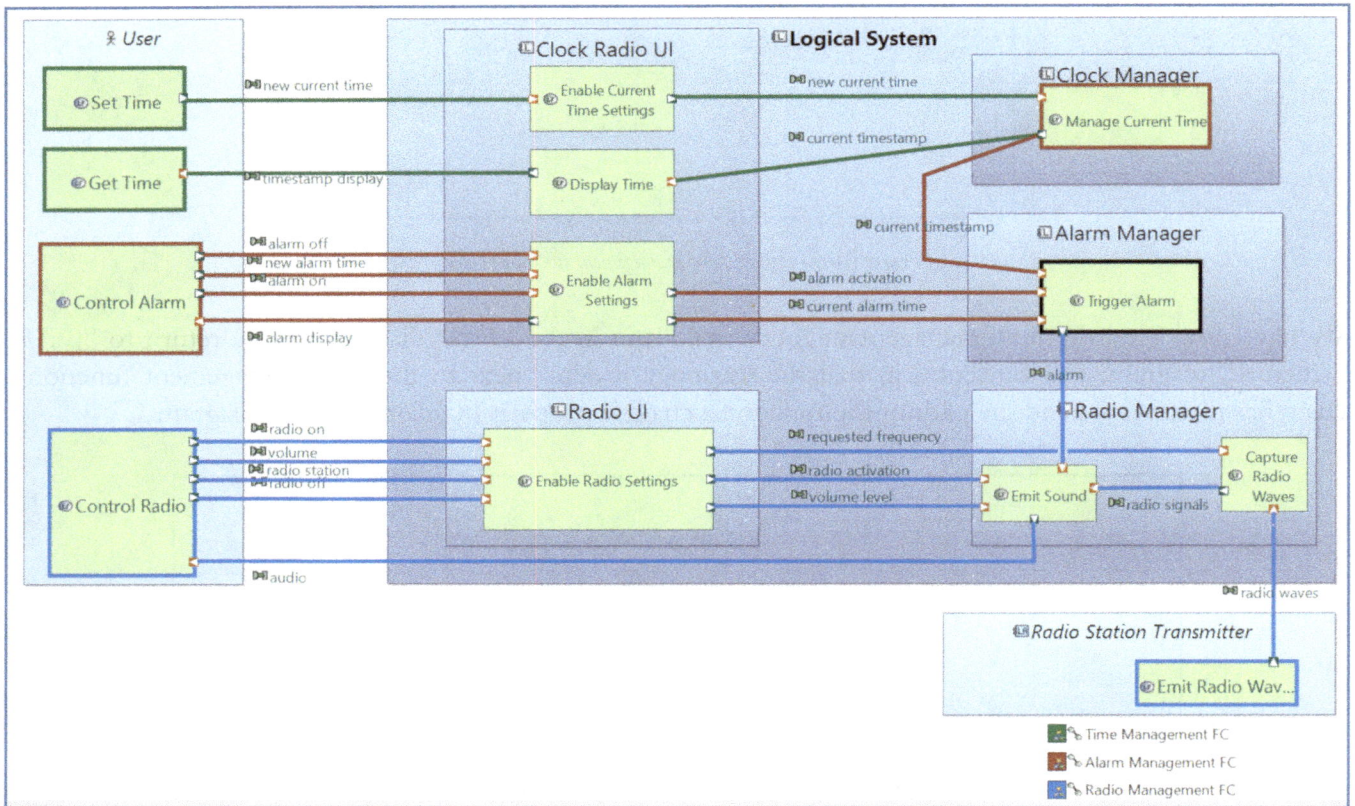

Figure 7-49 – Logical architecture blank with updated functional chains

Finally, we can see that the functional chains are all valid again. [22]

Further Study

Here is a list of further topics for more in-depth study:

---

[22] Sharp-eyed readers will notice that in this diagram, the paths through the functions are not shown. This seems to be a bug in *Capella*. If we examine Figure 6-16 on page 133, we will see that architecture blanks sometimes show the paths through the functions. We have not yet been able to determine when the tool displays functional chains which way.

- **System Subsystem Transition Add-on** – In this video, you will find a short illustration of the use of the System Subsystem Transition add-on on a real model.

  - https://url4ap.net/CH-7-LAB-Transition

- **System Subsystem Transition Add-on** – In this video, you will find a more in-depth explanation and illustration of the System Subsystem Transition add-on.

  - https://url4ap.net/CH-7-LAB-AddOn

- **LAB vs SAB Comparison** – In this short video, you will find a real example of a LAB diagram and also a comparison with the SAB of the same model.

  - https://url4ap.net/CH-7-LAB-Logical1

- **Commented Logical Architecture Model** – In this video, you will find a commented example of a Logical Architecture model, from the book of J.L Voirin.

  - https://url4ap.net/CH-7-LAB-Logical2

- **Replicating Elements** – In this video, you will find a short demo of the way to replicate elements in *Capella* using REC/RPL.

  - https://url4ap.net/CH-7-LAB-Replicate

- **Linking to Requirements** – In this advanced video, you will find an in-depth explanation and illustration of the Logical Architecture level in a complex real model. It shows also how to link to requirements using the dedicated add-on.

  - https://url4ap.net/CH-7-LAB-Reqs

Chapter 8 – Physical Architecture Blank Diagrams

In this chapter, we will continue looking at the Architecture Blank Diagrams, by going to the physical architecture level. Architecture blank diagrams are useful at all *Arcadia* levels. The main goal at the physical architecture level is to refine again what the system will do, in terms of functions allocated to it, by allocating more precisely physical (internal) functions to physical components. At this level, the diagram is called a **Physical Architecture Blank (PAB)**.

The **physical architecture blank** diagram, or PAB, is very similar to the logical architecture blank diagram (LAB), with the difference that we can now show the real physical components inside the system, as well as the physical functions allocated to each of them.

> *Physical Architecture Perspective*
>
> *The Physical Architecture defines the solution at a level of detail sufficient to specify the developments of its subsystems or components, and to drive system IVV phases.*
>
> *It describes the designed solution at a degree of precision and completion far more significant than the previous Logical Architecture, with which it must however remain consistent. It integrates in particular technological and environmental issues among others, and hardware resources required to ensure the expected behavior.*
>
> *All components must have, at this point, a physical existence for engineering at all levels (development/ acquisition specification, configuration, contribution to integration).*
>
> [Voirin] – Chapter 8

Additional Concepts

There are a several additional concepts appearing at this level:

- **Node physical components** (yellow boxes) – which are hosting a number of behavioral components, providing them with the resources they require to function and interact with their environment.

- **Physical links** (red lines) – representing a real means of communication, transport or routing between two node components.

- **Physical ports** (yellow squares) – which are connection points of node components and are not directional.

The other familiar modeling concepts are still present: **physical functions** and **physical behavioral components** (to which physical functions can be allocated).

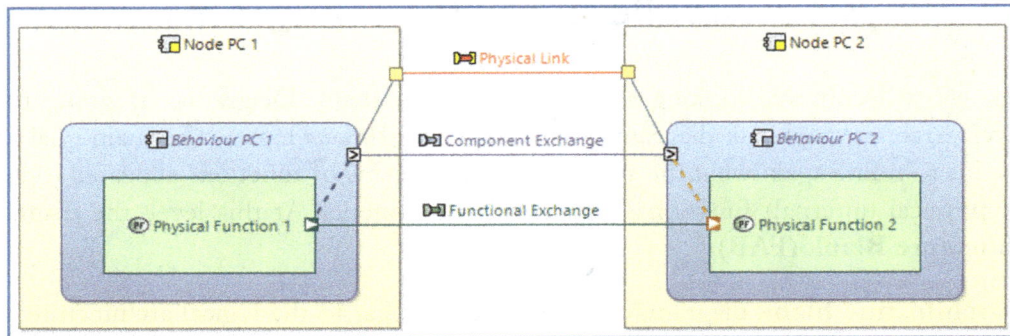

Figure 8-1 – Canonical representation of physical concepts

We will introduce and explain each of these physical concepts in the following sections.

Transition of Logical Functions and Actors

We can start again by transitioning all logical functions. *Capella* systematically creates one physical function for each logical function. It creates also similar function ports and functional exchanges. Finally, *Capella* adds a "realization" link between each physical architecture level model element and the corresponding logical architecture level model element.

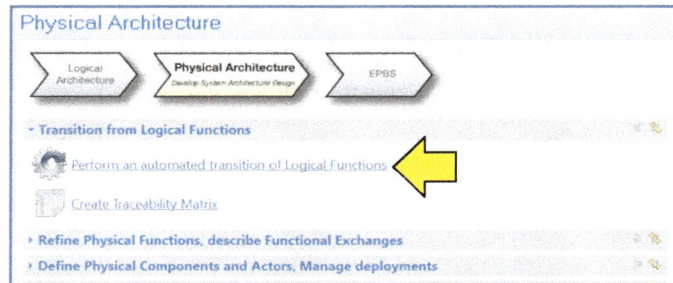

Figure 8-2 – Perform automated transition of logical functions

To perform this transition, simply go back to the activity explorer, and open the "Transition from Logical Functions" group. Then, select "Perform an automated transition of Logical Functions".

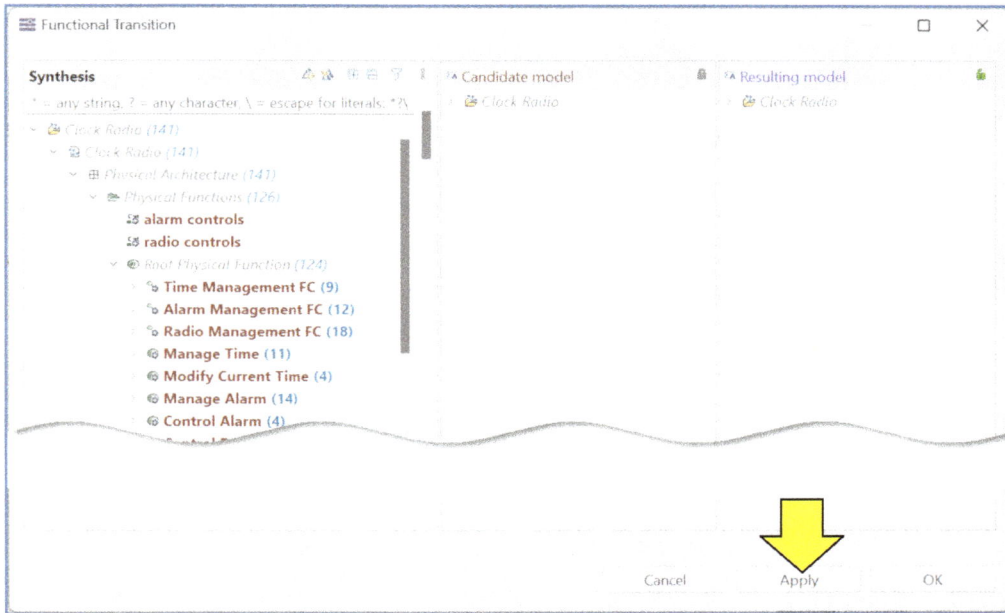

Figure 8-3 – Functional transition window

Capella opens the same kind of complex window as before, called "Functional Transition", where *Capella* shows what it intends to do. Once again, we will accept all of the elements proposed for creation by clicking on "Apply".

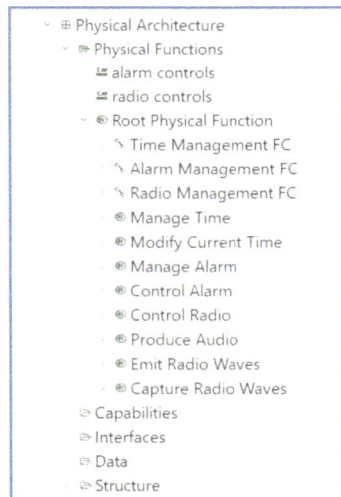

Figure 8-4 – Result of the functional transition in the project explorer

The (preliminary) physical architecture has been created and is now visible in the project explorer.

Next, we need to transition the actors.

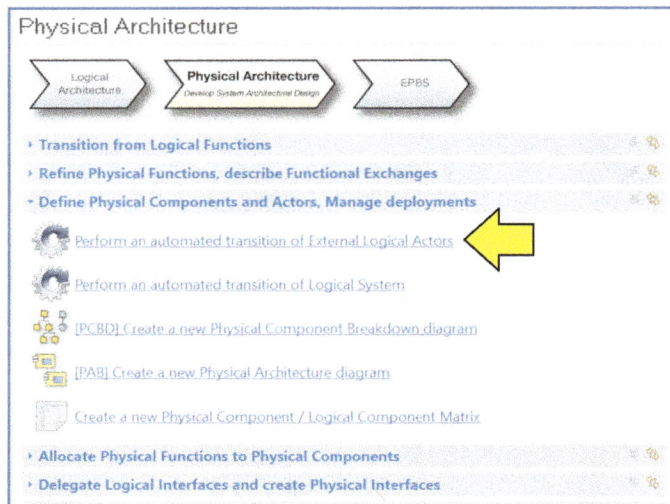

Figure 8-5 – Automated transition of logical actors

Return to the activity explorer. Open the "Define Physical Components and Actors, Manage deployments" group, then select "Perform an automated transition of External Logical Actors". Here also, *Capella* systematically creates one physical actor for each logical actor. It creates also similar component ports and component exchanges. Finally, *Capella* adds a "realization" link between each physical architecture level model element and the corresponding element at the logical architecture level. Notice that *Capella* defaults to keeping the same function allocations inside the actors.

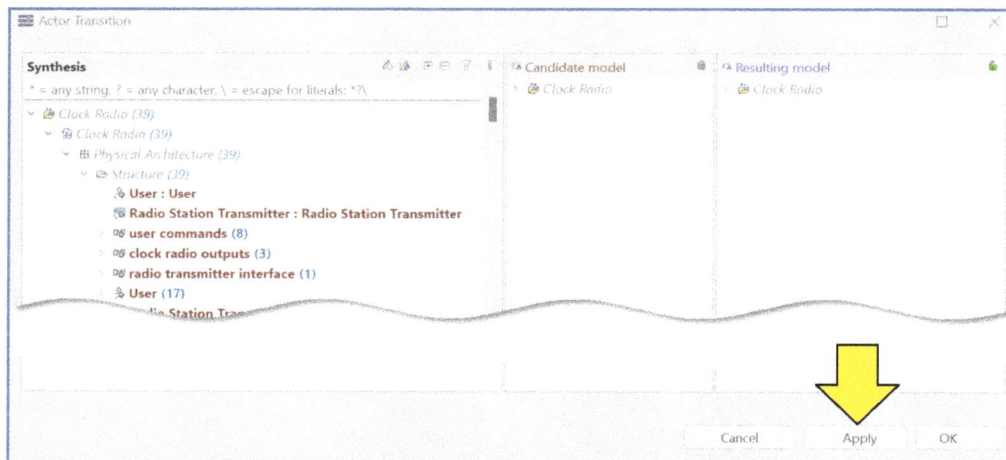

Figure 8-6 – Actor transition window

Once again, *Capella* opens the same kind of complex window as before, called "Actor Transition", where *Capella* shows what it intends to do. Once again, we will accept all of the elements proposed for creation by clicking on "Apply".

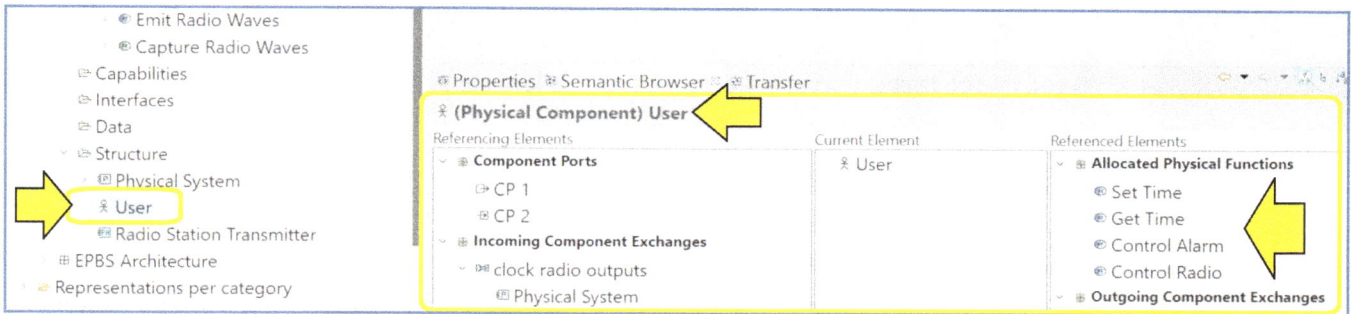

Figure 8-7 – Two physical actors are now visible

The two physical actors are now visible in the project explorer, and you can see their allocated functions in the semantic browser.

Creation of the PAB Diagram and the Physical Components

We can now proceed with the creation of the PAB diagram.

Figure 8-8 – Create new physical architecture blank diagram (PAB)

In the same group of the activity explorer, just click on "[PAB] Create a new Physical Architecture diagram". Accept the default diagram name of "[PAB] Structure" and click "OK".

Capella creates an empty (blank) diagram, with the most complex palette of all *Arcadia* diagrams!

In the palette you will find the three main physical concepts through the first three groups: Node Components, Behavior Components, and Functions, as well as the usual accelerators and common groups. Notice that even though the palette is stuffed full, there are multiple little grey bars in each section that allow you to pull down even more tools!

.

Once again, we can use an accelerator, as we did for the LAB.

.

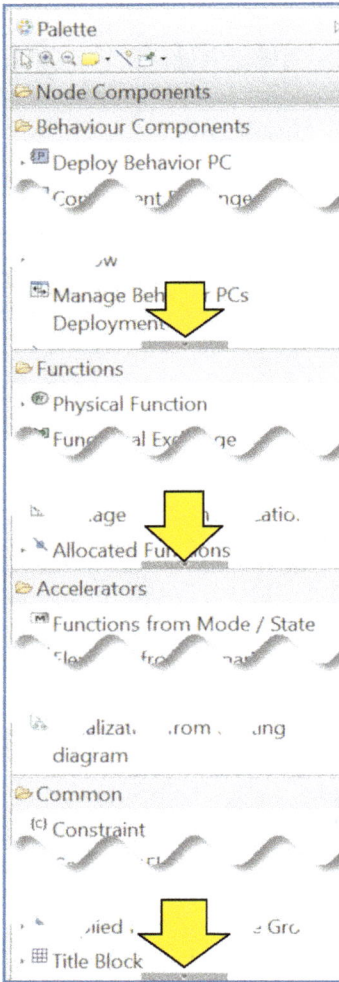

Figure 8-9 – [PAB] Palette

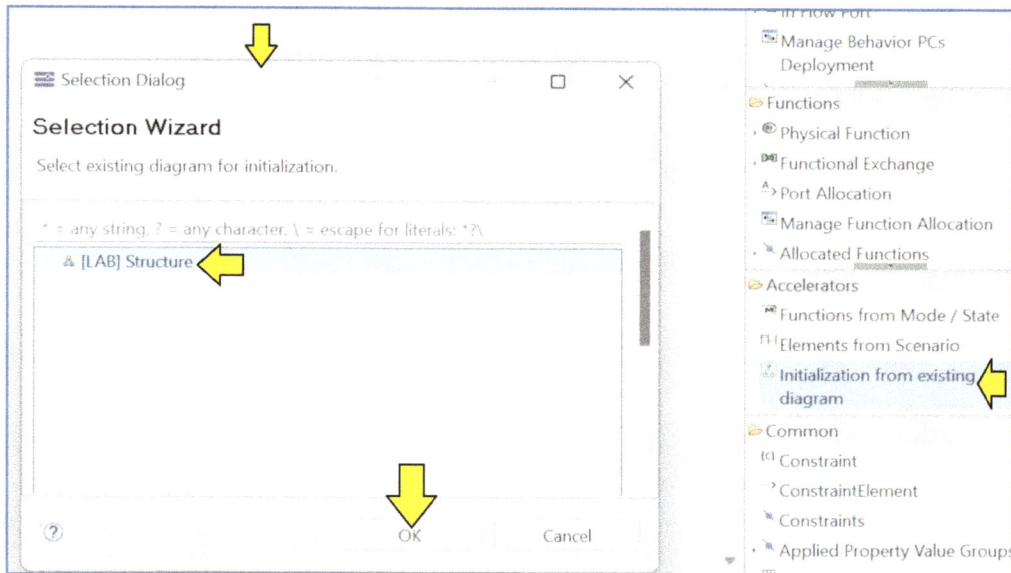

Figure 8-10 – Initialize from existing diagram

1) In the palette expand the "Accelerators" group.

2) Select "Initialization from existing diagram".

3) Click in the background of the diagram.

4) Select the only available logical architecture blank (LAB) "[LAB] Structure".

5) Click "OK".

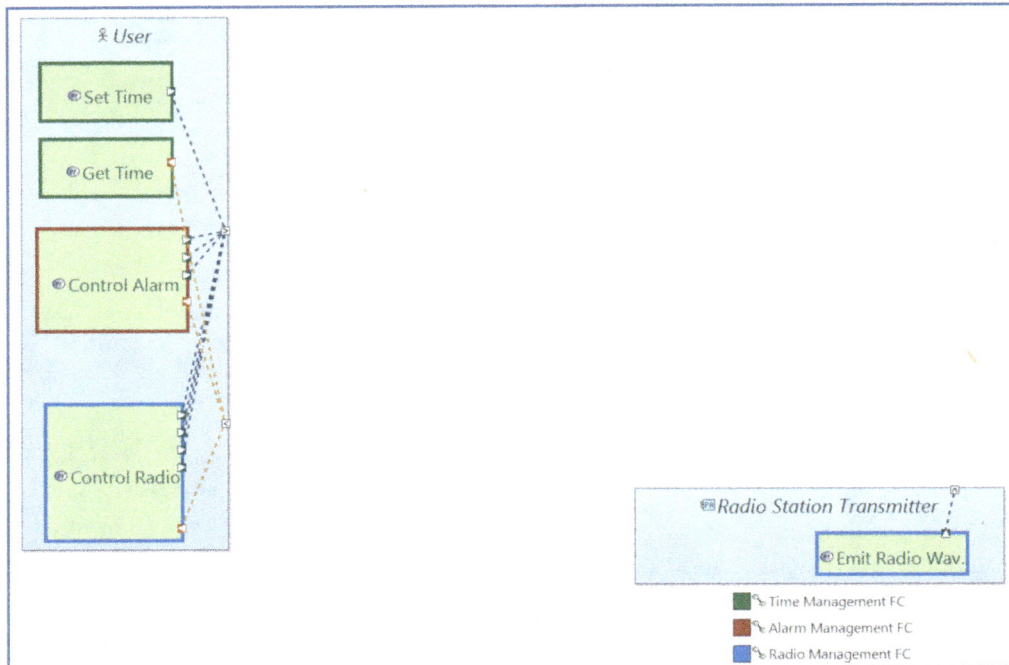

Figure 8-11 – Automatically constructed diagram

The result is quite immediate, showing the two physical actors, as well as the allocated functions, function ports and even component ports (with dashed lines indicating allocating function ports).

The only thing missing is the actual physical system that all of this belongs to. Next, we need to add that physical system and create the node and behavioral physical components inside of it.

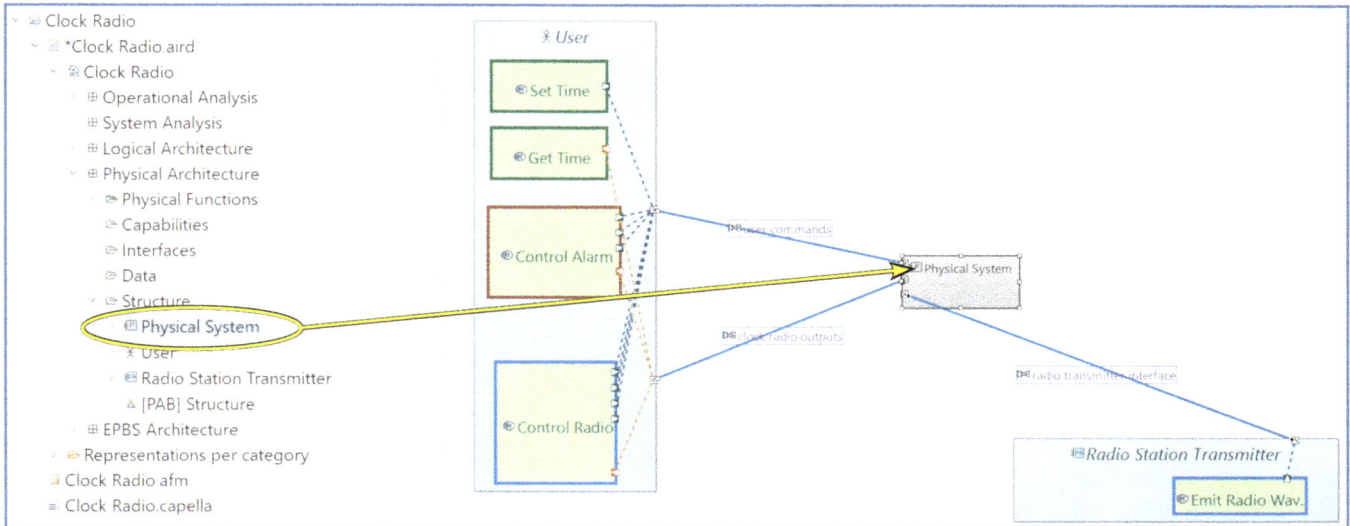

Figure 8-12 – Drag the physical system from the project explorer

Select "Physical System" in the project explorer and drag it to the diagram. Notice that the component exchanges with the actors will appear automatically.

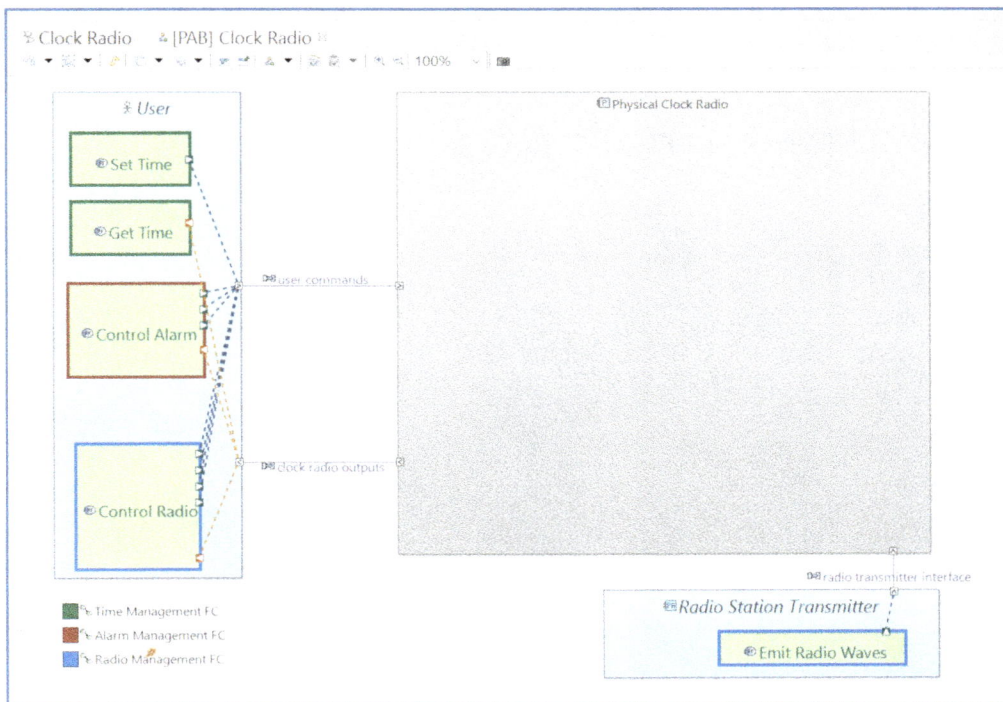

Figure 8-13 – Reorganize the diagram

1) Make the "Physical System" quite a bit larger.

2) Drag the component ports around to align the component exchanges.

3) Change the name of "Physical System" to "Physical Clock Radio".

4) Change the name of the diagram from "[PAB] Structure" to "[PAB] Clock Radio".

We can now start with the creation of node components.

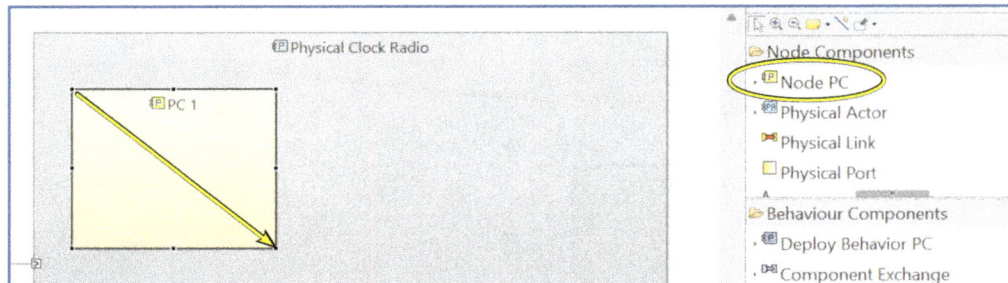

Figure 8-14 – Create a node physical component

Select "Node PC" in the first group of the palette "Node Components" and drag to create your first physical component.

Note that there is a mandatory order of creation in the PAB. First, you need to create node components. After that, you can deploy behavioral components inside the node components. Finally, you can allocate physical functions inside the behavioral components. You can remember the order by the default colors of the elements: "yellow, blue, and green".

Rename this first node "Clock Radio Display".

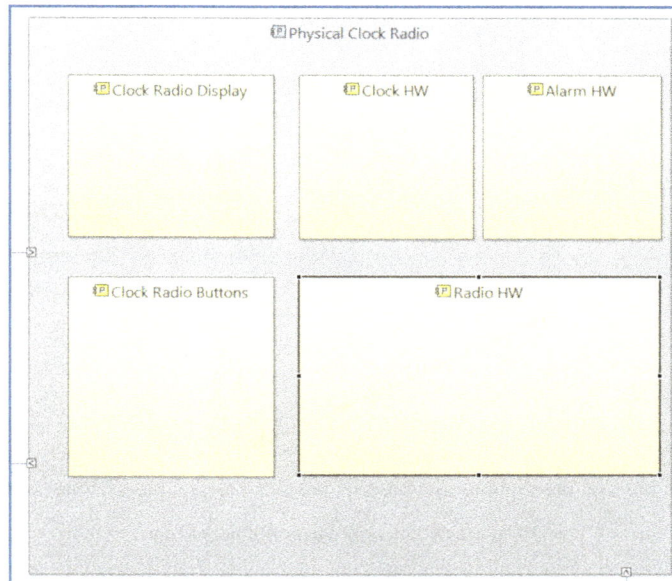

Figure 8-15 – Create four additional node physical components

Create four additional node physical components:

- "Clock Radio Buttons"
- "Clock HW"
- "Alarm HW"
- "Radio HW"

Once the node components are created, we can create behavioral components inside of them.

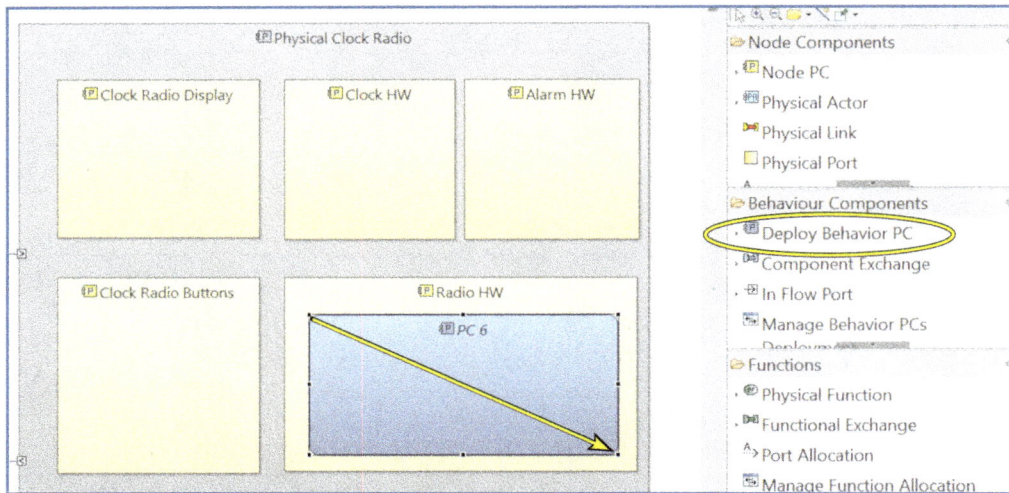

Figure 8-16 – Create behavioral component

Select "Deploy Behavior PC" [23] in the second group of the palette "Behaviour Components" [24] and drag to create your first behavior physical component inside the "Radio HW" node. Rename the new behavior physical component: "Radio Electronics".

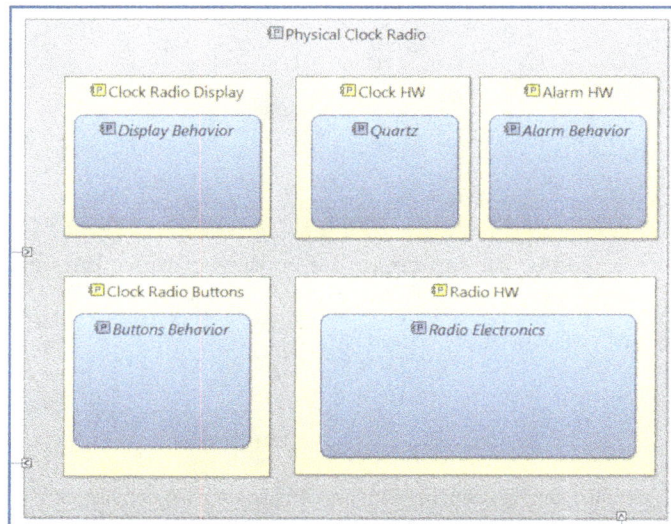

Figure 8-17 – Behaviors Created

Create four more behavioral components:

- "Display Behavior" inside "Clock Radio Display"

---

[23] This is slightly odd English. "Deploy" is a verb used for things that already exist. "Create" would have been better here.

[24] The French spelling of "Behaviour" seems to have slipped into the group name, but not into the entries which use "Behavior".

- "Buttons Behavior" inside "Clock Radio Buttons"

- "Quartz" inside "Clock HW"

- "Alarm Behavior" inside "Alarm HW"

This example has been kept very simple. In a real system, there might be multiple behavioral components (created and) deployed in any given node.

Allocation of the Physical Functions

Next, we will allocate the new physical functions created by the transition process to the physical behavioral components created in the previous section.

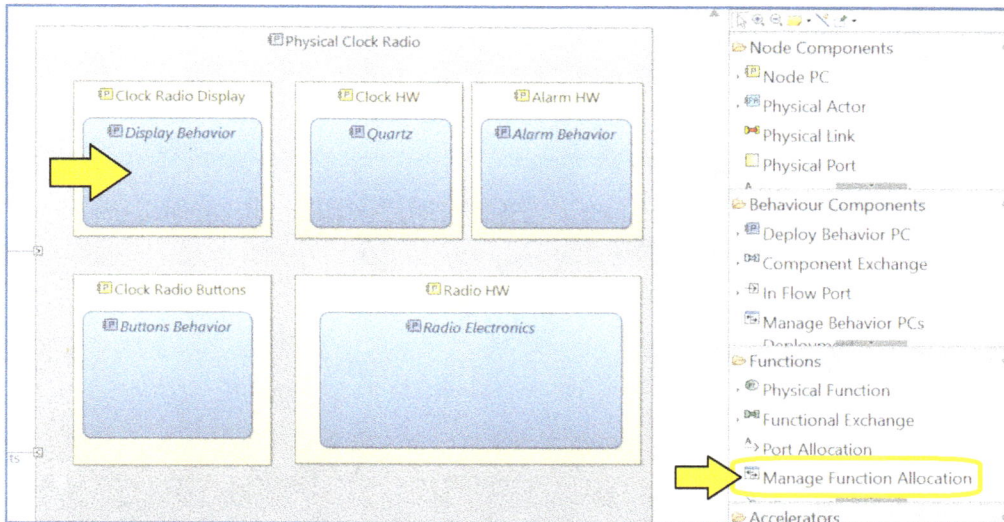

Figure 8-18 – Prepare to allocate functions to the display behavior component

In the palette, select "Manage Function Allocation" in the third group (Functions), and click inside the "Display Behavior" blue component.

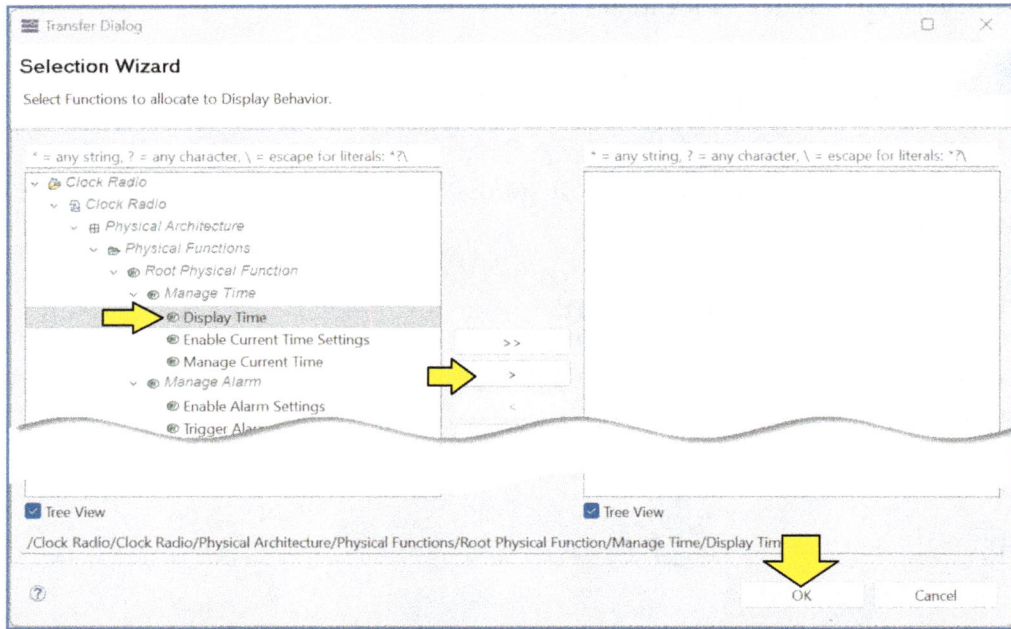

Figure 8-19 – Allocate the display time function

Select "Display Time" function and click on the single arrow (>) to allocate it to the selected component. Click "OK".

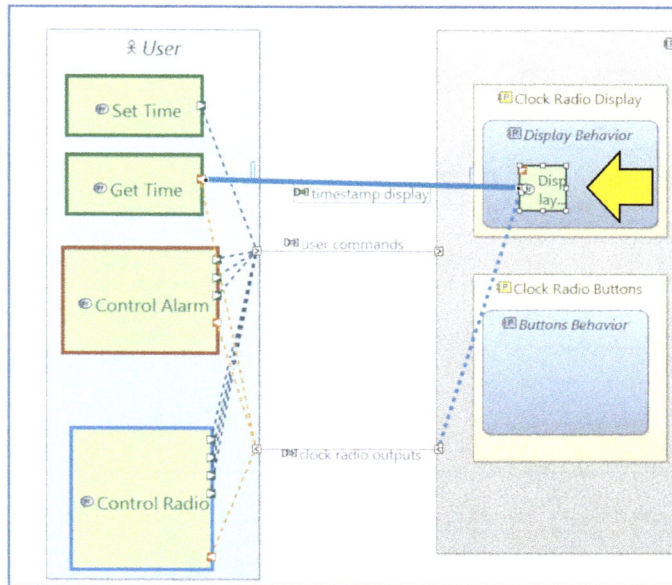

Figure 8-20 – Green display time function visible in blue display behavior component

The "Display Time" physical function then appears inside the "Display Behavior" behavioral component. The existing functional exchanges connected to already displayed functions in the actors (here only the "Get Time" function) are added to the diagram as well. [25]

Resize the function to make the diagram readable. Use the filter function to "Hide Component Exchanges" and "Collapse Component Ports". [26]

---

[25] These physical architecture level functional exchanges were created automatically when we transitioned the model from the logical architecture level.

[26] See Figure 6-39 on page 151 and Figure 6-42 on page 152.

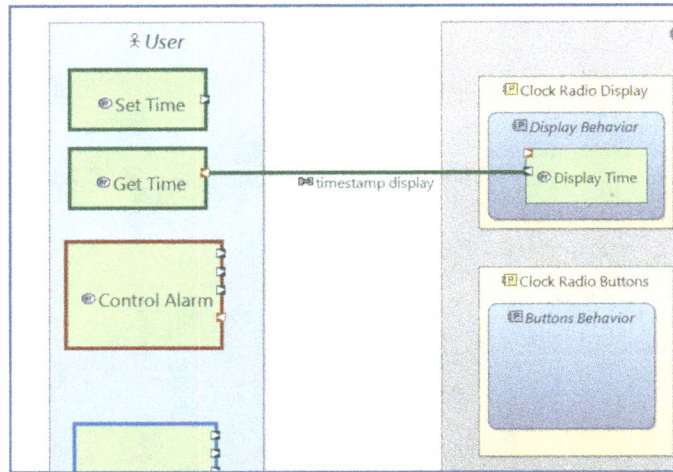

Figure 8-21 – Cleaned up physical architecture blank diagram (PAB)

Having cleaned up the diagram, we can now continue to allocate functions to components. Remember functional exchanges are automatically displayed on the blank diagrams, as soon as both source and target functions are present in the diagram.

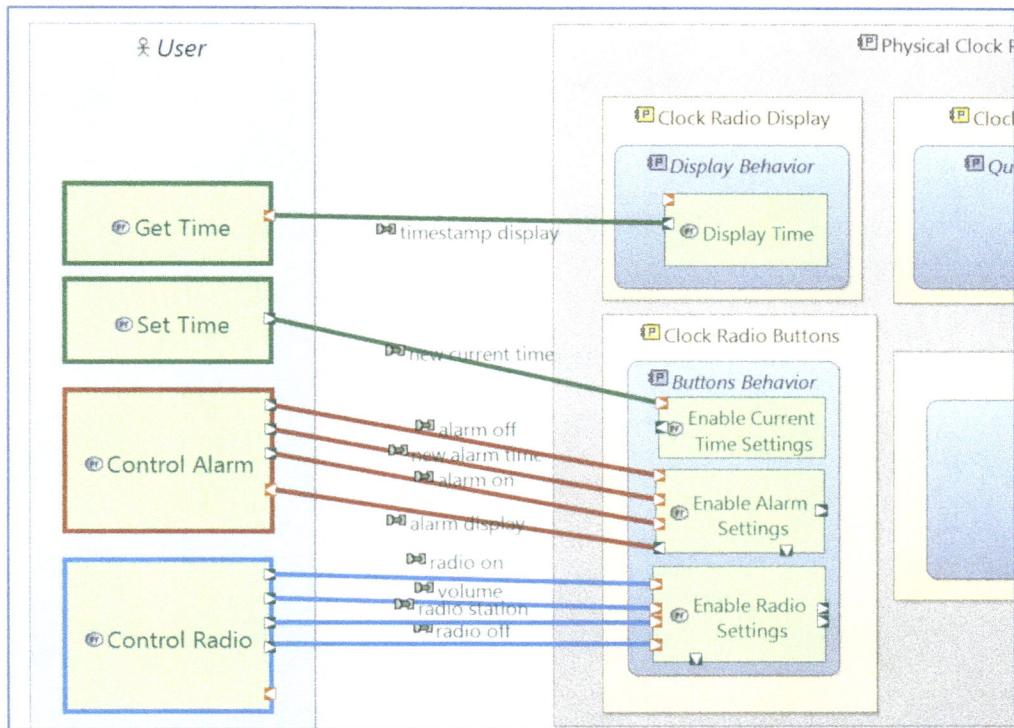

Figure 8-22 – Continue to allocate functions

Continue to allocate functions: "Enable Current Time Settings", "Enable Alarm Settings", and "Enable Radio Settings" to the "Buttons Behavior" behavioral component. Drag things around to make the diagram readable.

So far, so good. Taking a close look at our model, [27] we realize that the "alarm display" functional exchange should be produced by something in the "Display Behavior" component, not by the "Enable Alarm Settings" function in the "Buttons Behavior" component.

Fortunately, this sort of problem is very easy to fix on the fly in a modeling tool!

_____

[27] Something every modeler should do frequently!!

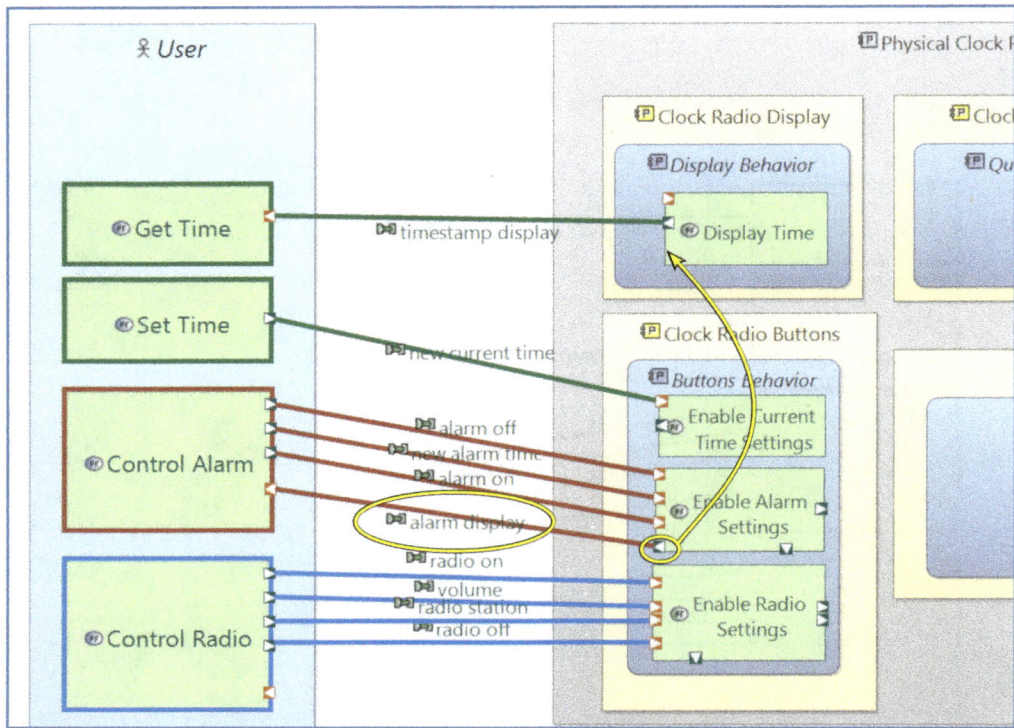

Figure 8-23 – Move the function output port of alarm display

1) Find the "alarm display" functional exchange that runs from "Enable Alarm Settings" to "Control Alarm".

2) The port on the "Enable Alarm Settings" end of the functional exchange is the output port.

3) Grab that port with the mouse and drag it up to the "Display Time" function.

4) Rename the "Display Time" function to "Display Time & Alarm" to indicate the new function we have assigned to it by giving it the new output port.

5) Drag things around as needed to neaten up the diagram.

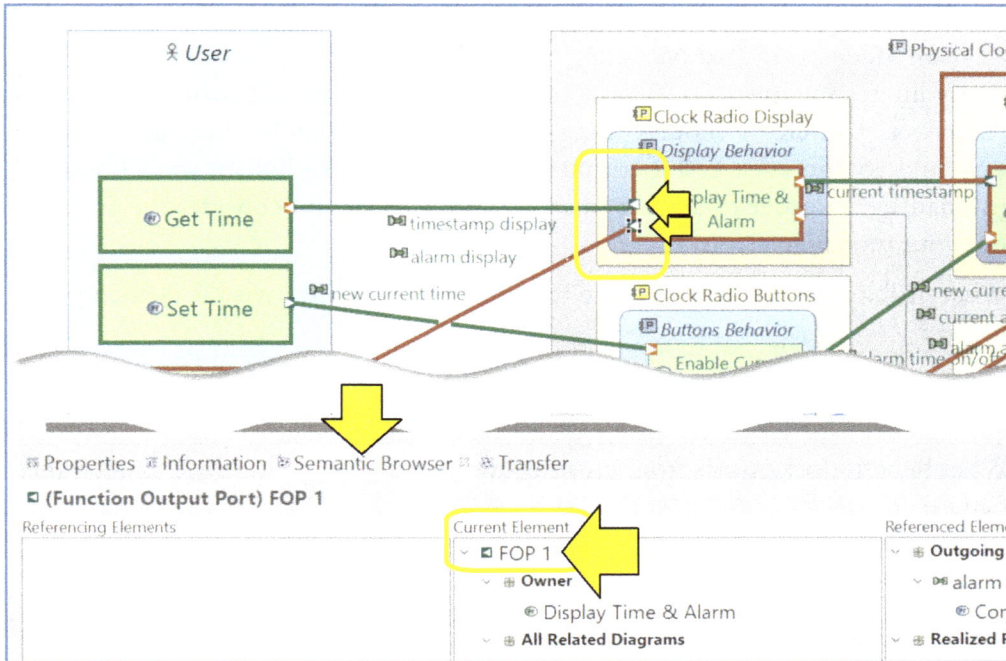

Figure 8-24 – Conflicted port names

The previous operation has left us with a small model integrity problem. Moving the ports around has left the function "Display Time & Alarm" with two ports named "FOP 1".

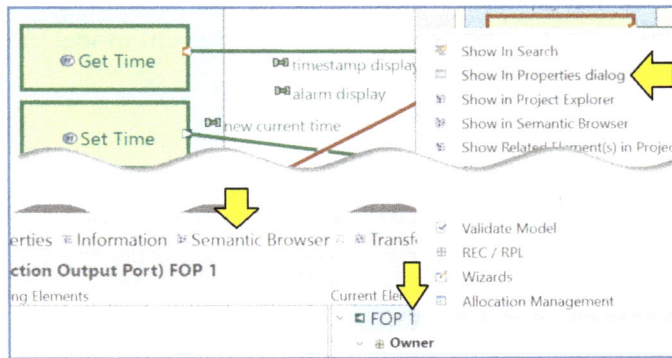

Figure 8-25 – Open the properties dialog for the port and fix its name

With one of the conflicted ports still selected in the diagram, in the semantic browser right-click on the port and select "Show in Properties dialog". Change the port name from "FOP 1" to "FOP 1**a**" in order to deconflict its name. It really doesn't matter which of the conflicted ports you rename. You simply need to be sure that they don't have the same name.

Should I Create Meaningful Port Names?

This is a very interesting question. *Capella* automatically generates all sorts of hidden modeling element names as you create models using the diagramming tools. Port names are only one example. Should you go back and meticulously give all of these hidden model detail elements meaningful names? In order to answer this question, you should consider the following questions about your modeling effort:

1) What problem am I trying to solve by giving these mostly invisible elements meaningful names?

2) How much will it cost me to create (and maintain!) all of these hidden meaningful names?

3) What benefit do I expect from creating and maintaining all of these hidden meaningful names?

The answer will usually depend on what sort of system you are trying to model.

- Are you modeling a system to administer lifesaving medications in an emergency helicopter rushing injured patients to a hospital? In this case: Yes, you should sweat every single detail of the model, name every hidden element, check it front to back, check it back to front, and have three different independent reviewers inspect every single detail of the model.

- Are you modeling a machine to dispense ice cream in the executive lunchroom? In this case, the automatically generated default names are probably good enough. However, you may need to fix up a name here-and-there to prevent model integrity problems.

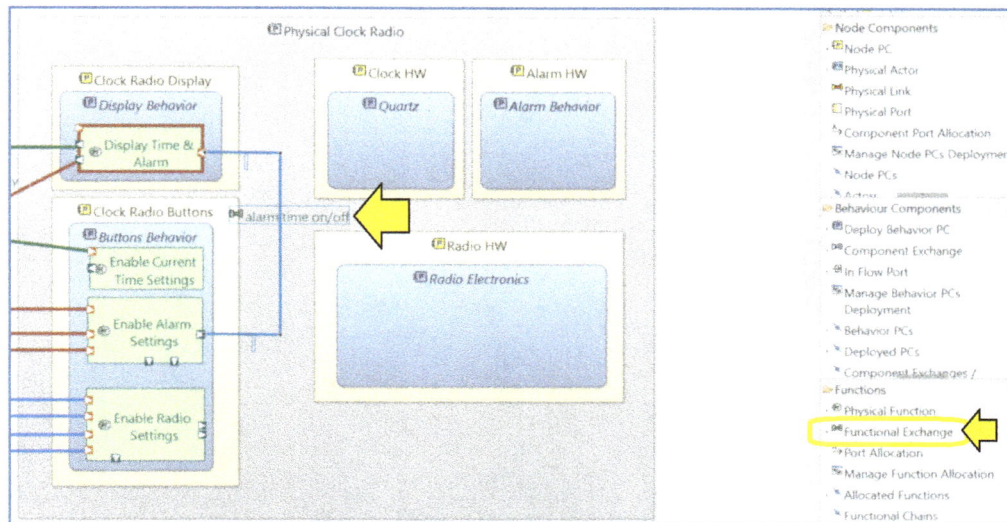

Figure 8-26 – Create a functional exchange for alarm time

We probably need also to create a functional exchange from "Enable Alarm Settings" to "Display Time & Alarm" so that the later function knows the alarm time and whether the alarm is on or off.

It is very common that the physical architecture differs from the logical one, because we choose specific technical solutions and modify in consequence the dataflow, allocation of functions, and so on.

We can now complete the function allocation process:

- "Manage Current Time" to the "Quartz" behavioral component,
- "Trigger Alarm" to the "Alarm Behavior" behavioral component,
- "Emit Sound" and "Capture Radio Waves" to the "Radio Electronics" behavioral component.

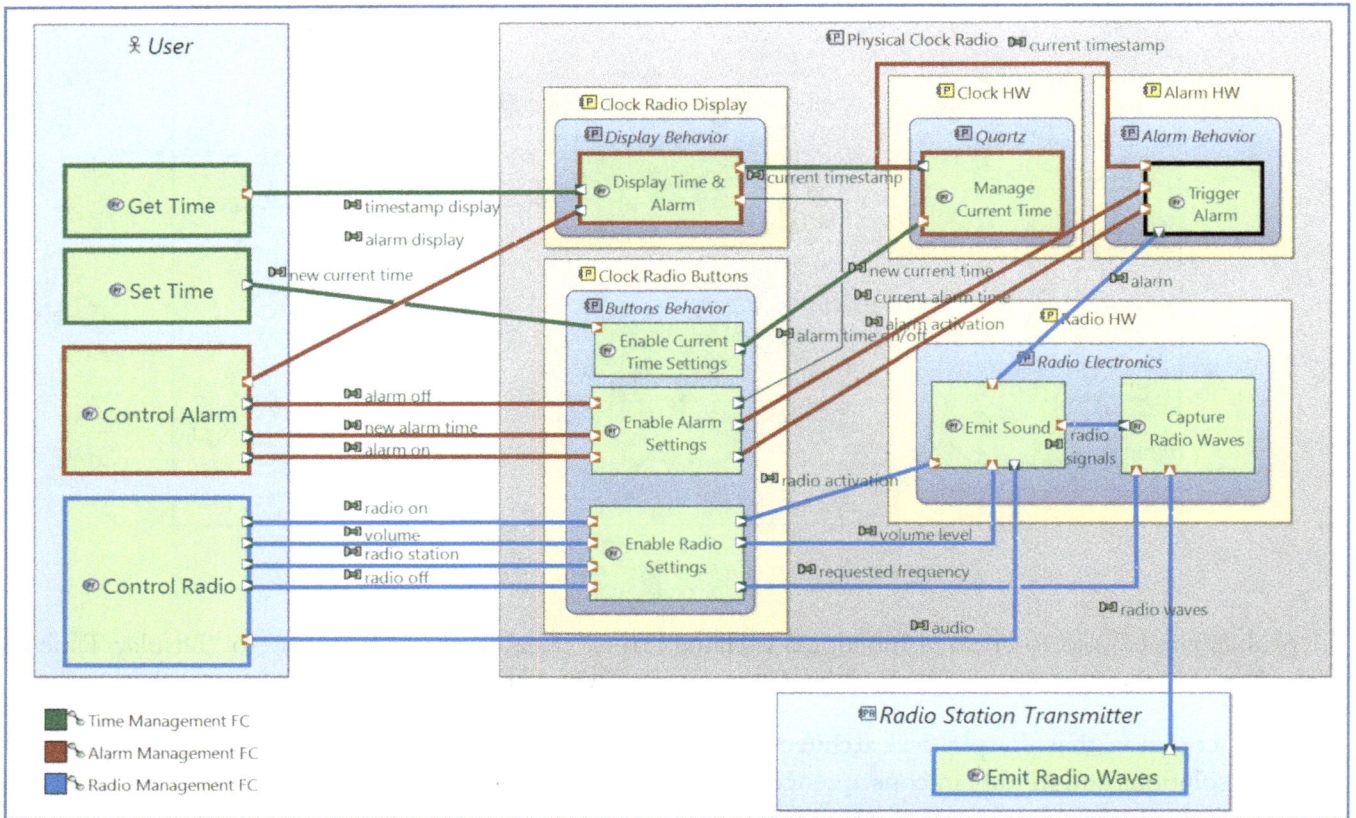

Figure 8-27 – Allocation complete

Neaten up the diagram as needed. Notice that the new functional exchange that we added is not yet part of any functional chain. We might consider adding it to the alarm management functional chain.

Notes Images

Of course, the "Radio Electronics" component is much more complex that showed in our PAB. However, we know that it can be purchased off-the-shelf. As such, we do not want to spend time modeling its internals. It is good practice to document this sort of assumption by adding a note to the relevant diagram.

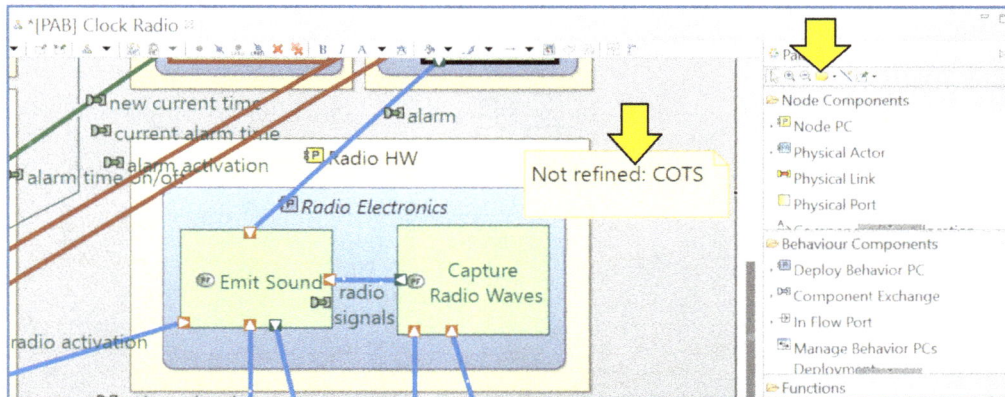

Figure 8-28 – Add a note to a diagram

At the top of the palette, you will find a yellow icon similar to a post-it note that represents a note. A note is just a graphical decoration for one and only one diagram; it is not a model element. A note cannot be reused in another diagram.

You can write text inside the note, for instance: "Not refined: COTS". [28]

It is possible to move it near the concerned element, or even to create a graphical attachment represented by a dashed line. Here we will just place it partially inside the component boxes.

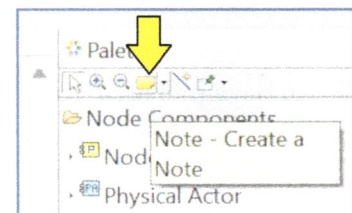

Figure 8-29 – Closer look at the icon

To make *Arcadia* diagrams easy for stakeholders to understand it is often very useful to replace component rectangles by images. Such images are especially helpful for communicating physical-level diagram content to the types of experts who are interested in physical details!

We would like here, for instance, to display a picture of a clock radio inside the grey box representing the physical system. It is quite easy with *Capella*. We just need to have some images in the current workspace. Either we can prepare this before working with *Capella* by copying files to the workspace with the usual operating system commands at our disposal. Or we can do it directly inside *Capella* by drag and dropping the selected image in the *Capella* project explorer on the root of our model.

---

[28] "COTS" = "Commercial-Off-The-Shelf"

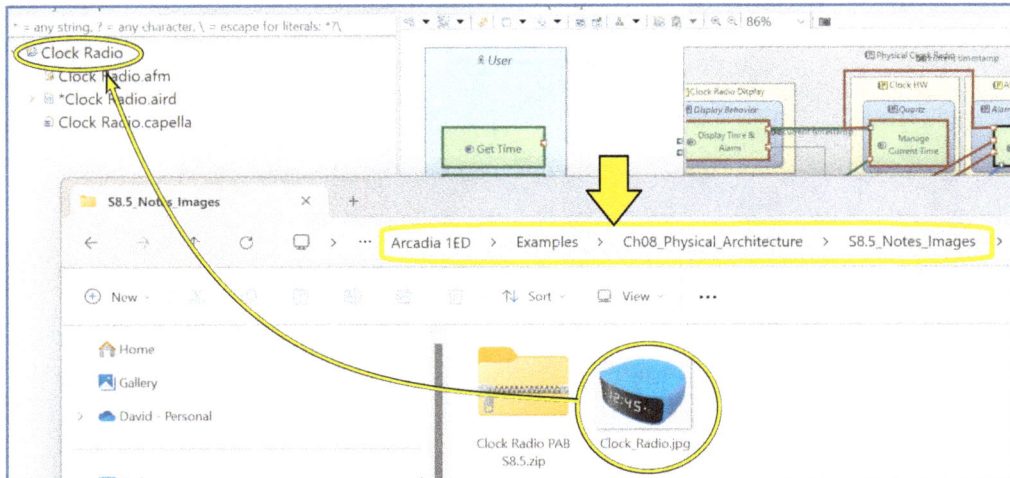

Figure 8-30 – Drag image into project explorer

If you have prepared an image file in a folder on our computer, you can now just drag and drop it onto the model root in the *Capella* project explorer. [29]

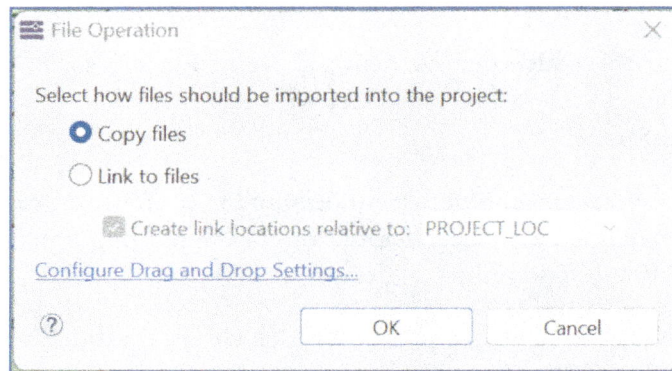

Figure 8-31 – Copy or link

Capella asks us whether we prefer to copy it physically or just create a link. In this case we can accept the default option to copy the file into the work folder.

---

[29] Notice in Figure 8-30 that we have provided the image file in the examples package.

Figure 8-32 – Image added to model

As a result, the image now appears in the project explorer and will be available to be used inside our diagrams.

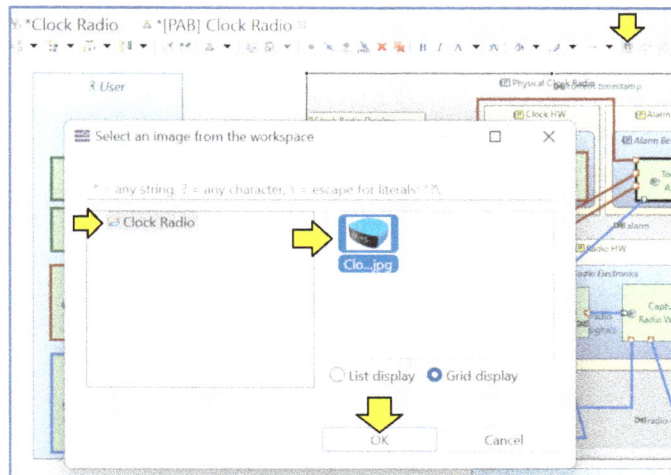

Figure 8-33 – Set background image

1) Select the "Physical Clock Radio" grey box.

2) Select the image icon in the diagram toolbar.

3) *Capella* will open a selection dialog.

4) In the left panel, navigate to the image. (Just one click actually).

5) Select the image.

6) Click "OK".

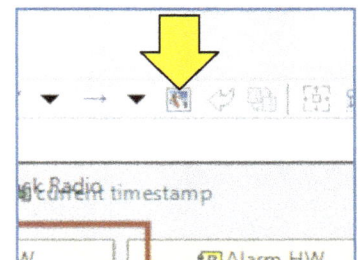

Figure 8-34 – Closer look at the icon

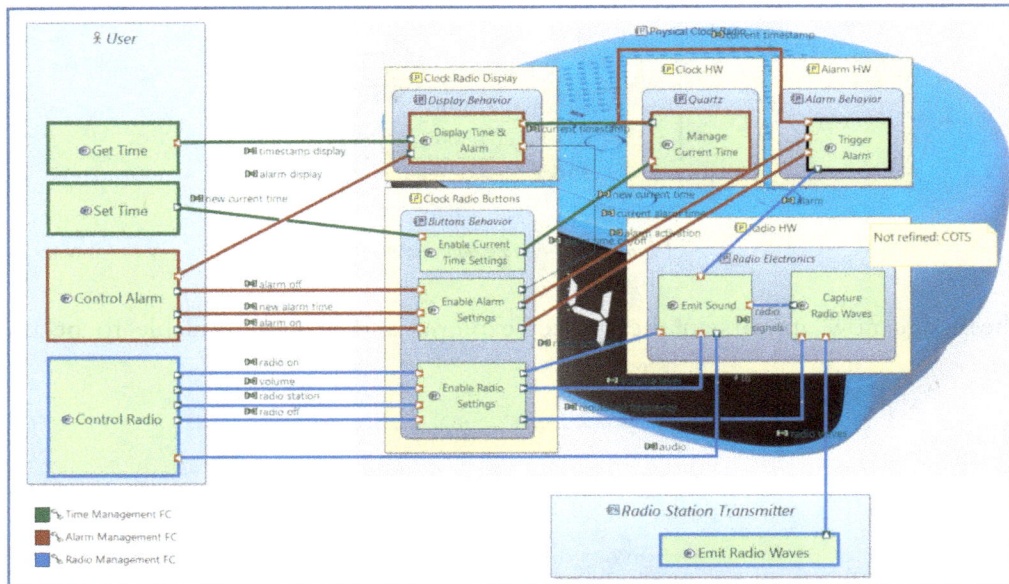

Figure 8-35 – Diagram with image background

As soon as we select one available image file, it appears in the background of the selected box. You will very likely have to adjust the size of the image by dragging one corner of the image as needed. The diagram is now ready to be communicated to domain experts.

Physical Links

The last of the three concepts are **physical links** and **physical ports**.

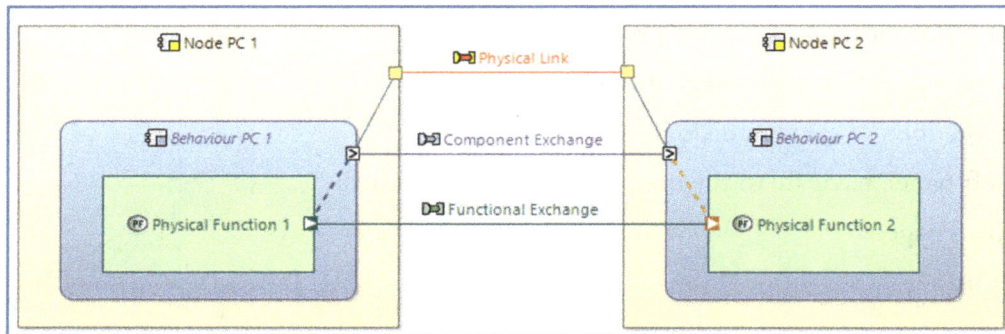

Figure 8-36 – Physical links with physical ports (review)

The allocation approach recommended by *Arcadia* is to allocate all functional exchanges to component exchanges, and then all component exchanges to physical links.

Let's start with a very simple example. If we focus on the interface between the radio components and the external radio station transmitter, we may want to express the fact that the radio waves functional exchange is allocated to a component exchange that we can call radio transmitter interface. [30]

In principle, we could continue adding to Figure 8-35. However, we would quickly run into several practical problems.

1)   The diagram is already quite intricate. If we also attempt to show the multiple layers on this diagram, it will become very difficult to read.

2)   *Capella* does have a "visibility mode" to display hidden elements on a diagram. However, this mode has a number of limitations:

   •   The color of the newly "visible" elements is very faint.

   •   *Capella* does not allow you to restore elements one-by-one. As soon as you select "show" for any one element, it turns off the filter for the entire diagram.

   •   Hidden elements tend to not merely be turned off, but physically positioned underneath visible elements, making them difficult to manipulate.

3)   The limitations of "visibility mode" are further compounded by the fact that we have replaced the system element with an image that has a white background. With that image in place, it is quite difficult to determine where the boundary of the system element is.

Fortunately, we are engaged here in *model-based systems engineering* – there is absolutely no need to cram everything into one diagram!

We can simply make some additional physical architecture diagrams to show the allocation of functional exchanges to component exchanges, and the allocation in turn of component exchanges to physical links – and nothing else!

---

[30] You may wish to briefly return to the section *Component Exchanges* on page 140 to review the procedure for allocating function exchanges to component exchanges.

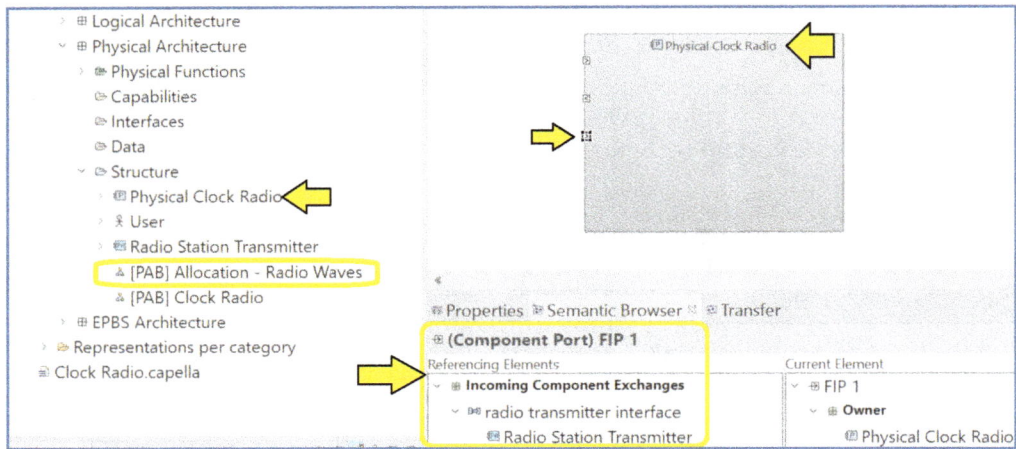

Figure 8-37 – Create a physical link

1) Return to the activity explorer and create another physical architecture diagram called: "[PAB] Allocation - Radio Waves".

2) Drag the physical clock radio to the diagram.

3) Notice that the physical clock radio already has three component ports, one of which leads to the radio station transmitter.

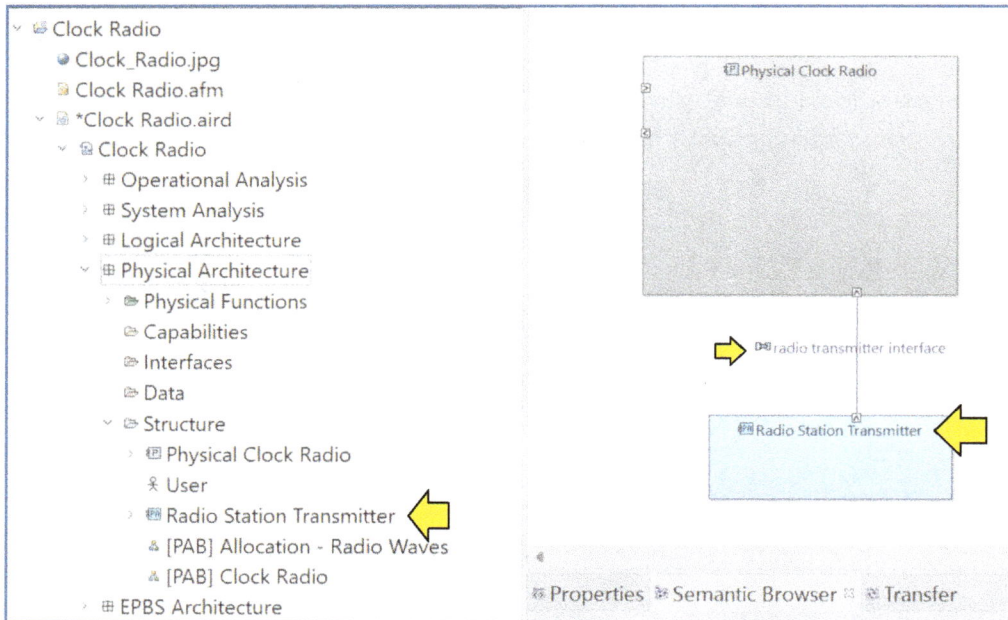

Figure 8-38 – Add the radio station transmitter

Drag the radio station transmitter to the diagram. The component exchange will appear automatically. Drag the component ports around as needed to make the diagram look neat.

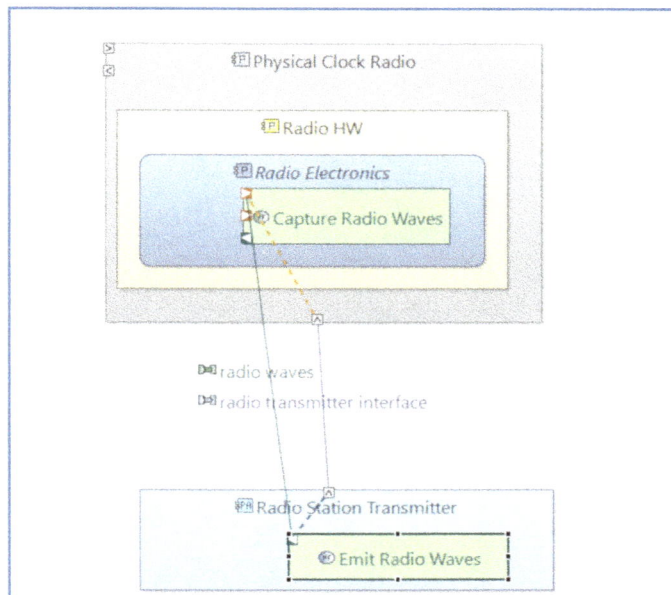

Figure 8-39 – Add the node, component, and functions

1) In the project explorer, under "Structure", and then "Physical Clock Radio", you will find the "Radio HW" node.

2) Drag the "Radio HW" node to its correct place inside of the "Physical Clock Radio" element in the diagram.

3) Resize all elements to make more space.

4) In the project explorer, under "Structure", and then "Physical Clock Radio", you will find the "Radio Electronics" behavioral component.

5) Drag the "Radio Electronics" behavioral component to its correct place inside of the "Radio HW" node in the diagram.

6) In the project explorer, under "Physical Functions", and then "Root Physical Function", you will find the "Capture Radio Waves" function.

7) Drag the "Capture Radio Waves" function to its correct place inside of the "Radio Electronics" behavioral component in the diagram.

8) In the project explorer, under "Physical Functions", and then "Root Physical Function", you will find the "Emit Radio Waves" function.

9) Drag the "Emit Radio Waves" function to its correct place inside of the "Radio Station Transmitter" actor in the diagram.

Notice that the functional exchange is automatically added to the diagram as well.

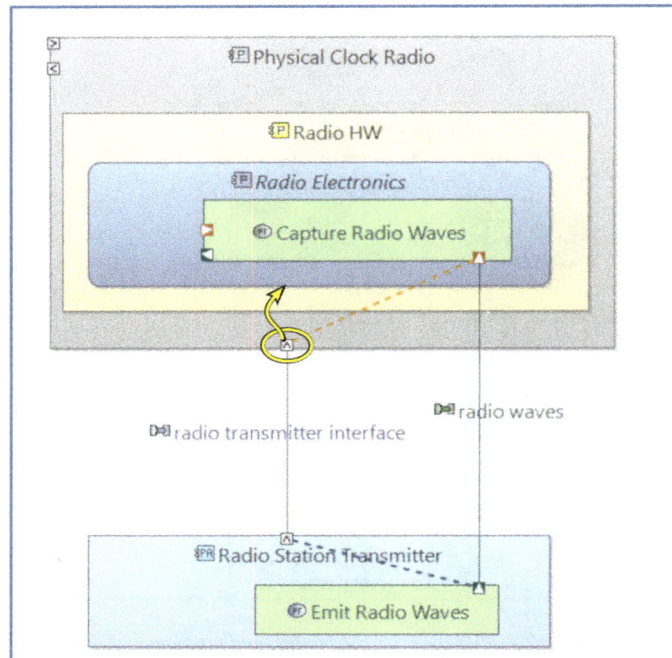

Figure 8-40 – Neaten diagram and drag component port to radio electronics component

Continue to resize and drag ports to make the diagram look neat.

Currently the component port is attached to the system. Drag it to the "Radio Electronics" behavioral component.

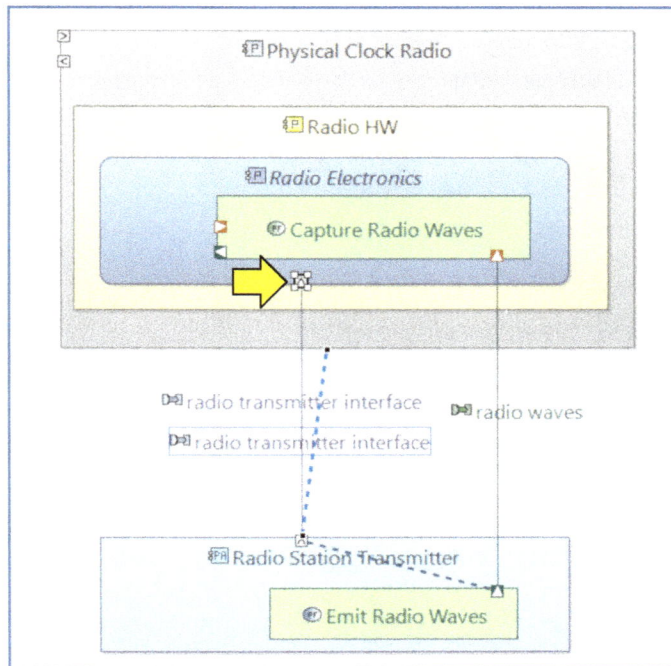

Figure 8-41 – The diagram looks confusing

The component port is now in the right place, but the diagram looks confusing. This confusing appearance is an indication that *Capella* has not moved the allocations around appropriately. We will need to fix them manually.

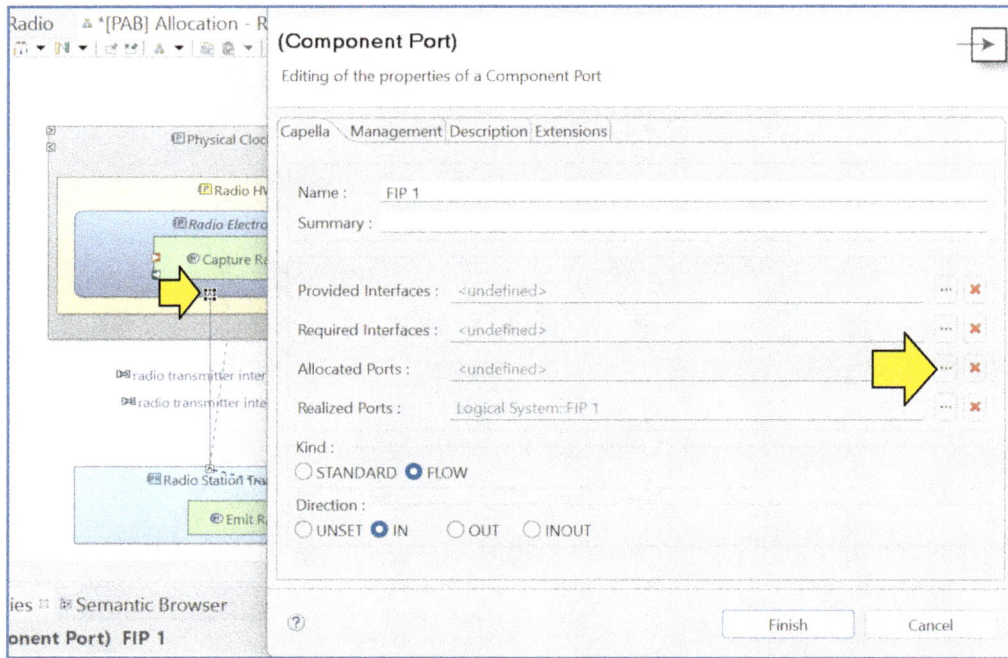

Figure 8-42 – Open property sheet and allocated ports panel

Double-click on the component port to open its properties sheet. In the "Allocated Ports:" field, click on the three dots to open the allocate ports panel.

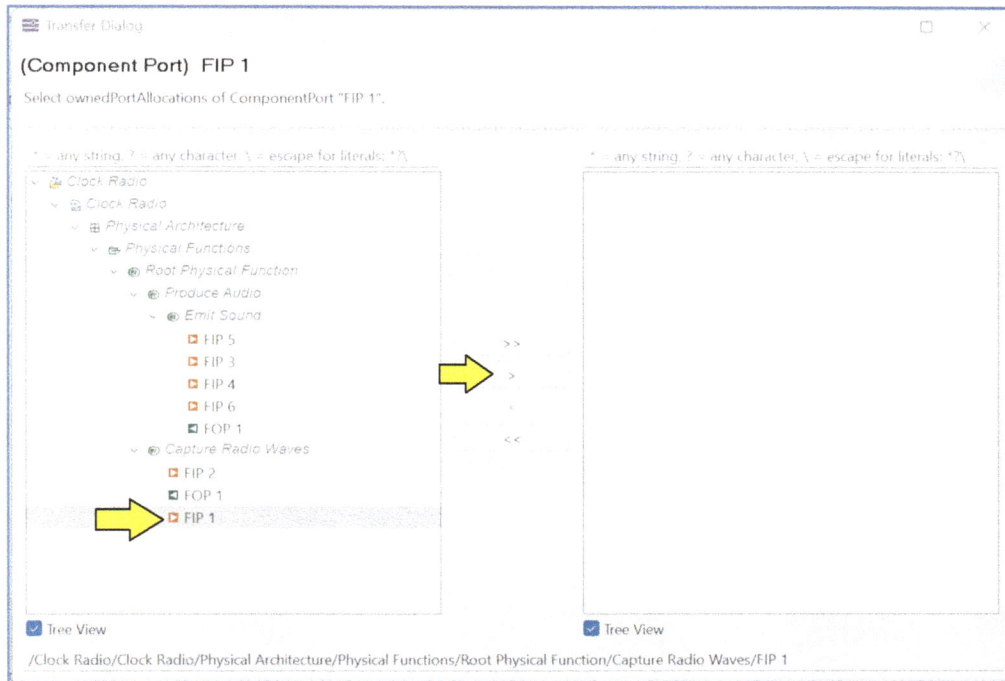

Figure 8-43 – Allocate FIP 1 from the capture radio waves function

Allocate the FIP 1 functional port from the capture radio waves function. Click "OK" and then "Finish".

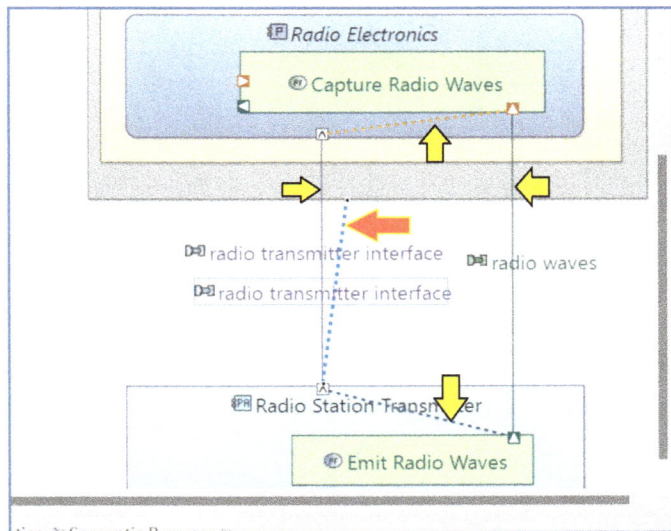

Figure 8-44 – Looks great, except that Capella may show ghost link

The function exchange, the component exchange, and the allocations at each end, now look great. You can select each link in turn and carefully review the information in the semantic browser – it is all exactly as it should be.

However, *Capella* tool may be showing a "ghost" of the original connection of the component exchange. If you select this ghost link and examine its information in the semantic browser, you will see that the diagram rendering is incorrect. This is simply a ghost of the correct component exchange being displayed in the wrong place.

You can make the ghost disappear by selecting it, right-clicking, selecting "Show/Hide" and then "Hide Element".

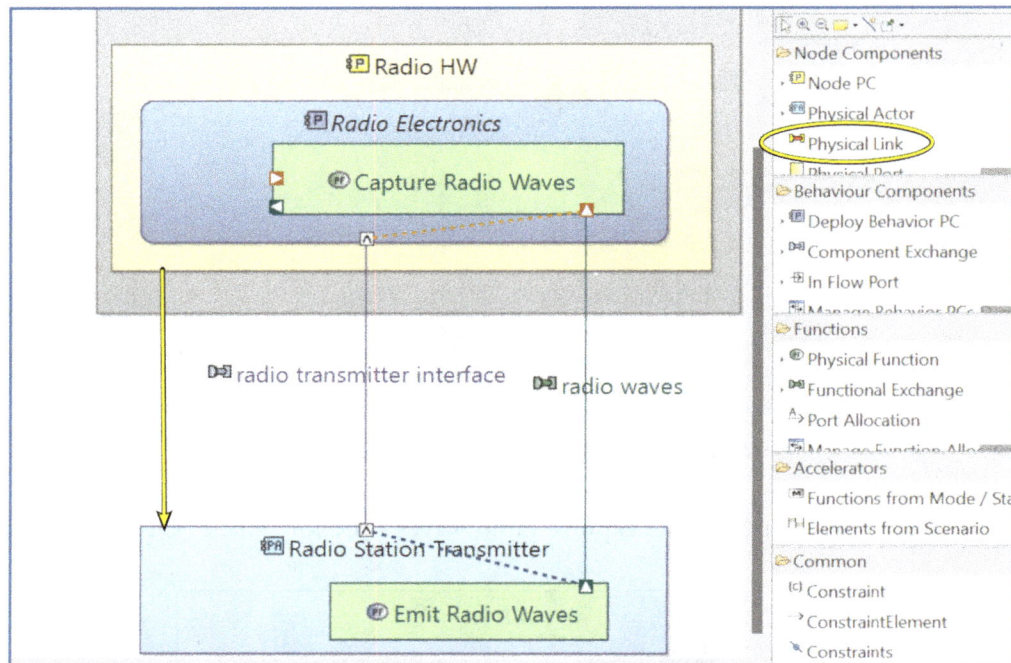

Figure 8-45 – Add a physical link

In the palette, select the "Physical Link" tool. Draw a physical link from the yellow "Radio HW" physical node to the "Radio Station Transmitter" physical actor.

Rename this physical link "air".

Double-click on the physical link to open its properties sheet. In the "Component Exchange Allocations:" field, click on the three dots to open the allocate ports panel.

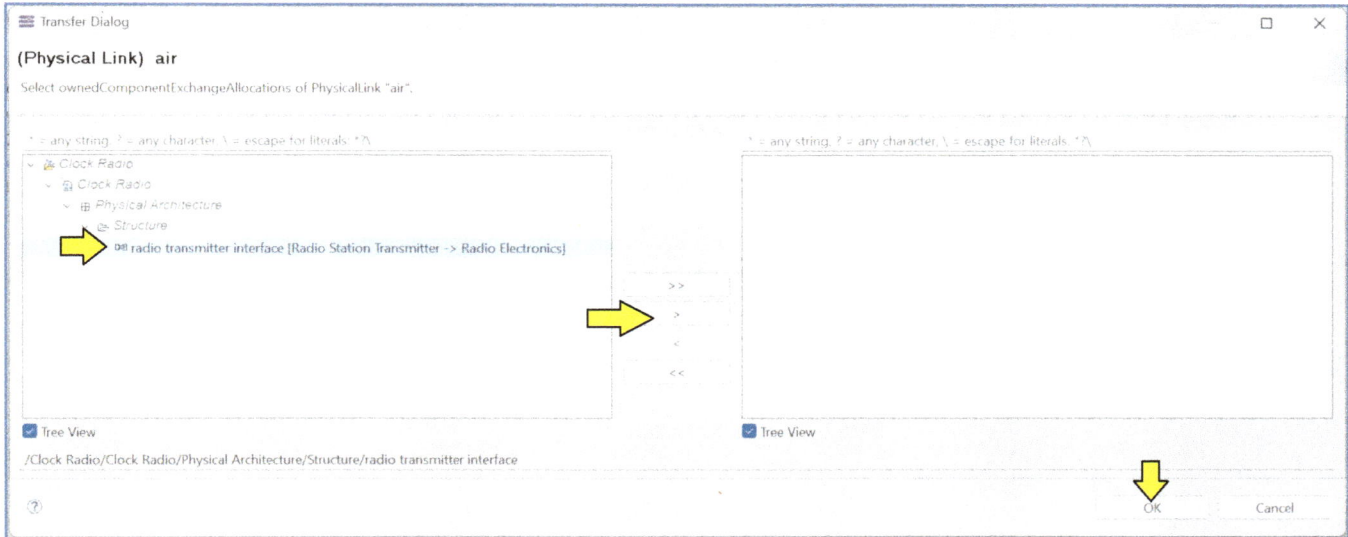

Figure 8-46 – Allocate component exchange to the physical link

Capella is smart enough to figure out that only one component exchange is a candidate for allocation to this physical link. Select that component exchange, use the single arrow to allocate it, click "OK" and then "Finish".

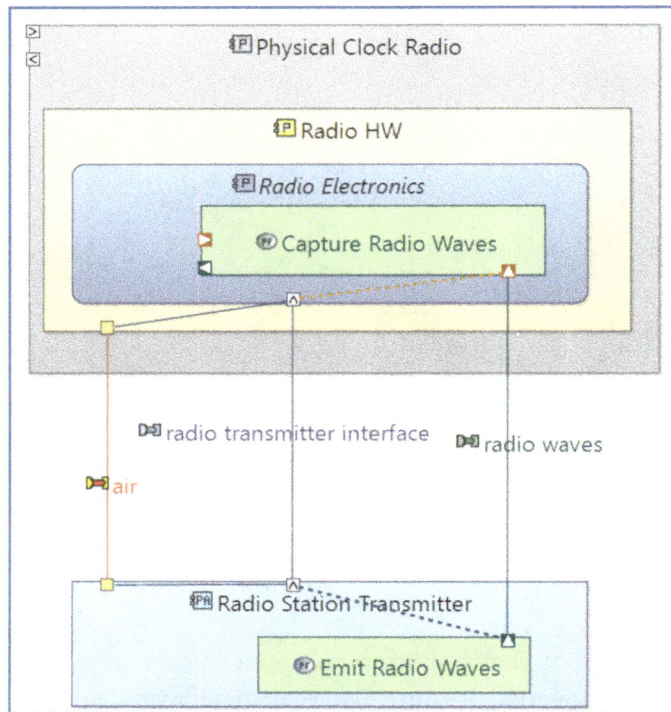

Figure 8-47 – Finished allocation diagram

We can now see the finished diagram that shows the allocation of functional exchange to component exchange which is in turn allocated to a physical link.

Traceability Matrices

One of the most useful features supported by almost all MBSE tools (not just *Capella*) is the **traceability matrix**. This type of matrix shows the relationships between one category of elements in the rows and another in the columns.

Clock Radio New Physical Components - Logical Components	Clock Radio UI	Radio UI	Clock Manager	Alarm Manager	Radio Manager
Physical Clock Radio					
Clock Radio Display	X				
Clock Radio Buttons	X	X			
Clock HW			X		
Alarm HW				X	
Radio HW					X
Radio Electronics					X
Display Behavior	X				
Buttons Behavior	X	X			
Quartz			X		
Alarm Behavior				X	

Figure 8-48 – Example of a traceability matrix

Construction of such a matrix is a little beyond the scope of a beginner's book. However, Pascal has prepared a helpful video to explain this topic in more detail:

https://url4ap.net/CH-8-Matrix

Further Study

Here is a list of further topics for more in-depth study:

- **Interface Control Document Generation** – In this advanced video, you will find an in-depth explanation and illustration of the use of Capella model elements to produce an Interface Control Document and connect to an external PLM tool thanks to a specific add-on.

  - https://url4ap.net/CH-8-PAB-ICD

- **Arcadia/Capella for a Large Complex Mechanical System** – In this video, you will find an advanced explanation and illustration of the use of the System Subsystem Transition add-on for a large complex mechanical system.

  - https://url4ap.net/CH-8-PAB-Mech

- **PVMT add-on** – In this short video, you will find a real example of PAB diagram. And also, an example of the use of the PVMT add-on to enhance the model by adding property values.

  - https://url4ap.net/CH-8-PAB-Physical1

- **Physical Architecture model** – In this video, you will find a commented example of Physical Architecture model, from the book of J.L Voirin.

  - https://url4ap.net/CH-8-PAB-Physical2

- **Reverse Engineering Physical Architecture Model** – In this advanced video, you will find an in-depth explanation and illustration of how to reverse engineer a *Capella* model from the physical architecture level and which add-ons can be useful.

  - https://url4ap.net/CH-8-PAB-Reverse1

- **Physical Architecture of a Smart Robot Car** – In this advanced video, you will find an in-depth explanation and illustration of the development of the physical architecture of a smart robot car.

  - https://url4ap.net/CH-8-PAB-Reverse2

- **System of Systems Modeling** – In this advanced video, you will find an in-depth explanation and illustration of the use of the System Subsystem Transition add-on to perform system of systems modeling with *Capella*.

  - https://url4ap.net/CH-8-PAB-SOS

- **TitleBlocks** – In this video, you will find an in-depth illustration in a PAB of the use of TitleBlocks in *Capella*, in order to add meaningful information extracted from the model.

  - https://url4ap.net/CH-8-PAB-Title

- **Traceability Matrices** – As mentioned previously, this video covers the mechanics of creating a traceability matrix in *Capella*.

  - https://url4ap.net/CH-8-Matrix

Chapter 9 – Breakdown Diagrams

In this chapter, we will take a closer look at the so-called **Breakdown Diagrams**, representing hierarchies of either Functions (Activities) or Components (Entities) at all *Arcadia* levels. This type of diagram is probably the simplest of all *Arcadia* diagrams. However, it behaves according to very specific rules that we need to understand.

In the preceding chapters, we looked at different types of "Blank" diagrams, and the general rule was that they are empty at creation time and update automatically only for connectors under precise circumstances (both source and target boxes of the connector should be already present in the diagram). For the "Breakdown" diagrams, it is nearly the exact opposite! They are automatically populated with all existing relevant elements at creation time and update automatically as soon as any corresponding model element is created in any other type of diagram. By default, their objective is to be always complete.

Of course, the modeler can still "hide" model elements. However, the "Delete from Diagram" command is never available in breakdown diagrams, as graphical objects would be automatically created again the next time we opened the diagram.

The example used in this chapter is again the simple Clock Radio for domestic use. We will continue using the model we developed in the previous chapters.

Function Breakdown

Introduction

As we already mentioned in Chapters 5 to 8, a function can be refined into subfunctions or child functions, which detail and clarify its content. Leaf functions are functions which are not further broken down.

In the same way, an operational activity can be refined into several subactivities. As the notion of operational activity shares most of the properties of the notion of a function, with the exception of owning ports, we will not cover the breakdown of operational activities in this chapter. The techniques that we present for function breakdown diagrams also work for the corresponding **Operational Activity Breakdown** diagram [OABD].

Arcadia has strong rules for function decomposition, such as:

- A non-leaf function shall not own function ports (refer to Chapter 5).
- A non-leaf function shall not be allocated (refer to Chapters 6 to 8).

About Functional Decomposition

Several functions can be grouped into a mother function (they are then called subfunctions, or daughter functions, of this function). Symmetrically, a function can be refined into several functions. This grouping is not a strong structural decomposition relationship; **function grouping only forms a synthetic representation of these, essentially for documentary purposes. Generally, in a finalized model only the leaf functions (without subfunctions) refer to and carry the functional description expected.**

...

As indicated above, function grouping is only one synthetic representation of functions, essentially for documentary purposes. **At the end of model building, the exchanges should link only the leaf functions, and only the leaf functions should host ports.** *On the one hand, this is to avoid leaving any ambiguity or imprecision in the model, as if two mother functions are linked by an exchange, there is nothing to indicate which of their subfunctions should take it into account; on the other hand, it is to avoid redundancies or incoherencies between exchanges that would be defined both on the functions and their subfunctions.*

[Voirin] – Chapter 17

System Function Breakdown (SFBD)

We will create our first function breakdown diagram at the system analysis level. Open the activity explorer and select the "System Analysis" activity.

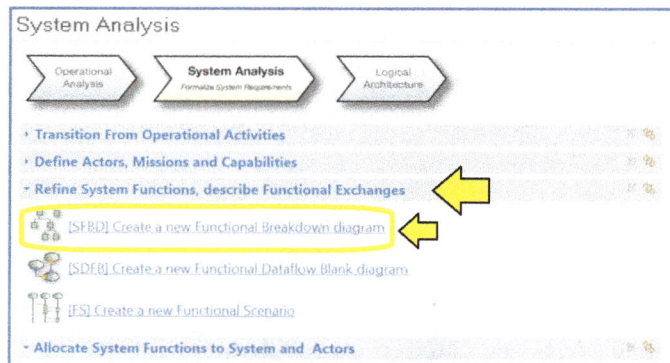

Figure 9-1 – Create a SFBD

Open the topic "Refine System Function, describe Functional Exchanges" and select "[SFDB] Create a new Function Breakdown diagram".

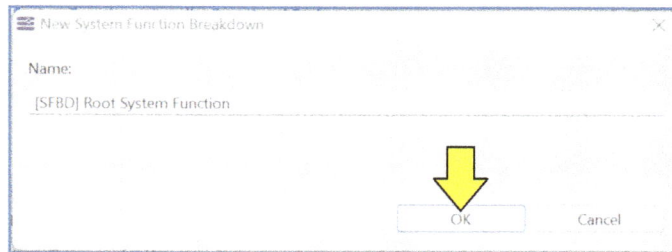

Figure 9-2 – Name the diagram

A panel appears with a field to set the diagram name. The default diagram name matches the name of the model element under which the diagram will be located. In this case, *Capella* will automatically place the diagram inside the "[SFDB] Root System Function" within the "System Functions" folder. Accept the default name and click "OK".

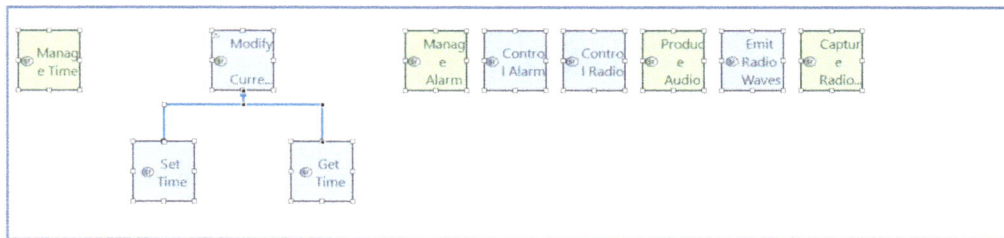

Figure 9-3 – Resulting SFBD

The resulting diagram is automatically filled with all existing system-level functions, along with their decomposition. All functions are displayed as squares with a default size and selected. You may need to adjust the size of some of the boxes.

Notice that *Capella* computes the light blue color from the allocation to an external actor.

If you add a new system-level function to the model (somewhere) it will automatically be added to this diagram.

Logical Function Breakdown (LFBD)

Next, we will move to the logical architecture. At this level we broke down several functions to allocate them precisely to logical components. As a result, the functional breakdown diagram should be more interesting!

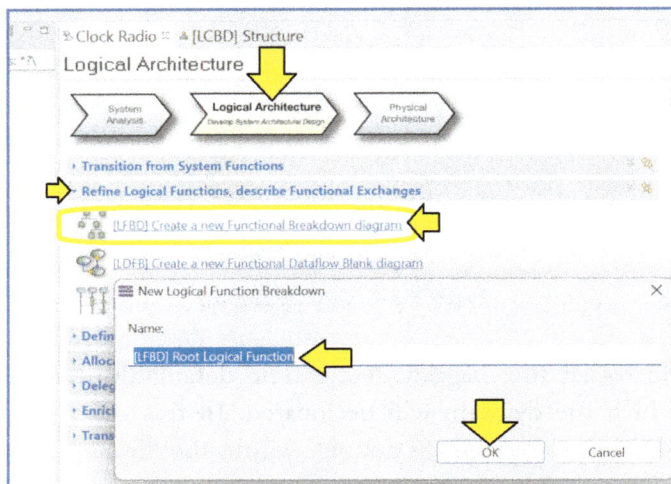

Figure 9-4 – Name the diagram

Again, we can create a "Logical Function Breakdown (LFBD)" diagram from the activity explorer, by returning to the logical architecture activity, selecting the topic "Refine Logical Function, describe Functional Exchanges", and clicking on "[LFBD] Create a new Functional Breakdown diagram". A panel appears with a field to set the diagram name. The default diagram name matches the name of the model element under which the diagram will be located. In this case, *Capella* will place the diagram "[LFBD] Root Logical Function" inside the "Logical Functions" folder.

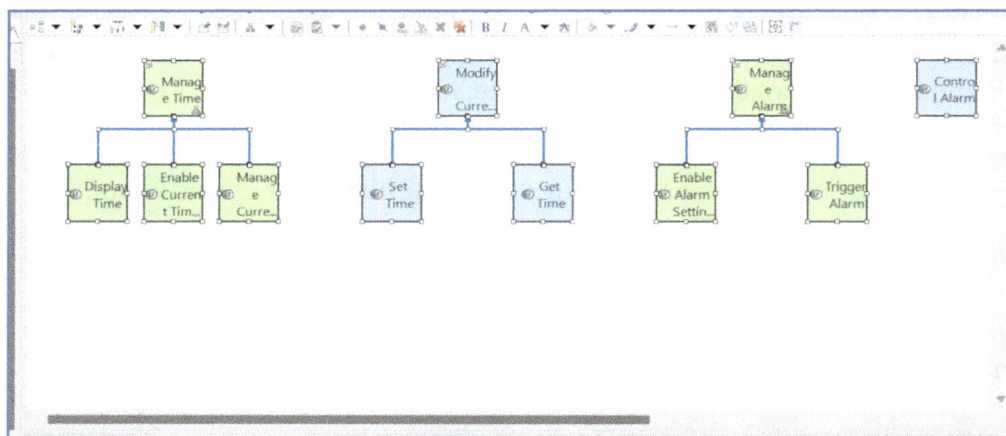

Figure 9-5 – All functions are displayed

Once again, all functions are displayed as squares with a default size and selected. This time, we have more functions which are broken down, so the diagram will probably be wider than the screen. Once again, you may need to manually adjust the size of the boxes.

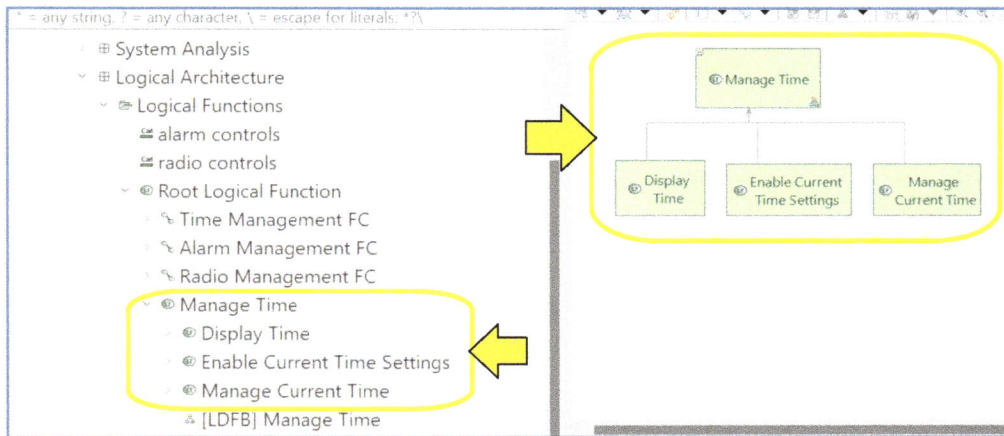

Figure 9-6 – Functional breakdown in the project explorer

It is interesting to compare the hierarchy shown in the diagram with the hierarchy shown in the project explorer. In the project explorer, function decomposition is visible by means of indentation, and the use of arrowheads to open or close the breakdown.

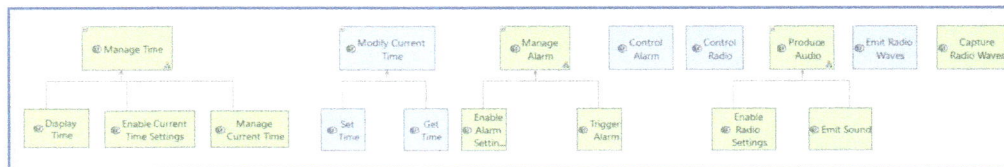

Figure 9-7 – Use zoom and rearrange

We can change the zoom ratio of the diagram, or manually rearrange functions to save space. We recommend keeping the vertical alignment of parent functions and child functions, and only moving squares horizontally.

Imagine we have a much more complex model with several levels of functional decomposition. The diagram would become quite complicated and difficult to view. *Capella* provides a tool to help control this complexity.

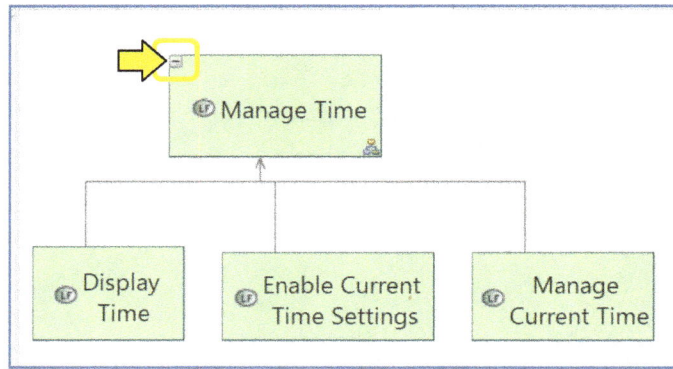

Figure 9-8 – Control to collapse and expand hierarchy

You can see a little " - " sign on the top left corner of decomposed functions. This minus sign is a control that allows you to "collapse" and "expand" the decomposition, to simplify the diagram at will.

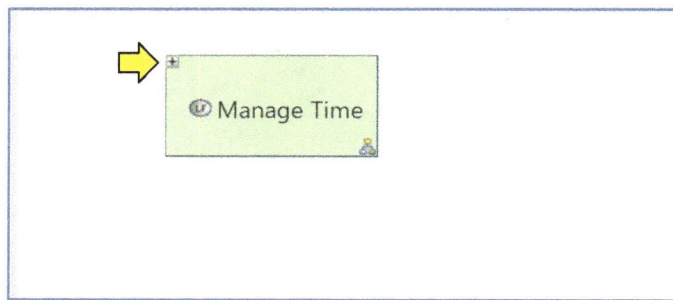

Figure 9-9 – Manage time collapsed

We can do it on "Manage Time" first to see the result. All three subfunctions have been hidden.

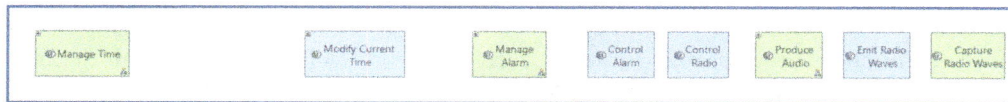

Figure 9-10 – All functions collapsed

If we do the same on all four parent functions, we only see the first level of decomposition, and a " + " sign has replaced the " - " sign on the top left corners.

We can also "clone" the functional breakdown diagram in order to keep different versions of the diagram, with or without subfunctions, for different readers. [31]

If we have a lot of decomposed functions, the breakdown diagram can become very wide and difficult to read as a whole. In this case, we can create a partial FBD for a specific function.

---

[31] See: *Cloning Diagrams to Provide Different Views* on page 149.

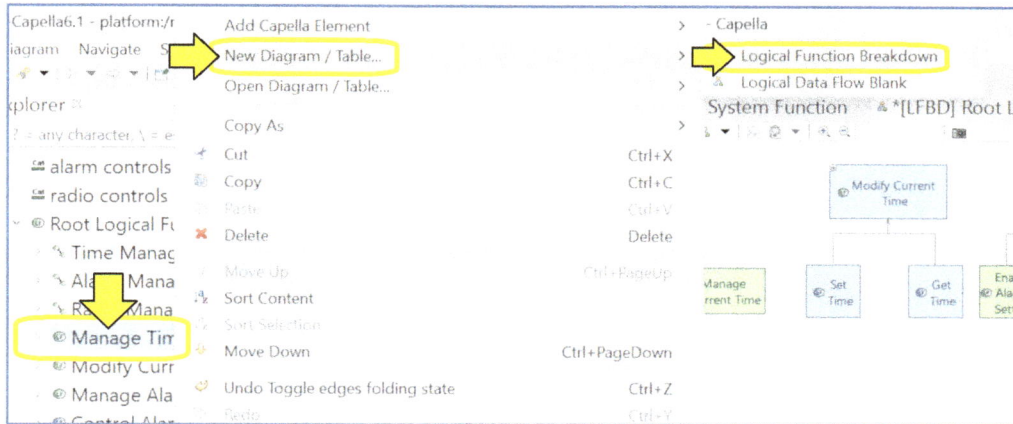

Figure 9-11 — Create partial FBD

In the project explorer right-click on the "Manage Time" logical function, select "New Diagram/Table...", and then "Logical Function Breakdown".

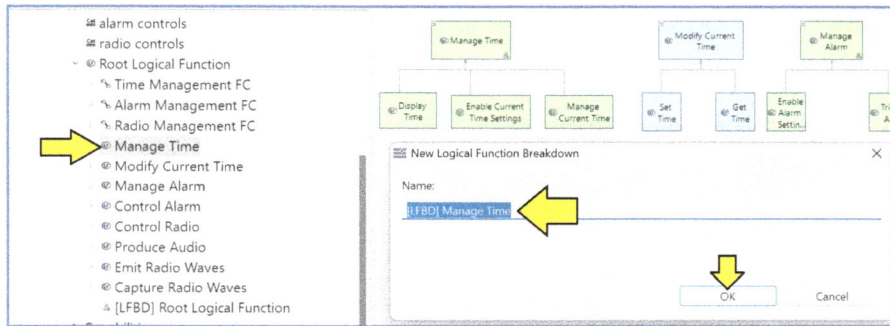

Figure 9-12 — The name of the selected function becomes the diagram name

Capella automatically puts the name of the selected function as the diagram's name. Click "OK".

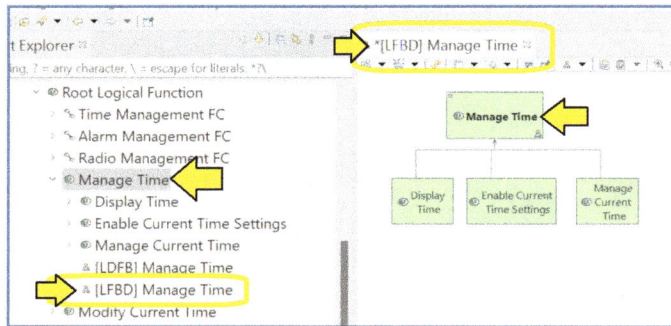

Figure 9-13 – Chosen function appears in bold

The result is a simple tree where the chosen function is the root and appears in bold. Notice that the breakdown diagram itself is stored under the function in the project explorer instead of being at the end of the folder.

We can create as many individual function breakdown diagrams as we need.

Physical Function Breakdown (PFBD)

We can create the same sort of diagram using the same procedure in the physical architecture.

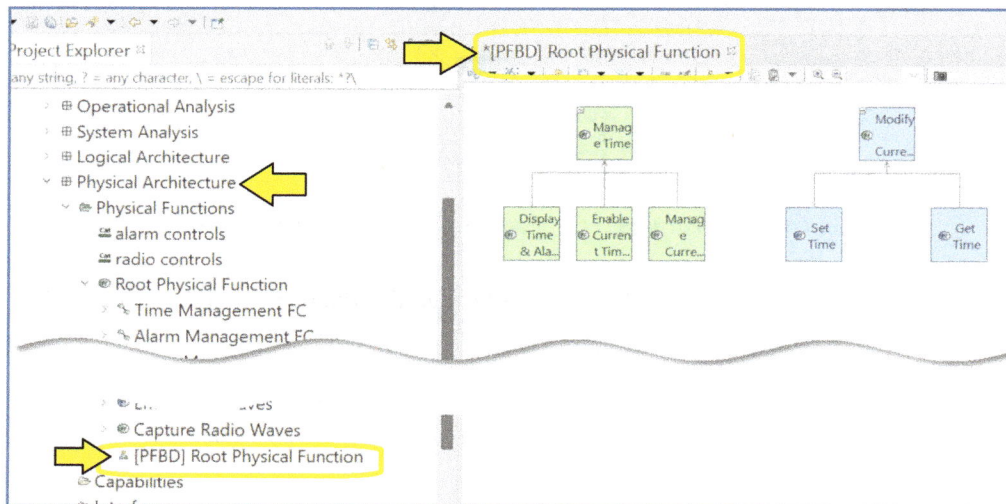

Figure 9-14 – Physical function breakdown is similar to logical function breakdown

As we did not refine functions any more at physical architecture level in Chapter 8, the "Physical Function Breakdown (PFBD)" diagram looks exactly the same as the LFBD, except that it shows physical functions instead of logical functions.

Component Breakdown

Introduction

As we already mentioned in Chapters 7 and 8, a component (logical or physical) can be refined into subcomponents or child components. We also define **leaf components** as components which are not broken down.

Operational entities (but not operational actors) can also be broken down. We will not be covering them here, but Operational Entity Breakdown diagrams [OEBD] at operational analysis level work pretty much the same way as the component breakdown diagrams presented in the following sections.

Arcadia treats components and functions in a very different way, as parent components have the same "rights and duties" as leaf components (unlike functions):

- A non-leaf component can have functions allocated to it.

- A non-leaf component can own ports.

- A component port belonging to a (father) component should be delegated to a port belonging to one of its (children) subcomponents by a delegation link.

Logical Component Breakdown (LCBD)

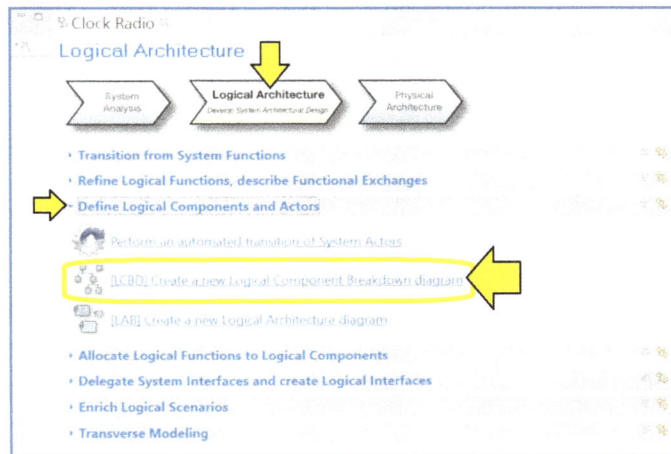

Figure 9-15 – Create a Logical Component Breakdown (LCBD)

Create a "Logical Component Breakdown (LCBD)" diagram from the activity explorer, by selecting the topic "Define Logical Components and Actors".

Remember that there is no similar diagram at the system analysis level, as there is one and only one "system" model component element, which is a black box and is not structurally decomposed...

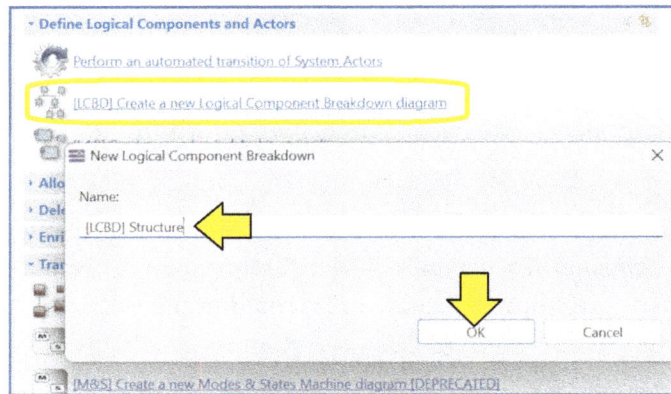

Figure 9-16 – The name defaults to the element under which it is created

Click on "[LCBD] Create a new Logical Component Breakdown diagram". A panel appears with a field to set the diagram name. The default diagram name matches the name of the model element under which the diagram will be located. In this case, the "Structure" folder, along with the logical components and actors.

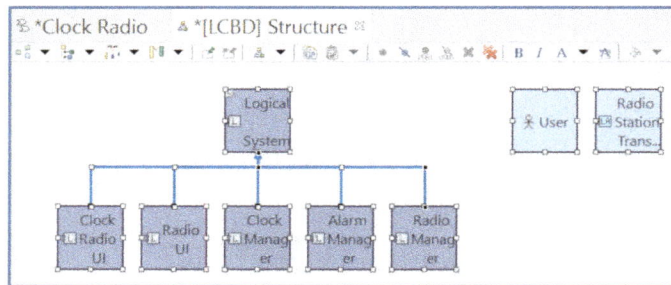

Figure 9-17 – The resulting diagram is automatically filled

The resulting diagram is automatically filled with all existing logical components or actors, along with their decomposition. All components and actors are displayed as squares with a default size and are selected by default. Notice that *Capella* uses the light blue color for the external actors, and the darker blue for the components, as in the LAB diagram.

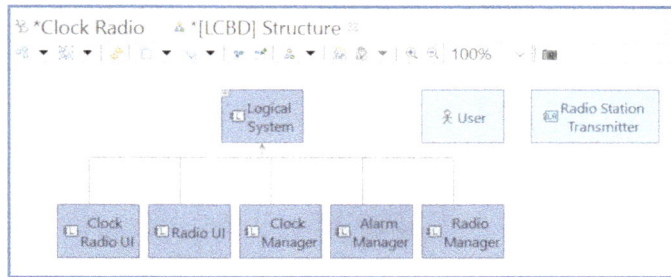

Figure 9-18 – Resize the elements as needed

This time, the default size of boxes does not fit the names of all actors, so we have to work a little on the layout of the diagram. We can change the zoom ratio of the diagram, and manually resize the last actor.

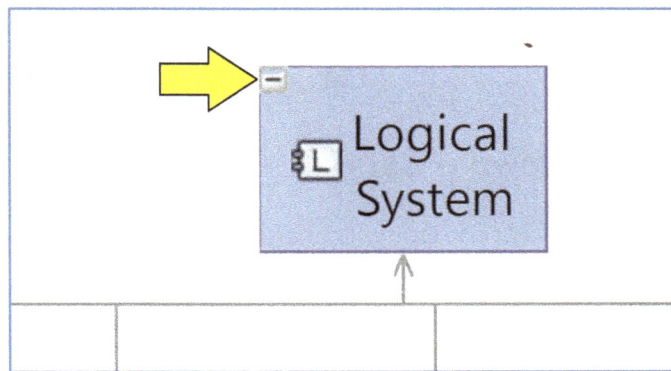

Figure 9-19 – The logical system can be collapsed

Once again, you can see a little " - " sign on the top left corner of the logical system. It allows you to "collapse" and "expand" the decomposition, to simplify the diagram at will.

Figure 9-20 – Expand the logical system again

The external actors are not decomposed in our example, so we can either keep them, or choose to hide them from the diagram.

We can do this easily by applying the "Hide element" button of the diagram toolbar on the selected actors. Now, only the breakdown of the logical system remains on the diagram.

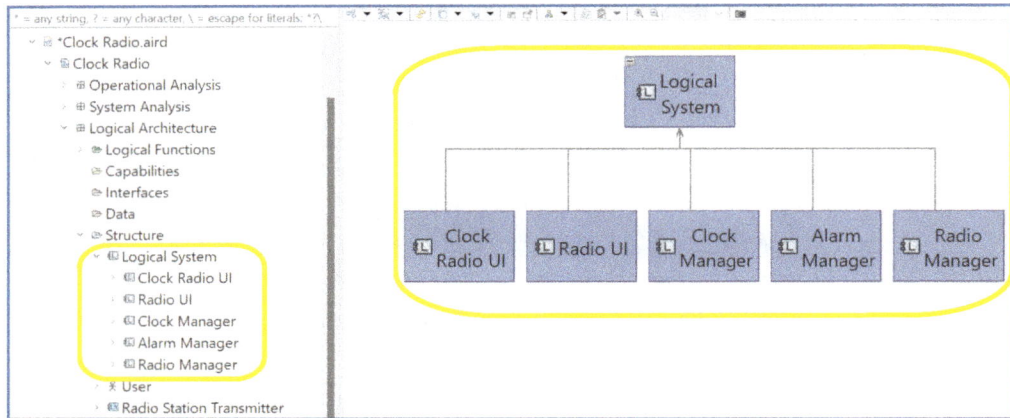

Figure 9-21 – Decomposition in the project explorer

Again, we can compare the contents of the project explorer with the diagram to see the different methods of representing component breakdown.

Physical Component Breakdown (PCBD)

Finally, we can take a look at how physical components get broken down.

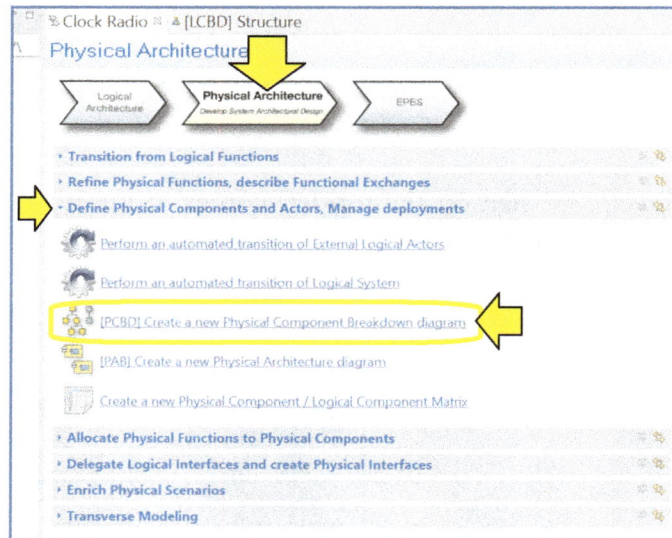

Figure 9-22 – Create a new physical component breakdown diagram (PCBD)

Again, we can create a "Physical Component Breakdown (PCBD)" diagram from the activity explorer, by returning to the physical architecture activity, selecting the topic "Define Physical Components and Actors, Manage Deployments", and clicking on "[PCBD] Create a new Physical Component Breakdown diagram".

A panel appears with a field to set the diagram name. The default diagram name matches the name of the model element under which the diagram will be located. In this case, *Capella* will place the diagram "[PCBD] Structure" inside the "Structure" folder.

Figure 9-23 — Behavior and node components are in separate hierarchies

The resulting diagram is automatically filled with all existing physical components or actors, along with their decomposition. You may need to manually adjust the sizes of the boxes.

Note that the relationship between behavior (blue) and node (yellow) physical components is called deployment and is not the same as containment. As such, behavior and node components are always in separate hierarchies.

Further Study

Here is a list of further topics for more in-depth study:

- **Delegation of Component Exchanges** – Delegation of component exchanges is an important topic related to encapsulation and modularity of your model. It is, however, a bit tricky in that allocation of component exchanges to function ports has to be done carefully when delegation is used.

    - https://url4ap.net/CH-9-Delegate

- **Adding Elements from a Breakdown Diagram** – Breakdown diagrams can also be used to add elements to the model.

    - https://url4ap.net/CH-9-BD-Add

- **Actor Decomposition in a Breakdown Diagram** – Breakdown diagrams can be used to decompose actors.

    - https://url4ap.net/CH-9-BD-Actors

Chapter 10 – Scenario Diagrams

In this chapter, we will take a closer look at the so-called Scenario Diagrams, showing the vertical sequence of messages passed between model elements at all *Arcadia* levels. This type of diagram is probably one of the most widely used of all *Arcadia* diagrams and is directly inspired by the UML/SysML sequence diagram.

Different types of Scenario Diagrams

Capella provides several types of scenario diagrams at each *Arcadia* level:

- **functional scenarios** – (the vertical lines are functions, and the sequence messages are functional exchanges),

- **exchange scenarios** – (the vertical lines are components/actors, while the sequence messages are functional exchanges or component exchanges),

- **interface scenarios** – (the vertical lines are components/actors, while the sequence messages are exchange items).

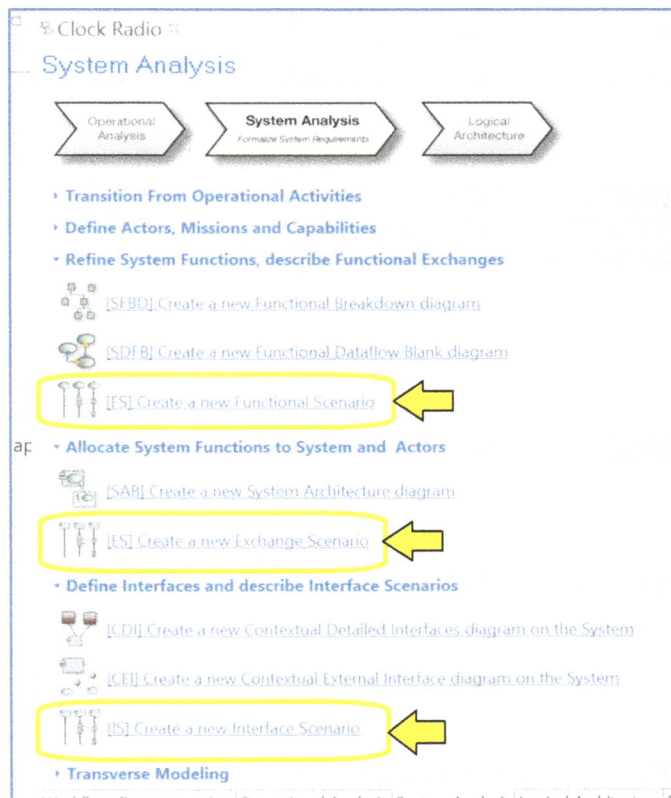

Figure 10-1 – Three types of scenario diagram are available

In the activity explorer in the "System Analysis" section we can see these three types of scenario diagram.

Functional Scenario –

A **functional scenario** is a time-ordered dynamic flow, on a temporal axis (conventionally vertical from top to bottom), of functional exchanges between different functions in the context of implementing a capability.

Such a scenario is formed of a set of references to functions and functional exchanges that link them, but unlike functional chains, these exchanges (in fact, their references) are ordered in relation to one another on a single temporal axis.

[Voirin] – Chapter 17, section 10

Exchange Scenario –

An **exchange scenario** is a time-ordered dynamic flow, on a temporal axis (conventionally vertical from top to bottom), of component exchanges between different behavioral components in the context of implementing a capability.

Such a scenario is formed of a set of references to components and functional exchanges that the functions implemented by these components. These exchanges (in fact their references) are time ordered in relation to one another on a single temporal axis. Occurrence of a function implemented by a component can be specified on the temporal axis associated with this component. A reference is then added in the scenario description.

[Voirin] – Chapter 20, section 6

We will not cover **interface scenarios**, which can only use **exchange items** (which we will cover in more detail in Chapter 12) as messages and are more related to UML and the object-oriented "interface" concept.

If you are modeling according to the core *Arcadia* concepts, in most cases you will be best served by focusing on **exchange scenarios**.

These three types of scenarios also exist at logical and physical levels. Note that only the first two types exist at operational level and are named differently: **Operational Activity Scenario (OAS)** and **Operational Entity Scenario (OES)**.

Basic Sequences

Introduction

We already briefly introduced the scenario diagram in Chapter 2. In this chapter, we will cover scenario diagrams in detail.

The example used in this chapter is again the simple Clock Radio for domestic use. We will continue using the model we developed in the previous chapters.

Arcadia on Scenarios and Capabilities

Arcadia scenarios are tightly tied to the previously presented capability concept. (See *Capabilities* on page 61). Scenarios explain the behavior of model elements (system, logical components, physical components, actors, etc.) in the context of a given capability. In most system designs, we will create several different scenarios for each system capability, such as "Main Success Scenario", "Alternate Scenario", and "Error Scenario".

Capabilities exist at all levels of *Arcadia*. *Capella* requires scenarios at each architecture level to directly belong to capabilities at that level and to be stored in the capability section of the project explorer for that level.

Our introduction to scenario diagrams in *Adding a Scenario Diagram* on page 31 in Chapter 2 explained that if you were not working from an existing scenario, *Capella* would create a capability for you when you created the scenario diagram. That explanation was correct and sufficient for the quick start section, but not entirely complete:

- *Capella* will only create a new capability for you on the fly if no capabilities have been previously defined.

- If any capabilities have already been defined at the relevant *Arcadia* level, *Capella* will force you to select one of the already defined capabilities for your new scenario diagram.

While we have already defined a few capabilities at the system architecture level, none of them fit the "main success scenario". Before we can create the scenario diagram for the "main success scenario", we will need to create a suitable capability to own it. Fortunately, it is quite easy to create a capability directly from the project explorer.

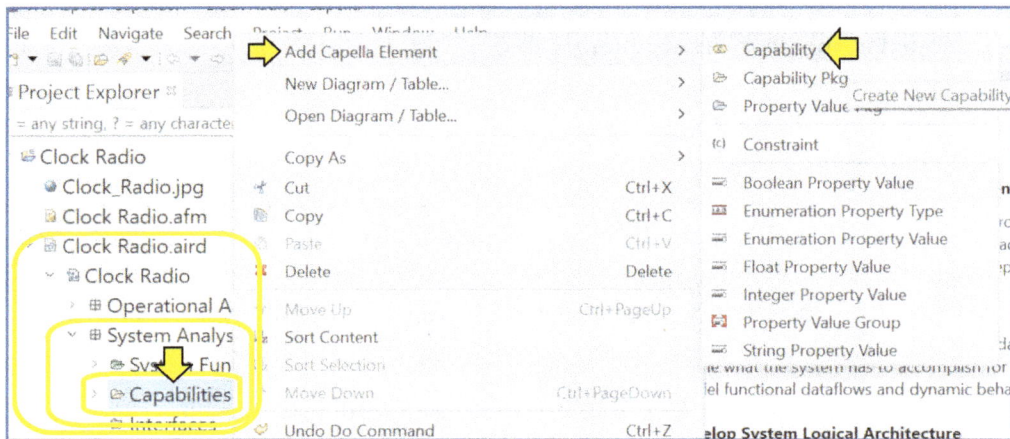

Figure 10-2 – Add a capability from the project explorer

1) In the project explorer, find the capabilities folder within the system analysis section.

2) Right-click on the capabilities folder, select "Add Capella Element", and then select "Capabilty".

3) Rename the capability to "Wake User in Morning".

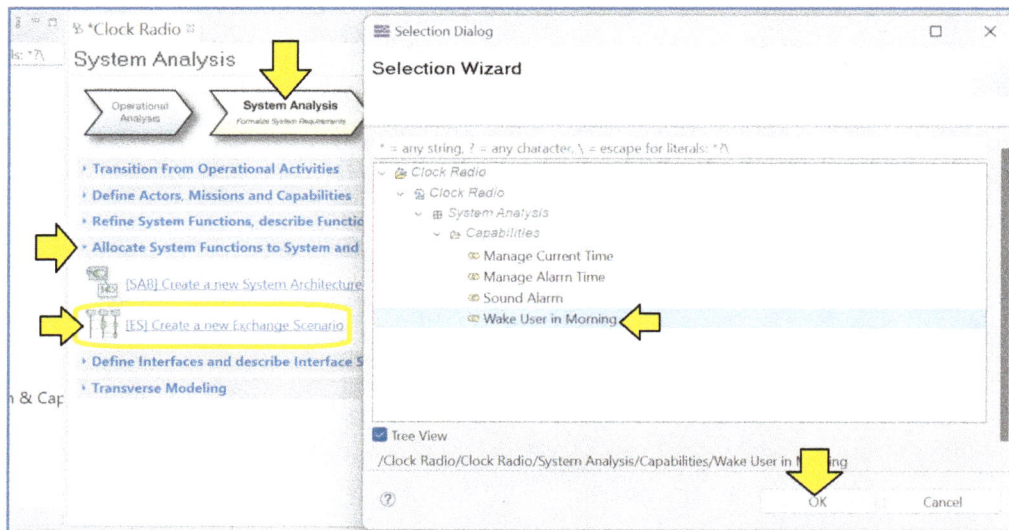

Figure 10-3 – The name defaults to the capability

1) In the activity explorer, open the system analysis section.

2) Expand the "Allocate System Functions to Systems and Actors" section.

3) Click on "[ES] Create a new Exchange Scenario".

4) The selection wizard will appear. Select the capability "Wake User in Morning" that we just created. Click "OK".

5) A panel appears with a field to set the diagram name. The default name shown is "[ES] Wake User in Morning". While that is a very appropriate name for a system capability, it is not a good name for a scenario. Replace "Wake User in Morning" with "Main Success Scenario".

6) Experienced *Capella* users recognize the "[ES]" prefix. In fact, we recommend enhancing the prefix by adding a first letter for the *Arcadia* level, as in many other diagram types. [32]

7) Altogether, the diagram name should be "[SES] Main Success Scenario".

8) Click "OK".

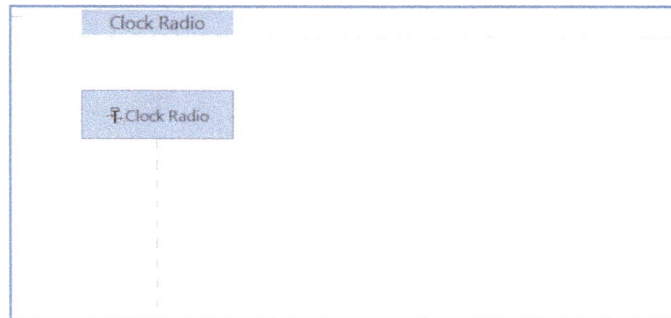

Figure 10-4 – At the system level, the system itself is always included

At system level (and only at the system level), the scenario diagram is not empty because *Capella* "knows" you need to represent the system in each scenario. This *Capella* behavior is similar to the behavior when creating system architecture blank diagrams that we discussed in Chapter 6.

In fact, if you look closely at the "Allocate System Functions to Systems and Actors" section in the system analysis activity in the activity explorer, you will see that creation of the exchange scenario is just below creation of the architecture blank.

We recommend placing the system as a black box in the center of the diagram and the actors to the left and right with the which initializes the scenario on the left.

In order to insert the actors, we can either use the "Insert/Remove Actors" button in the palette or drag and drop them from the project explorer. We will use the palette.

---

[32] Strangely, *Capella* automatically adds the first letter only at operational level.

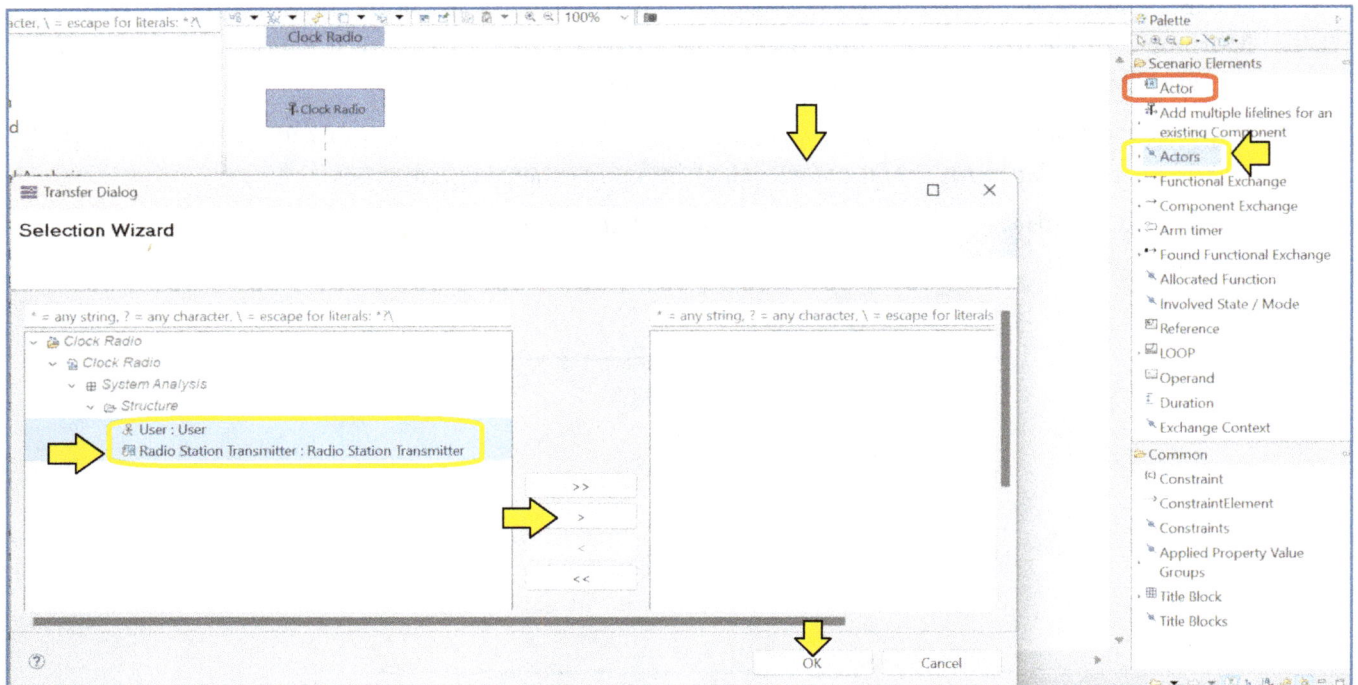

Figure 10-5 – Use the transfer dialog to add both actors to the diagram

1) Notice that the palette contains both a tool called "Actor" and another rather similar tool called "Actors". We want the second one.

2) Select "Actors".

3) Click anywhere in the diagram.

4) The transfer dialog will open.

5) Since we will want both actors in the diagram, select both actors and click on the single arrow (>) to add both to the diagram. [33]

---

[33] Or don't bother selecting the actors and instead click on the double arrow (>>) to "add all available actors" to the diagram.

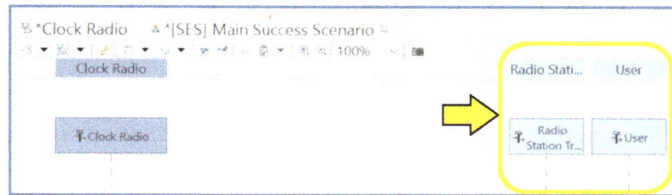

Figure 10-6 – Both actors appear in the diagram

Both actors appear in the diagram, as vertical dashed lines, similarly to the system element, except that the fill color of the top box is light blue instead of dark blue (as in architecture blank diagrams). (Depending on how your screen is laid out, the actors might appear in a slightly different position in the diagram).

Instance Role –

An **instance role** (**lifeline** in UML/SysML) is the representation of the existence of a model element that participates in the scenario involved. It has a name that reflects the name of the item of the model referenced and is represented graphically by a dashed vertical line. This dashed line indicates the passage of time in the "life" of the element, starting at the top and progressing downward in time.

Note: in this book we will be mostly referring to the dashed line portion of this diagramming element. We will be using the somewhat more intuitive UML/SysML term **lifeline** rather than the formally correct, but less understandable *Arcadia* term: **instance role**.

Figure 10-7 – Drag to adjust spacing of lifeline

You can resize the header box of the lifeline by selecting it and dragging the corners of the box around as needed. You can also change the horizontal position of the lifeline by dragging it back-and-forth as needed.

Sequence Messages

Now we are ready to start creating the actual interaction between these three elements. In the palette you will find icons for several different types of messages.

Sequence Message –

A **sequence message** is a unidirectional interaction between the element represented by the source lifeline and the one represented by the target lifeline. In *Capella*, according to the kind of Scenario, sequence messages can refer to Functional Exchanges, Component Exchanges, or Exchange Items.

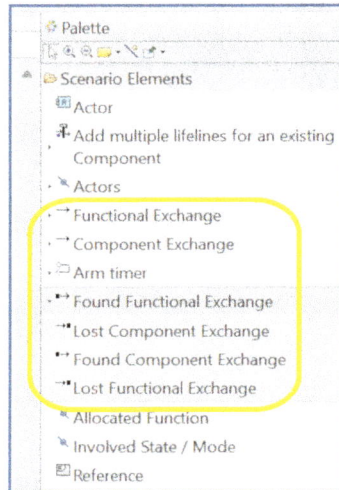

Figure 10-8 – There are lots of message types, but...

Capella proposes a lot of different types of messages. We will explain them later in this chapter, **even if we strongly recommend using only the first one: "Functional Exchange"!** Each message in this scenario diagram will refer to a functional exchange between system functions allocated to the system or an actor.

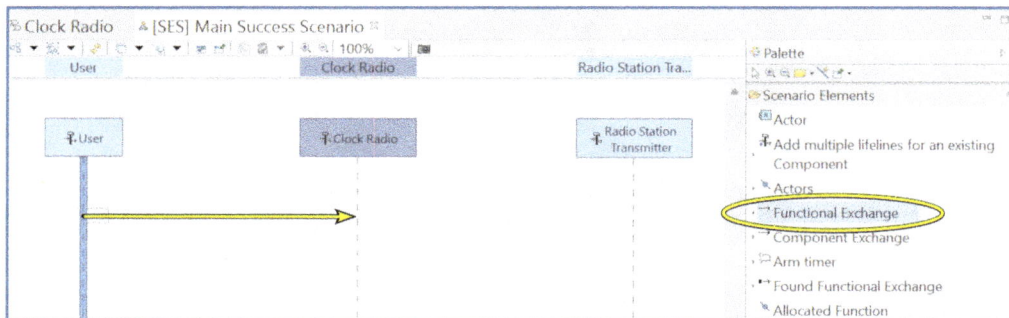

Figure 10-9 – Drag rightward to create functional exchange message arrow

1) Select "Functional Exchange" in the palette.

2) Hover the mouse pointer over the lifeline of the "User" actor until it turns thick and blue.

3) Depress and hold the left mouse button and begin to drag to the right towards the lifeline of the "Clock Radio".

4)   When the lifeline of the "Clock Radio" turns thick and blue, release the left mouse button and then immediately click once.

Capella automatically opens a window prompting you to select one of the existing functional exchanges. In our case, *Capella* has determined that there are eight functional exchanges connecting the user and the clock radio system and presented them to us so we can select one.

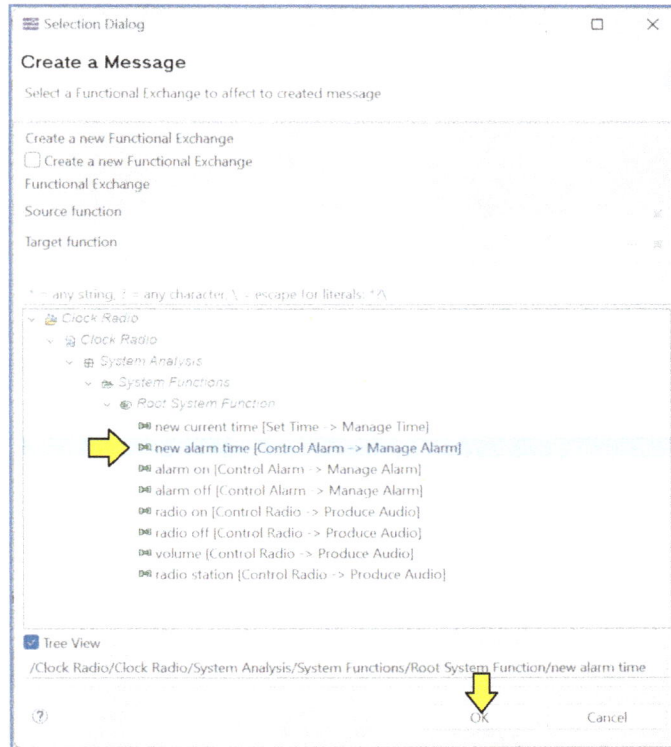

Figure 10-10 – Select new alarm time

Select the "new alarm time" functional exchange and click "OK".

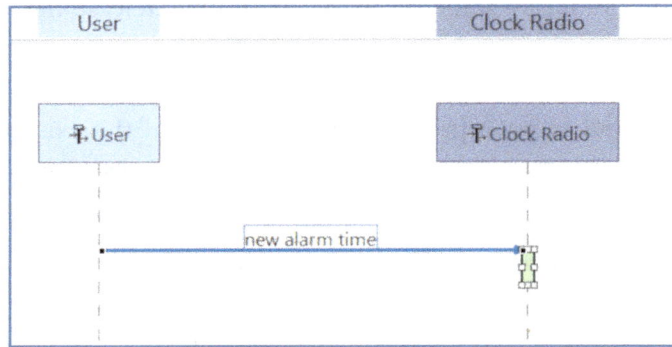

Figure 10-11 – Arrow from the user to the clock radio

Capella has created an arrow from the user to the clock radio. The tool has also created a green rectangle called "Execution" on the lifeline of the clock radio. [34]

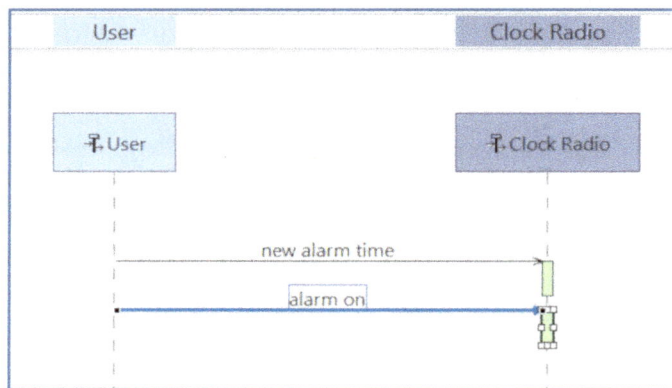

Figure 10-12 – Second message for alarm on

Create a second message in the same way by selecting "alarm on".

What if we want to model (and portray to the model reader) that the reception of this "alarm on" message, causes the clock radio to start continuously checking the current time to see if it is time to trigger the alarm?

The green box becomes really useful! This green box is similar to the "execution specification" concept in SysML and UML. However, *Capella* has a nice feature that is not supported in most SysML tools: we can link the green box to an actual function in the model.

Note: *Arcadia* does not seem to have a formal name for this green box. The UML/SysML term "execution specification" is a bit arcane. We will refer to this green box as the **execution bar**.

Let's link the "Manage Alarm" system function to this second execution bar.

---

[34] You can see the name "Execution" in the semantic browser.

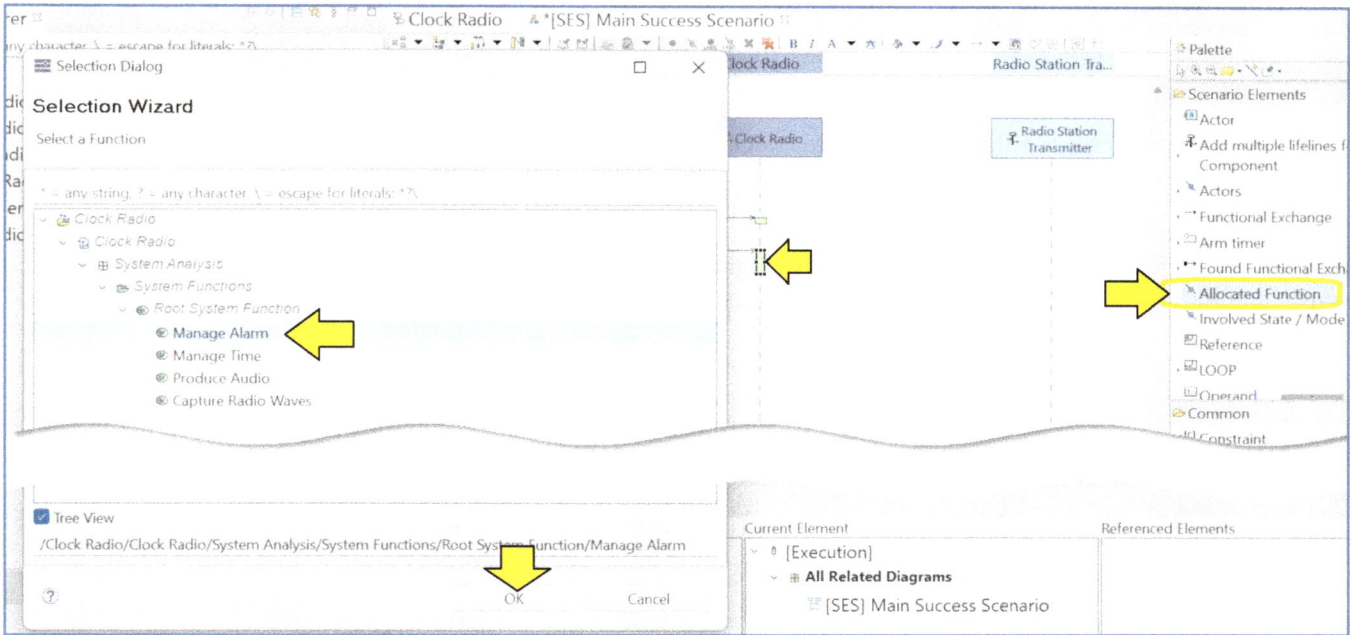

Figure 10-13 – Link system function to the execution bar in the lifeline

In the palette, select the "Allocated Function" tool. Click inside the execution bar. The selection dialog will appear. Select the "Manage Alarm" function. Click "OK".

Note that *Capella* will only insert functions that are allocated to elements corresponding to the selected lifeline.

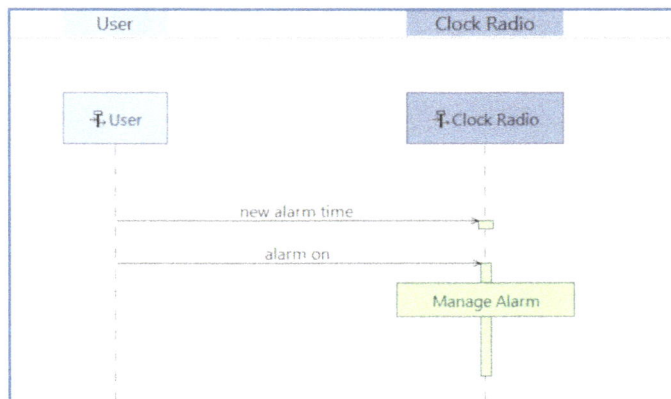

Figure 10-14 – Green rectangle appears now inside the execution bar

A green rectangle appears now inside the execution bar.

Notice that we have made a few manual format adjustments:

- We shrank the first execution bar after "new alarm time" because we have chosen not to add any detail or emphasize any particular action for this message.

- We have manually adjusted the size of the "Manage Alarm" box.

- We have increased the length of the execution bar after "alarm on" to make it clearly stick out from underneath the "Manage Alarm" box.

In fact, *Capella* created a so-called **State Fragment** model element which refers to the function. This strange name will be explained in the Chapter 11 *Mode or State Machine Diagrams* on page 287.

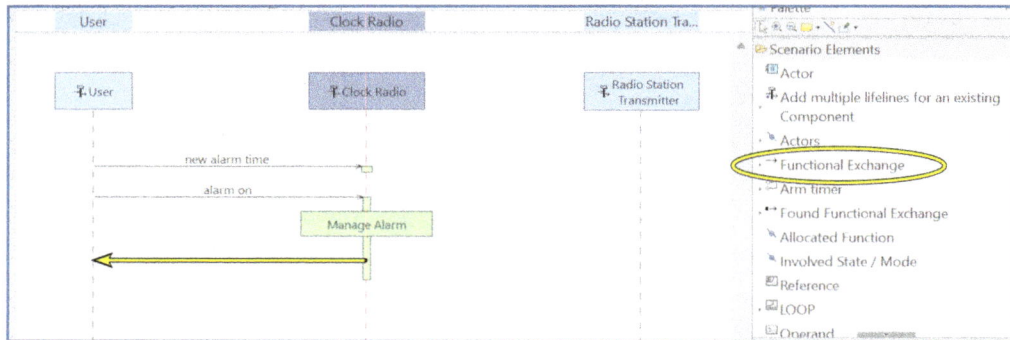

Figure 10-15 – Create message from inside the execution bar

Draw a third message, this time from the clock radio to the user. In order to stress the fact that this message is a result of the execution of the function, we will take care to make it start inside the execution bar and not below. Starting below the execution bar would not be a mistake, but the subtle difference will help the reader to understand our modeling intention.

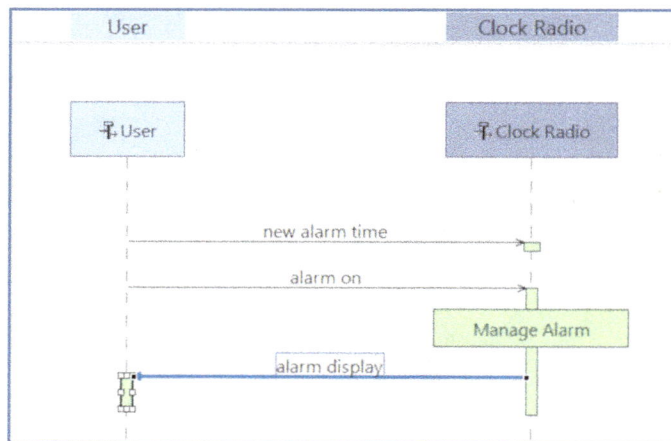

Figure 10-16 – Select alarm display

When *Capella* opens the selection wizard, select the "alarm display" functional exchange.

Figure 10-17 – Adjusting the execution bar to indicate the end of function execution

Notice that a new execution bar has been created on the user's lifeline, and that the source execution bar has been automatically resized by *Capella* to match the new one. If you want to express the fact that the clock radio has finished its job, you can manually shrink the execution bar under "Manage Alarm" to end just after the message is sent.

Next, we will explain how to vertically move messages. We may have made an error in the sequence, or we might simply want to rearrange them graphically to save space.

The tip & trick is the following: **do not move the arrows – move the execution bars!** We can illustrate this point simply by testing both possibilities.

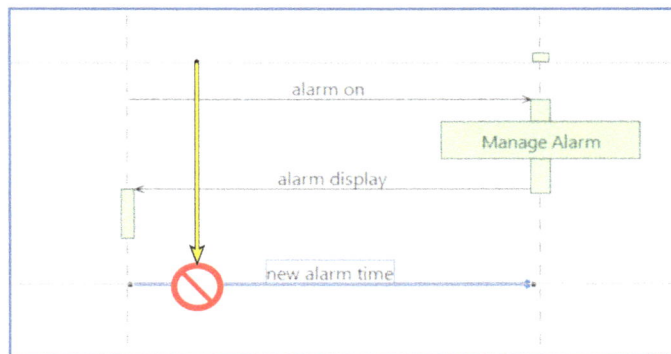

Figure 10-18 – Message arrow movement is forbidden

Using the mouse, grab the "new alarm time" message and try to drag it to the bottom of the sequence. *Capella* simply forbids it!

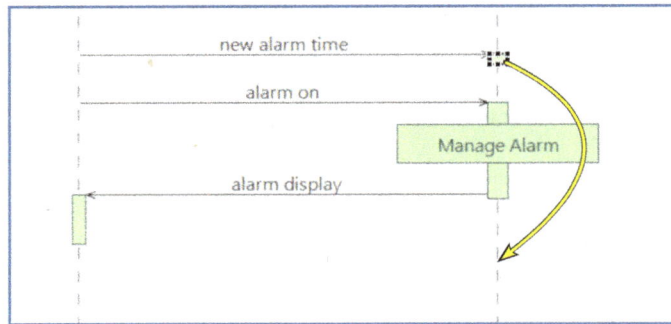

Figure 10-19 – Move the green rectangle to move the message arrow

To move the message down, simply move down the green rectangle of the execution bar.

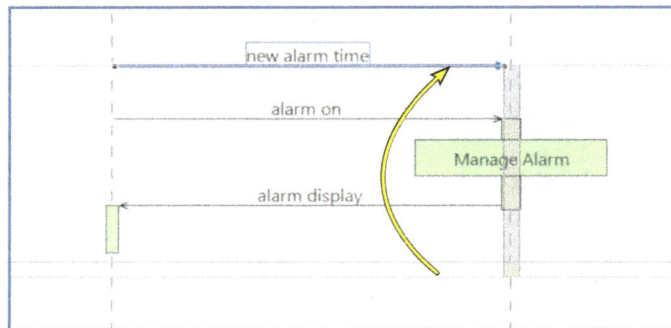

Figure 10-20 – Moving the arrow back will have unexpected effects

Moving the message down using the execution bar worked perfectly. However, if you want now to move back the message to its original position in the sequence and you try to directly move the arrow, *Capella* will move it up, but will resize the execution bar and put the existing one inside the resized one.

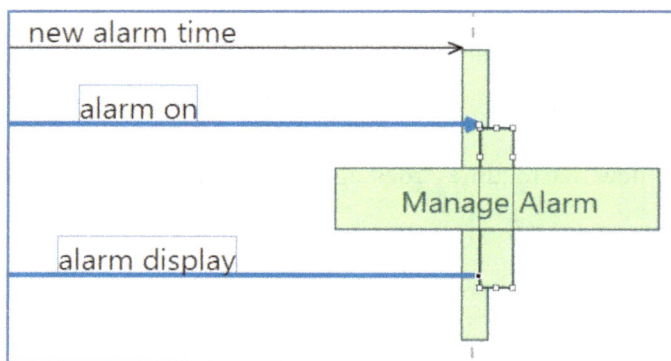

Figure 10-21 – We now have nested execution

This notation comes from the UML/SysML concept of parallel/nested executions. The meaning of this notation is probably not what you were trying to express in your model!

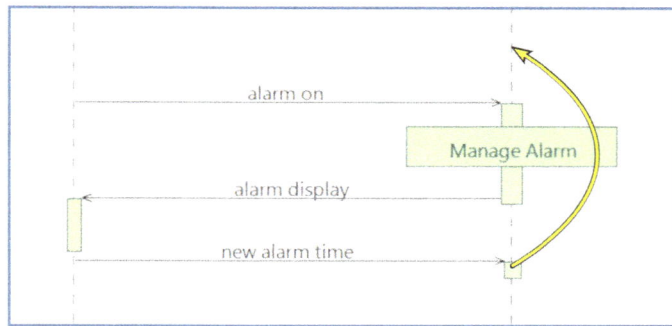

Figure 10-22 – Undo and move execution bar instead

Just undo the previous move and move the execution bar up instead, which works perfectly.

Let us go back to the different types of messages:

- Functional Exchange (+ Functional Exchange with Return Branch)
- Component Exchange (+ Component Exchange with Return Branch)
- Arm Timer (+ Cancel Timer)
- Found Functional Exchange (+ Lost Functional Exchange + Found Component Exchange + Lost Component Exchange)

The first two types are mutually exclusive: as soon as you use the "Functional Exchange" tool, the "Component Exchange" tool disappears from the palette, and vice-versa. As component exchanges are mostly groupings of functional exchanges, we have always found it strange to use them as messages.

The notion of "Return Branch" comes from UML/SysML and is interesting but unfortunately not handled very well by *Capella*.

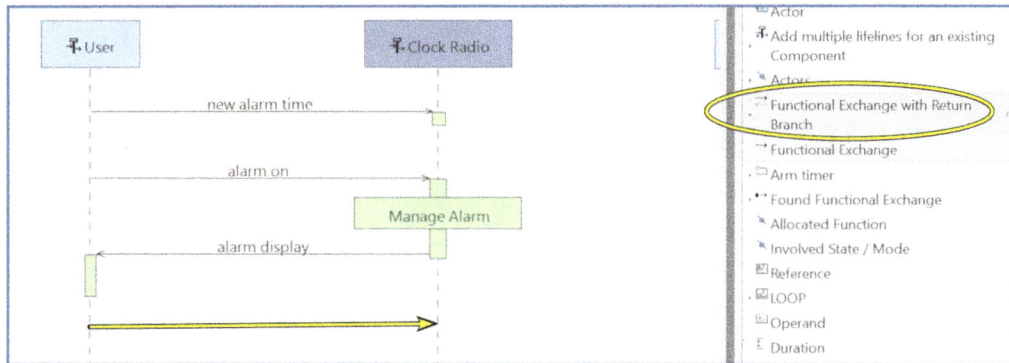

Figure 10-23 – Create a functional exchange with a return branch

Expand the "Functional Exchange" tool in the palette if needed to reveal the "Functional Exchange with Return Branch" tool and select the tool. Draw an arrow from the user's lifeline to the clock radio's lifeline. Select the functional exchange "alarm off".

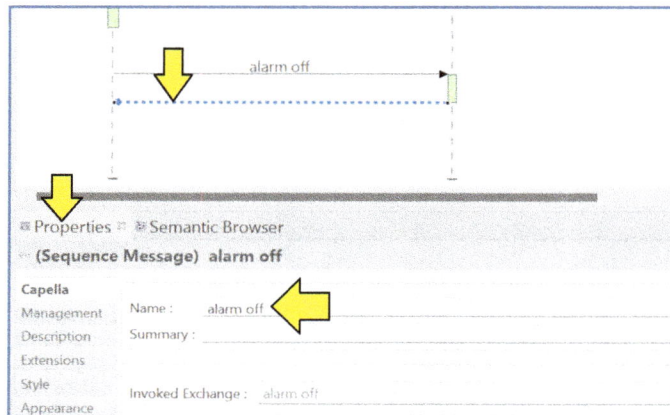

Figure 10-24 – By default, return branch has same name as message

By default, the name of the return branch is the same name as the initial message, and it does not appear on the diagram.

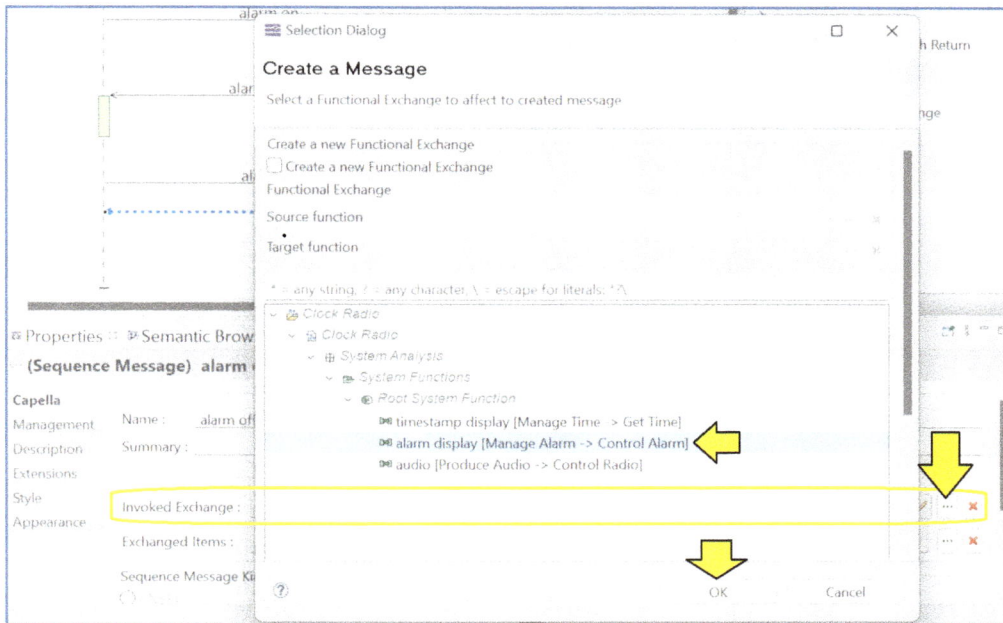

Figure 10-25 – Linking a message to a return branch

We would like to use a reference to the message "alarm display" for the return branch and it is possible to do so by modifying the "Invoked Exchange" field. Click on the three little green dots to open a selection dialog and select the desired message.

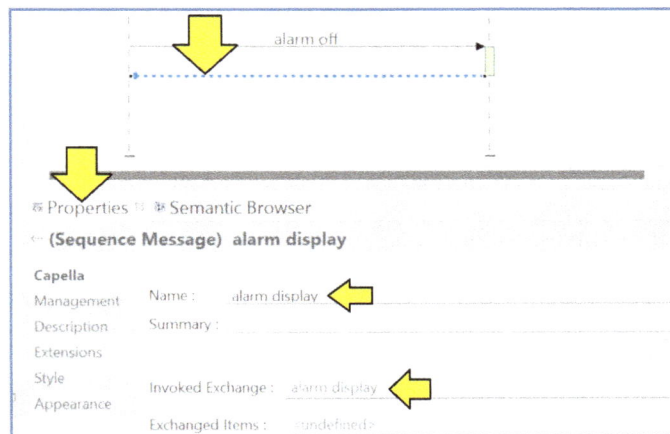

Figure 10-26 – Message linked, but message name not visible

Now the return message refers well to the desired message: "alarm display".

Unfortunately, there is absolutely no way to show the label in the scenario diagram…

The notions of Arm timer / Cancel timer are very specific, and do not conform to UML/SysML.

And the subtle notions of "Found …" and "Lost …" come from UML/SysML, when you do not want to declare in a sequence diagram from where a message comes, or to which element it goes. While the UML/SysML compliance is very nice, the basic concepts involved really do not fit with the *Arcadia* approach.

Let's review the different options for messages in the scenario diagram:

Message Type	Assessment
Functional Exchanges	**Recommended** – Functional exchanges fit the semantics and purpose of the diagram nicely.
Component Exchanges	**NOT Recommended** – Component exchanges are pipes between components that convey functional exchanges. Component exchanges themselves don't really do anything and are a poor semantic fit for the diagram.
Arm Timer	**NOT Recommended** – This concept is unique to *Arcadia*, is not well developed, and does not comply with UML/SysML usage.
Lost/Found Message	**NOT Recommended** – This concept is UML/SysML compliant but does not fit well with the *Arcadia* approach.

Table 10-1 – Messages in the scenario diagram

Self Messages

Self messages are useful for documenting the invocation of a function from within a component.

In order to keep the diagram simple, go ahead and delete the previous "alarm off" functional exchange with return message.

We can model the scenario that the clock radio sends itself an internal message from the "Manage Alarm" function to the "Produce Audio" function. We can represent this internal message as a "reflexive" arrow on the scenario diagram.

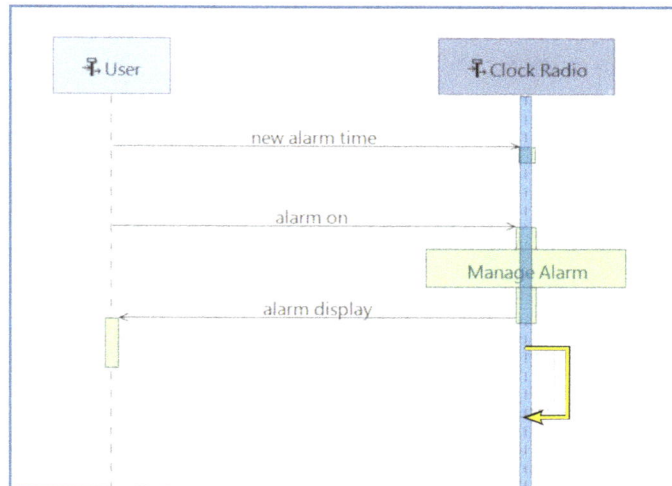

Figure 10-27 – Create a self message

1) Select "Functional Exchange" in the palette as usual.

2) Hover over the lifeline of the clock radio until it turns blue.

3) Press the left mouse button down and drag to the right – don't release the mouse button yet.

4) Continue to drag downward slightly – don't release the mouse button yet.

5) Drag back to the lifeline of the clock radio.

6) Release the mouse button.

Capella will open the selection dialog.

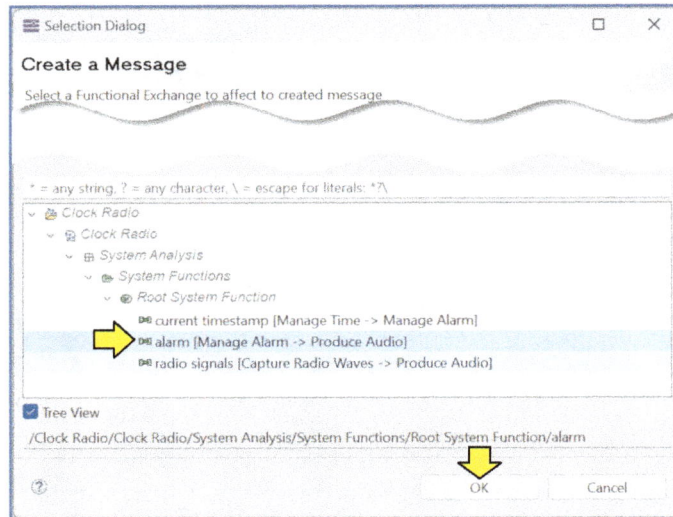

Figure 10-28 – Select an internal function

Only the internal functional exchanges of the selected element (clock radio) are shown. Select "alarm" and click "OK".

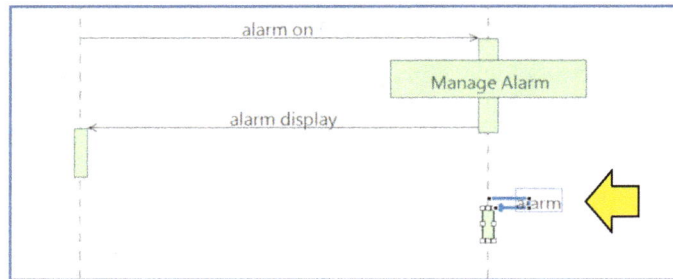

Figure 10-29 – Reflexive arrow added

Capella has created a reflexive arrow. The format may not be exactly what we want yet. In particular, we may want to adjust the execution bars.

Figure 10-30 – Moving the self message

For instance, we may want to adjust the execution bars to emphasize that the "alarm" message occurs sometime after the "alarm display" message is sent but that the "Manage Alarm" function is still active.

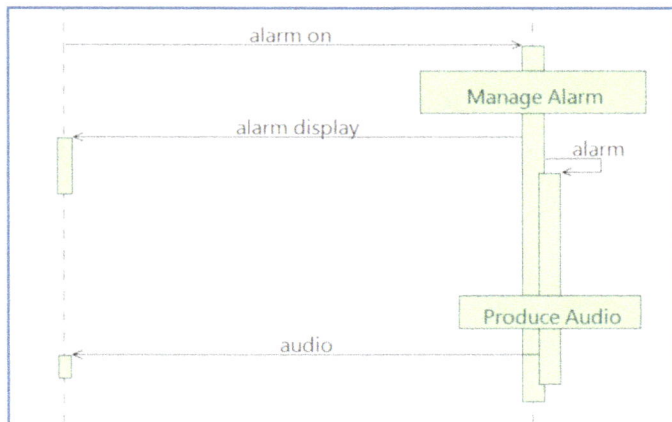

Figure 10-31 – We can now add the produce audio function

We can now easily add the "Produce Audio" function in the new execution bar, and the "audio" message to the user.

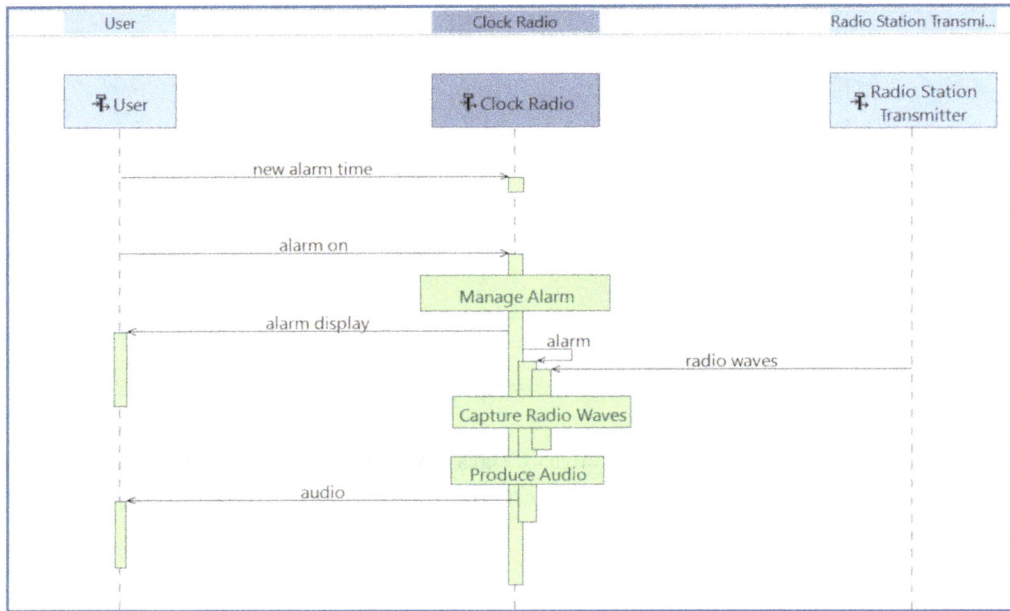

Figure 10-32 – We can add another level of function nesting

In the same way, we can decide to add the "radio waves" message coming from the radio station transmitter and the "Capture Radio Waves" function.

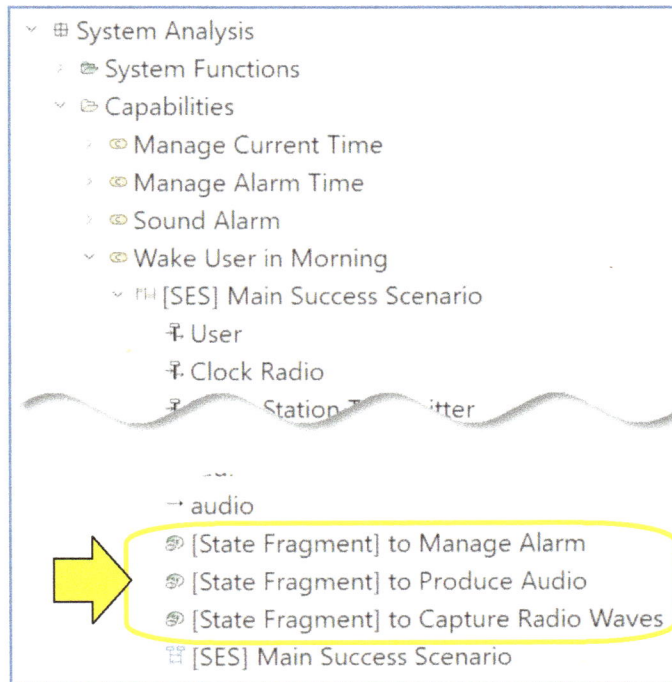

Figure 10-33 – State fragments in the project explorer

If we check the project explorer, we can find all important model elements: lifelines and messages. Functions appear in a strange way, as "State Fragments" (without a name until version 6.0 where it has been added). If you still use version 5.x of *Capella*, we recommend that you copy/paste the name of the function into the name of the State Fragment to clarify the project explorer.

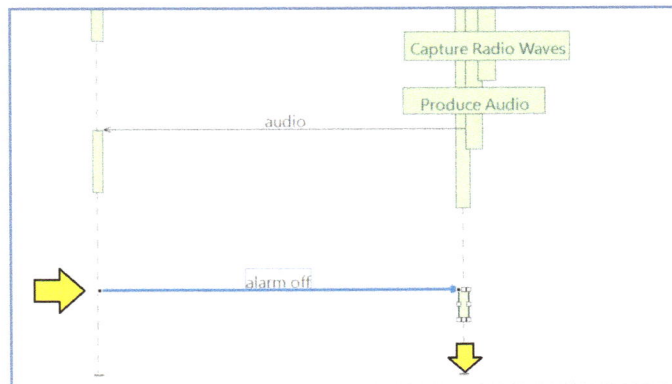

Figure 10-34 – Finish with alarm off message

Finish the scenario with an "alarm off" message. Note: there are some bugs in the *Capella* scenario diagram. You may find that the tool does not want to let you connect the message to the clock radio's lifeline. In this case, you should be able to get the tool to work by first dragging the lifeline down to make a lot of additional space.

Combined Fragments

In this section, we will introduce two important additional concepts: duration and combined fragment. The "Combined Fragment" concept comes from the sequence diagram of UML/SysML.

> *Combined Fragment –*
>
> A **combined fragment** consists of one or more interaction operands, and each of these encloses one or more messages, interaction uses (references), or nested combined fragments. Combined Fragments are mostly used to model loops, branches, and other kinds of alternative flows.

Suppose we now want to add an option for the user to allow the alarm to remain set for the next day. In this case, the final "alarm off" message of our previous scenario will become optional.

We will want to add an "OPT" combined fragment to our scenario. [35] Unfortunately, there are quite a few different types of combined fragment and the *Capella* path to the "OPT" combined fragment is a little complicated.

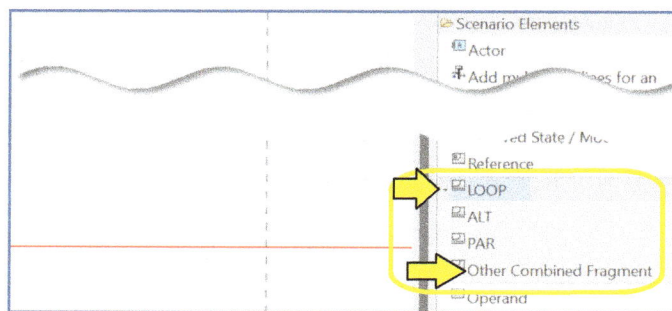

Figure 10-35 – Select Other Combined Fragment

In palette, first click on "LOOP" to expand the options. Then, select "Other Combined Fragment".

---

[35] Readers with any sort of programming experience will recognize the familiar "IF-THEN-ELSE" style of operation.

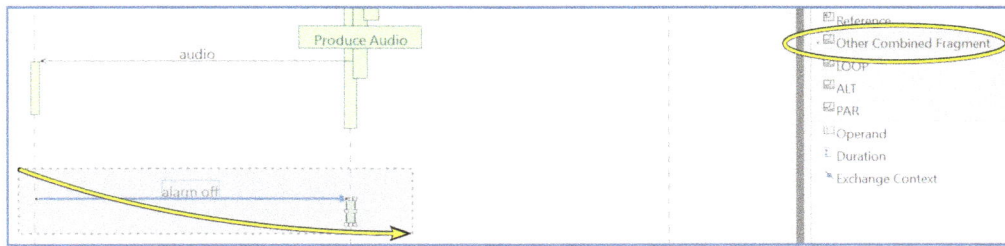

Figure 10-36 – Draw box around the message of interest

Draw a rectangle around the message of interest.

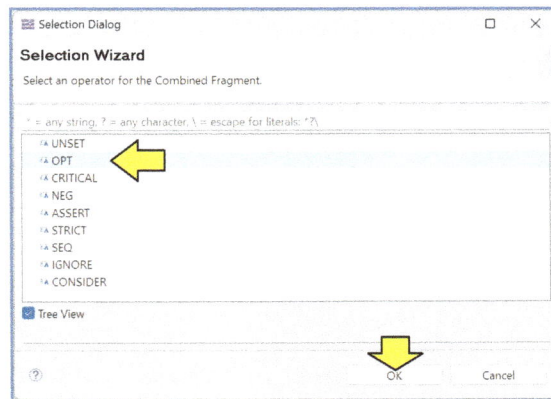

Figure 10-37 – Select OPT

Capella shows a list of possible types of combined fragments. Select "OPT" and click on "OK".

All these combined fragment types come from UML/SysML. The most popular are LOOP, ALT, PAR and OPT. LOOP and OPT only have one operand. ALT and PAR should contain at least two operands, representing either branches of alternatives, or parallel branches.

When you create a combined fragment, *Capella* only creates one operand. If you need to create a second operand, you will need to go back to the palette and select the "Operand" tool.

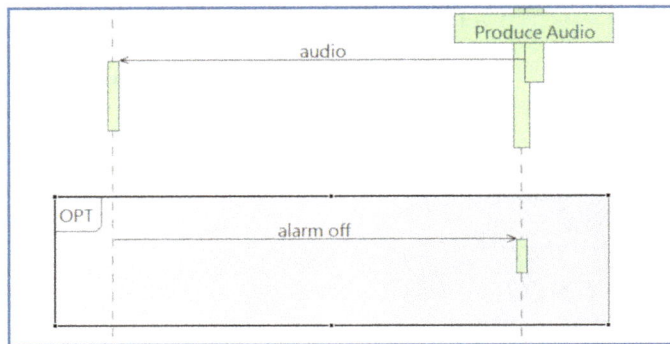

Figure 10-38 – OPT rectangle is created

In the case of the OPT combined fragment, a single operand is convenient. *Capella* creates a simple rectangle with the OPT keyword in the top left corner.

The initial appearance of this rectangle is deceptively simple. Without any further selection (or by selecting the top edge of the rectangle) *Capella* displays the combined fragment itself – the internal structure is not (yet) accessible from the properties panel.

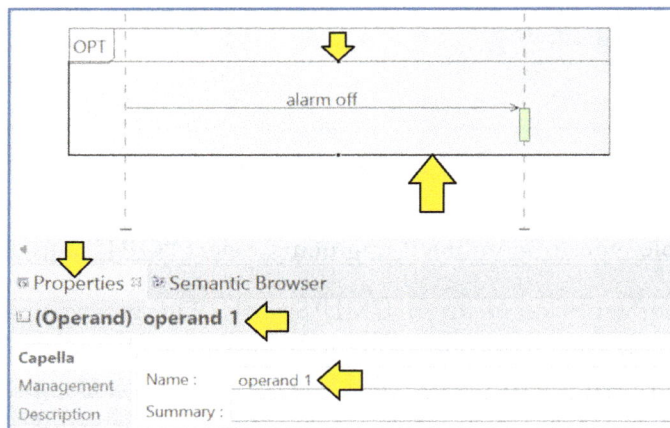

Figure 10-39 – Accessing the operand

In order to access the operand, click on the bottom edge of the combined fragment's outer rectangle. The *Capella* appearance will change slightly. An internal rectangle will appear, and the operand will become visible in the properties panel.

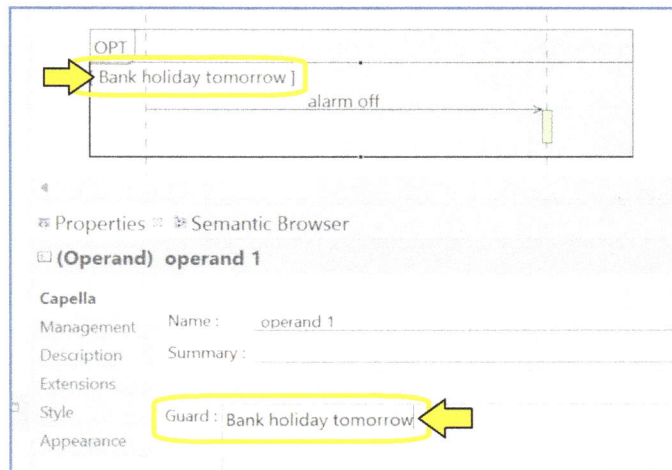

Figure 10-40 – Specify a guard condition

It is possible to specify a **guard**. The guard is a condition the controls execution of the combined fragment. The guard should be an expression that can (somehow) be evaluated as a Boolean variable. For instance, we could specify that only if the user has a bank holiday the next day, the system should send an "alarm off" message. The guard text appears at the top left of the operand rectangle enclosed in square brackets.

Should You Use Combined Fragments?

Combined fragments should be used carefully:

1) They can make your scenario diagrams difficult to understand for anyone except a UML or SysML expert.

2) The urge to use a combined fragment can be an indicator that you may have passed the point of diminishing returns in terms of modeling rather than coding.

For diagrams that need to be understood by a broad range of stakeholders, simpler is usually better. Split the diagram into multiple simpler diagrams or simply raise the level of abstraction so that the messy detail is not needed.

Combined fragments can be used successfully by certain kinds of teams, notably test teams who need to make intricate test procedures. However, such a team should be aware that complicated structures of nested combined fragments can make their sequence diagrams absolutely impenetrable to normal stakeholders outside of the team. Such models should be carefully positioned and consumed only within the team in question where all team members can be trained to understand the intricacies.

The last important concept in the scenario diagram is called **duration**.

The duration concept also comes from UML/SysML and allows us to specify a duration constraint (or timing constraint) between two messages or even on the execution of a function in *Capella*.

In our example, let's create such a duration constraint between the "alarm" message and the "audio" message. The meaning of the constraint will be to set a limit for the time between the triggering of the alarm and the production of audio.

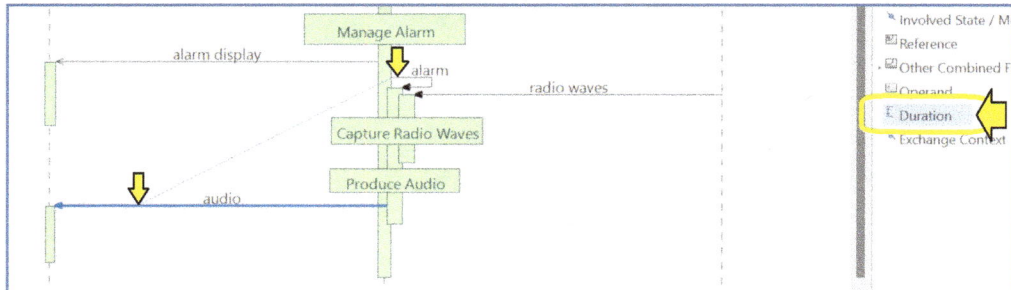

Figure 10-41 – Create a duration

To create this kind of constraint, we need to go to the palette and select the "duration" tool. Then we must first click on the source message and then click on the target message.

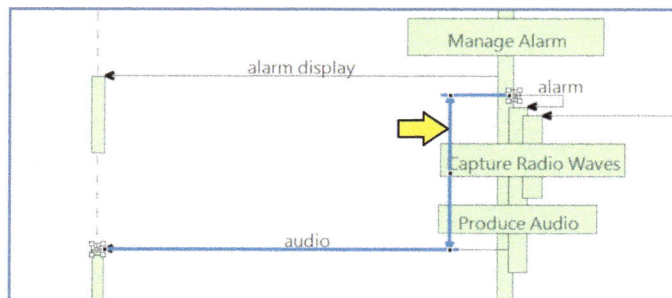

Figure 10-42 – A vertical double-arrow line has been added

Capella first adds blue horizontal continuation arrows for each message and then adds a vertical double-arrow line between them.

Figure 10-43 – Move the vertical line to make it more intuitive

Use the little dot in the center of the vertical double-arrow to move it horizontally to the left to make it easier to understand.

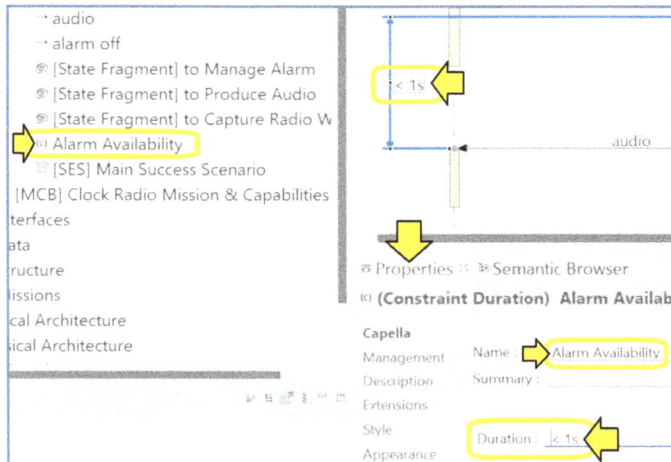

Figure 10-44 – The duration has a name

As a model element, the constraint has a name, which appears in the project explorer, and a specific property called "Duration" which appears in the diagram. Name the constraint "Alarm Availability" and specify that the duration must be less than 1 second. (You may need to manually adjust the position of the duration label in the diagram).

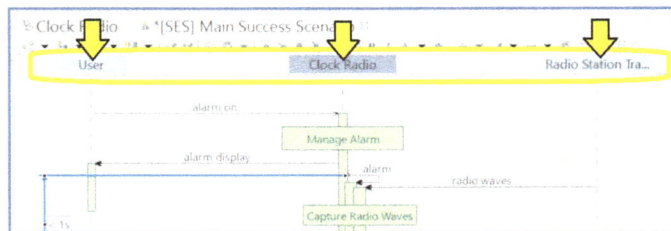

Figure 10-45 – Lifeline names at top of diagram

Our scenario diagram now contains quite a few messages and functions. As we start scrolling down or zooming in, we might not be able to see the tops of the lifelines, which could be quite confusing. Fortunately, *Capella* has a built-in feature to show the names of the lifelines above the top of the diagram.

The effect is somewhat similar to freezing the top row of a spreadsheet so that you can continue to see the column labels as you page down through the spreadsheet.

Logical Exchange Scenario (LES)

Next, we will look at scenario diagrams at the logical architecture level.

Of course, we can create completely new scenarios at this level, in the same way as was explained in the preceding paragraphs. However, we can also use again the very powerful *Capella* feature of "transition" between successive architecture levels.

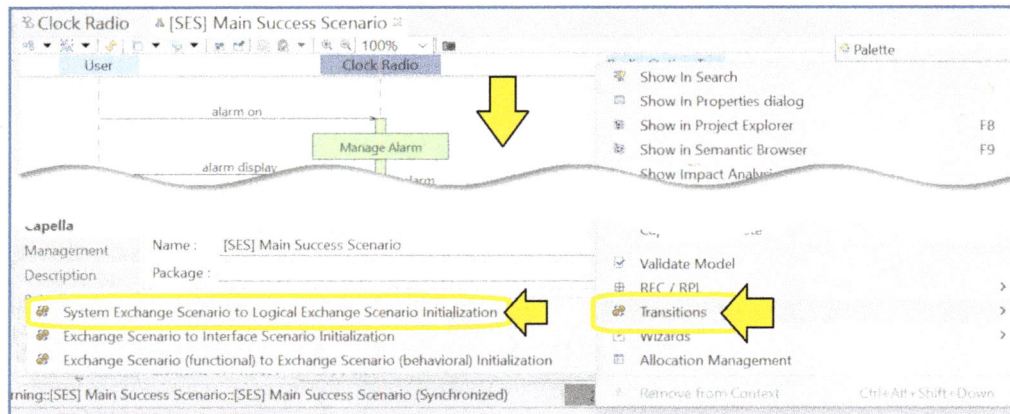

Figure 10-46 – Transition to logical exchange scenario

From the SES diagram (or from the scenario model element in the project explorer), we can right-click, select "Transitions" and then "System Exchange Scenario to Logical Exchange Scenario Initialization".

Capella computes the relevant logical actors, logical components, logical functional exchanges, and other related elements from the realization links that are present in the model.

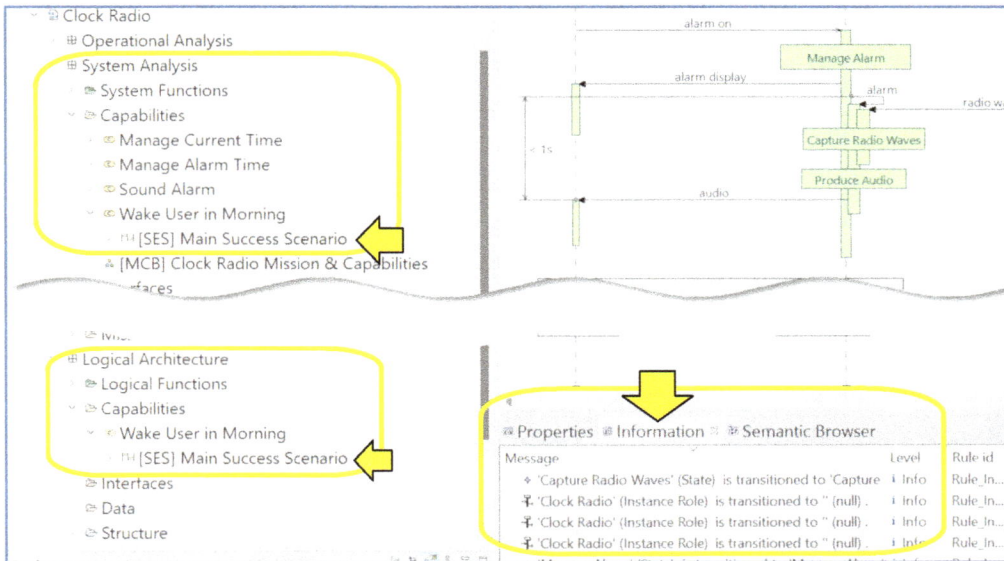

Figure 10-47 – Capabilities and scenarios were transitioned

In the project explorer, *Capella* created capability at the logical architecture level with the same name as the capability at the system analysis level: "Wake User in the Morning". Likewise, underneath this capability at the logical architecture level, it has created a scenario diagram with the same name as the diagram at the system analysis level: "[SES] Main Success Scenario".

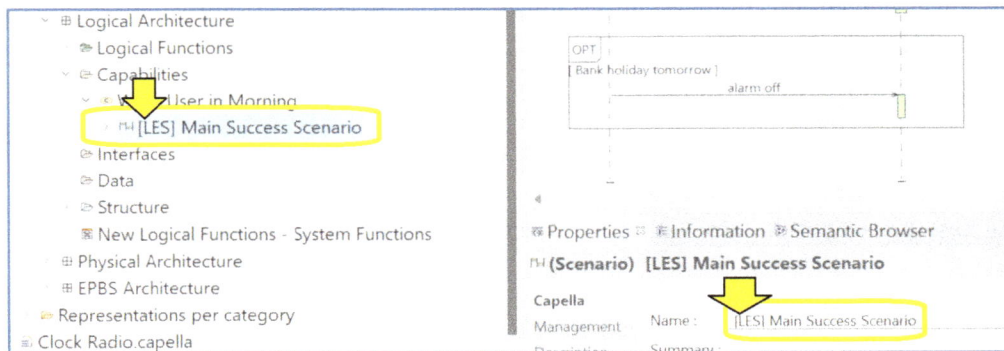

Figure 10-48 – Rename the scenario to reflect the level

Of course, "[SES]" does not make sense at the logical architecture level. Rename the scenario to "[LES]".

In order to avoid confusion as to what comes next, close the diagram from the system analysis level "[SES] Main Success Scenario".

Figure 10-49 – Lifelines were created automatically

Now, if we go back to the project explorer and expand the scenario model element, we can see that *Capella* has automatically created lifelines for us.

Note: these are only the model elements! ***The diagram has not been created yet.***

Continuing to explore the project explorer, we can see that *Capella* has created:

- Two lifelines for the logical actors involved: User, Radio Station Transmitter,
- Three lifelines for the relevant logical components which are involved in the scenario according to the functional exchanges and function allocation at logical level: Radio Manager, Alarm Manager, Clock radio UI,
- Six messages, in the same order as in the SES,
- Four state fragments.

Double-clicking on the scenario name in the project explorer will prompt *Capella* to create the scenario diagram.

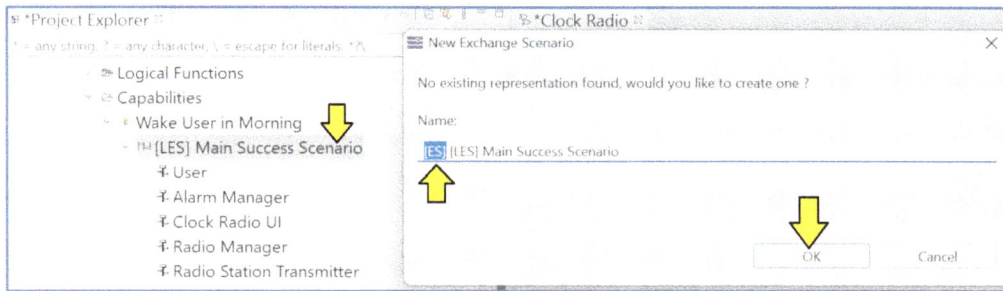

Figure 10-50 – Adjust the name

Capella will prepend "[ES]" to the diagram name. We don't need this extra notation. Delete the additional [ES] and click "OK".

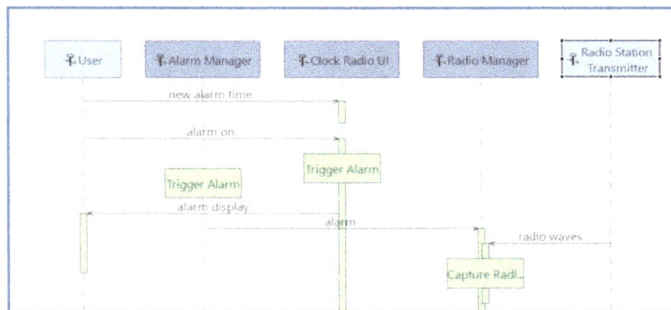

Figure 10-51 – The diagram is created automatically

The diagram is created automatically.

Capella does its best to guess the order of lifelines. However, the order will often be not quite what we want. Fortunately, it is quite easy to drag lifelines around. Reorganize the lifelines so that:

- The user is on the left and the clock radio UI is just to the right of the user.
- The radio station transmitter is on the right and the radio manager is just to the left of the radio station transmitter.
- The alarm manager is in the middle.

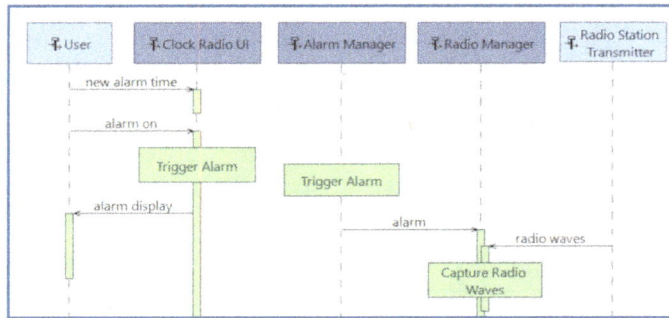

Figure 10-52 – The reorganized diagram is easier to read

The reorganized diagram is easier to read.

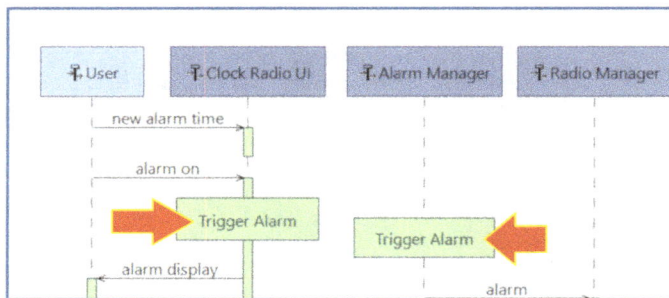

Figure 10-53 – Clean up erroneous state fragments

Two of the state fragment function allocations created by *Capella* don't actually makes sense [36] in their current location. Let's delete them temporarily as we fix up the model to reflect different structure and allocations at the logical architecture level.

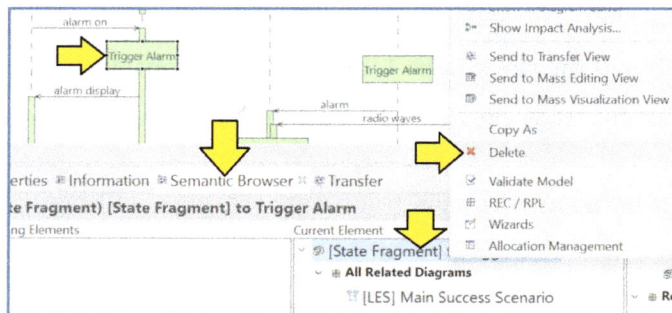

Figure 10-54 – Delete erroneous state fragments from semantic browser

Select each erroneous state fragment, open the semantic browser, right-click on the state fragment, and select "Delete", and confirm the deletion in the pop-up panel.

---

[36]and also fail model validation

We can now start working on the horizontal messages, and on the execution green rectangles. If we look back to the LAB diagram presented in Chapter 7, we can see which logical functions are dealing with the messages. For instance, the function "Enable Alarm Settings" deals with the first of the three messages.

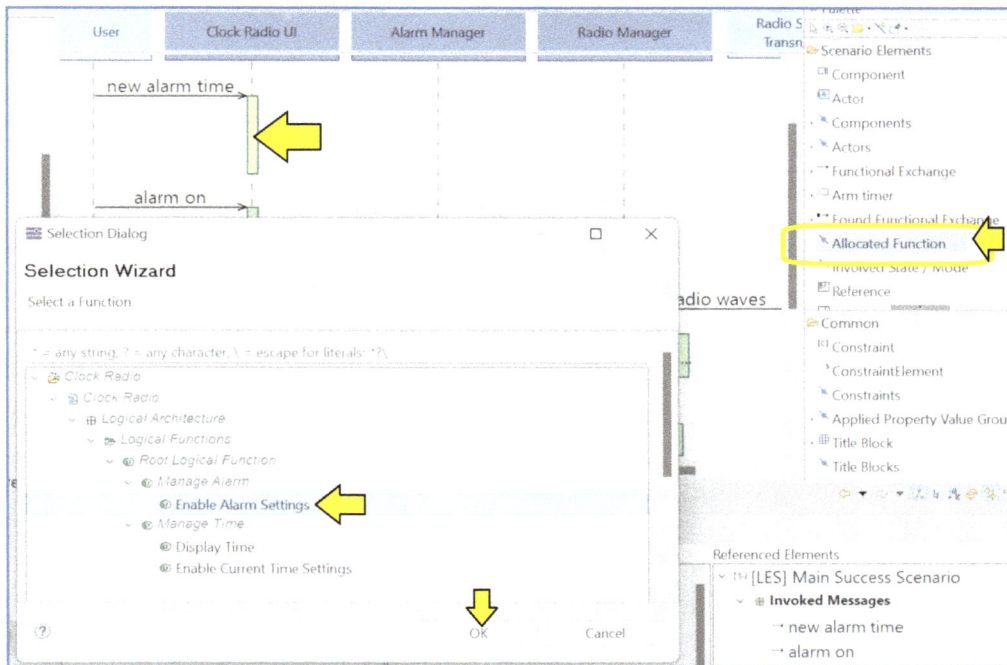

Figure 10-55 – Add allocated function to Clock Radio UI lifeline

Add "Enable Alarm Settings" to the clock radio UI lifeline using the "Allocated Function" tool.

What *Capella* could not invent were the messages needed between the logical components and which did not exist at system level. So, we need to add them manually at the right spot in the scenario.

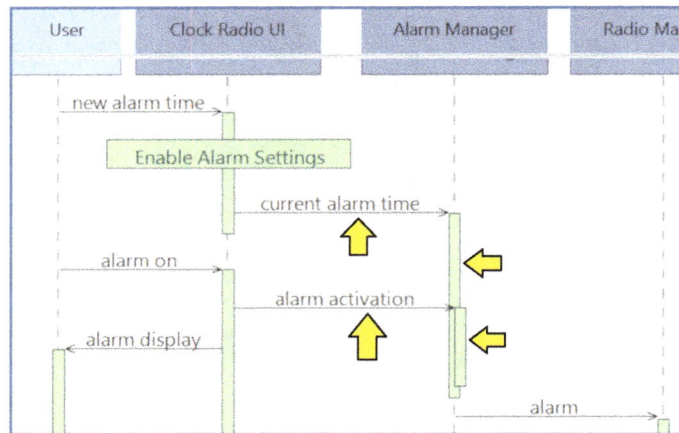

Figure 10-56 – Add messages between logical components

Add the "current alarm time" and "alarm activation" functional exchanges between the clock radio UI and the alarm manager. On the alarm manager, arrange the execution bars to prepare the "alarm activation" execution bar to receive an allocated function.

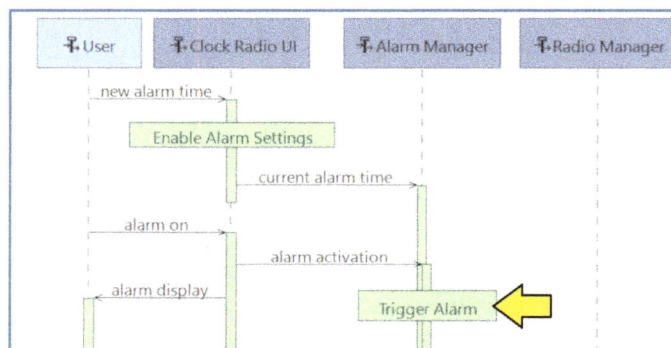

Figure 10-57 – Add the Trigger Alarm logical function

We also want to show the logical function "Trigger Alarm" as executed by the alarm manager. This function did not exist at system analysis level. This function was created as part of the functional analysis we performed at logical architecture level in Chapter 7.

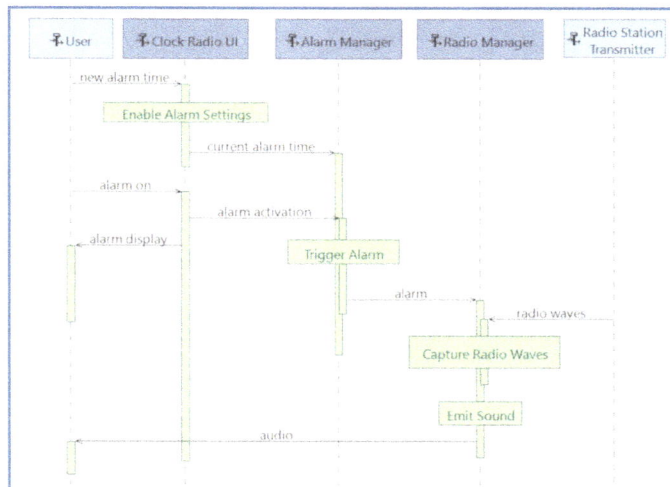

Figure 10-58 – Complete scenario diagram

Adjust the size of allocated function boxes and execution bars as needed.

Physical Exchange Scenario (PES)

Next, we will move to the physical architecture level. We created physical components and refined several functions in Chapter 8.

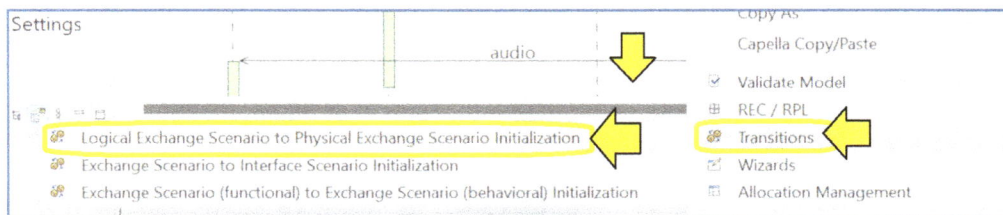

Figure 10-59 – Transition logical scenario to physical scenario

We can transition again the logical scenario to a physical one, using the same process as in the previous paragraph. From the LES diagram (or from the scenario model element in the project explorer), we can right-click, select "Transitions" and then "Logical Exchange Scenario to Physical Exchange Scenario Initialization".

Figure 10-60 – Information view

Note that the information view shows all of all the individual transitions that *Capella* performed. This information can be helpful for debugging problems.

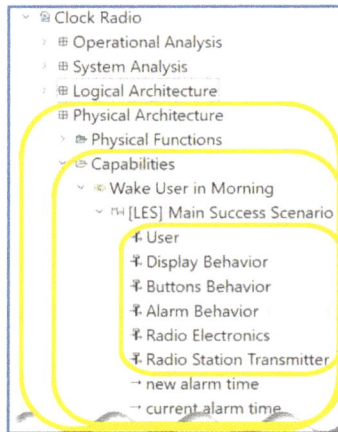

Figure 10-61 – Results in the project explorer

Of course, the results are easier to see in the project explorer.

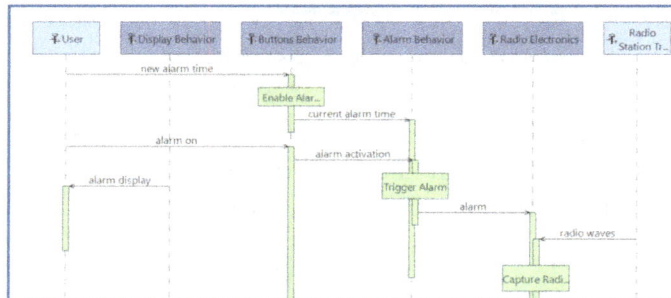

Figure 10-62 – Create the diagram manually

As before, clean up the transitioned elements and create the diagram.

1)   In the name of the scenario, change "[LES]" to "[PES]".

2)   Double-click on the scenario to create the scenario diagram.

3)   Delete the superfluous "[ES]" from the name of the scenario diagram and click "OK".

Instead of vertical logical components, we now find behavior physical components with allocated physical functions shown inside.

In scenario diagrams, we represent component lifelines as blue boxes at the system analysis, logical architecture, and physical architecture levels.

However, at the physical architecture level, *Capella* offers both "behavioral" and "node" components. We can also make a scenario from the yellow "node" components.

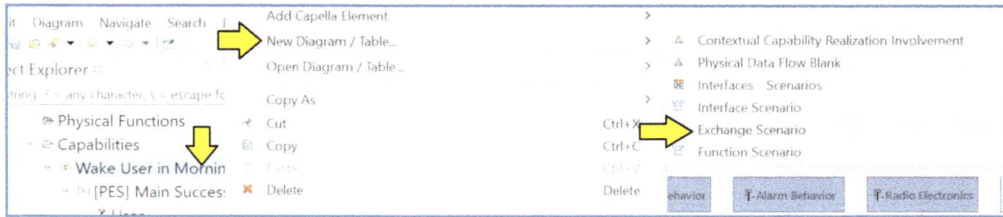

Figure 10-63 – Create new scenario from scratch

Node scenario diagrams cannot be made directly by the transition. We have to create a new scenario from scratch, by right-clicking on the capability. Select "New Diagram/Table…" and then "Exchange Scenario". When the naming panel pops up, rename this new diagram to "[PES] Main Success Scenario Node View".

Notice that we are adding a second scenario to the same capability. *Having multiple scenarios for a capability is completely natural, even recommended!* It is much better to have multiple, clear, easy to understand scenarios than to attempt to stuff everything into one scenario using a mess of combined fragments!

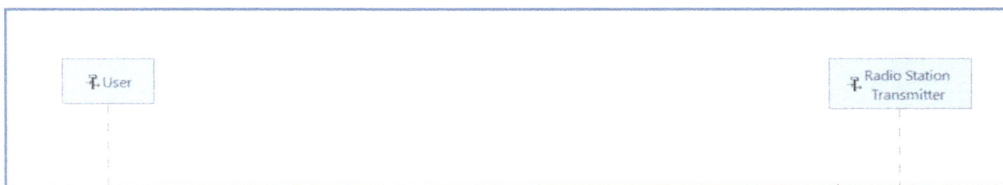

Figure 10-64 – Insert the two external actors

We can first insert the two external actors using the "Actors" tool in the palette.

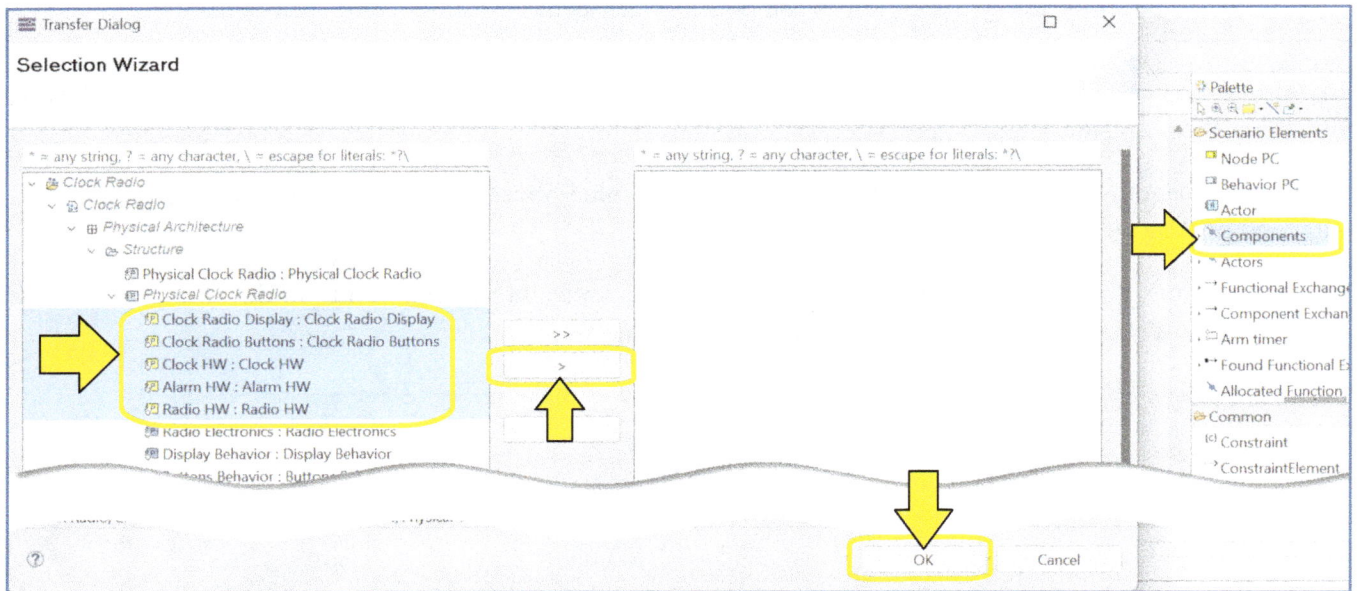

Figure 10-65 – Chose yellow node physical components

Using the components tool from the palette, insert the yellow node physical components.

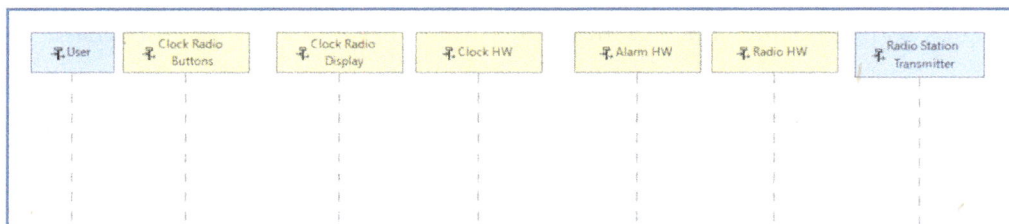

Figure 10-66 – Yellow node physical components are shown

The yellow node physical components are shown. Drag them around so that the radio hardware is to the right and the clock radio buttons are to the left.

Figure 10-67 – Add the messages

Again, we can create all of the messages using the functional exchange tool in the palette. As we draw messages from one lifeline to another, *Capella* will prompt use with the existing functional exchanges that are valid for the selected combination of lifelines.

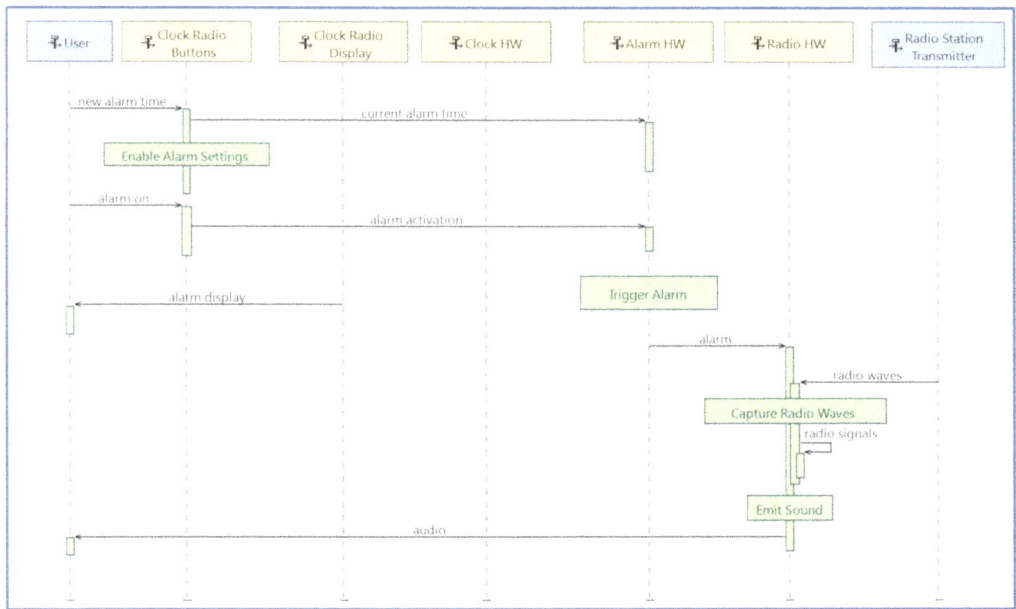

Figure 10-68 – Add function execution

We can even show the execution of the important functions, as if they were allocated directly to the node components. (The functions are actually allocated to behavioral physical components which are in turn allocated to node physical component. *Capella* "fudges" this small distinction for use and allows us to show the functions directly on the node physical components).

Further Study

Here is a list of further topics for more in-depth study:

- **Modeling Exchange Scenarios** – In this advanced video, you will find an in-depth explanation and illustration of exchange scenarios. This video also explains how to initialize exchange scenarios from functional chains.
  - https://url4ap.net/CH-10-ES-Exchange
- **Scenario Diagram in Capella (Part 1 of 2)** – This is the first of two videos that provide an in-depth illustration of the basic steps to create exchange scenarios.
  - https://url4ap.net/CH-10-ES-Scenario1
- **Scenario Diagram in Capella (Part 2 of 2)** – This is the second of two videos that provide an in-depth illustration of the basic steps to create exchange scenarios.
  - https://url4ap.net/CH-10-ES-Scenario2

Chapter 11 – Mode or State Machine Diagrams

In this chapter, we will take a closer look at the so-called Mode/State Machine Diagrams, showing the behavior of a structural element as it transitions between modes or states, at all *Arcadia* levels. Like the scenario diagram, the mode or state machine diagrams are inspired by a UML diagram. In this case, the inspiration is the UML/SysML state machine diagram. Mode and state machine diagrams do not behave quite like the architecture blanks that make up most of the core of *Arcadia*. Mode and state machine diagrams have their own specific rules that we will introduce in this chapter.

Introduction

Mode/State diagrams are graphical representations of State Machines inspired from UML/SysML.

Arcadia allows a Mode/State Machine to be associated with any structural element:

- Entity/Operational Actor
- System/Actor
- Logical Component/Logical Actor
- Physical Component/Physical Actor

Let us quickly present the main concepts used. *Arcadia* proposes two similar concepts: State and Mode. The use of one or the other is a methodological choice, in the context of the project or the company. A fairly widespread difference among system engineers involves considering that the Mode depends on a choice, often of an operator, while the State is the result of that which has happened to the structural element.

> *Mode –* A **mode** is a behavior *expected* of the system, a component or also an actor or operational entity, in some chosen conditions. A mode can be broken down into submodes. [Voirin] – Chapter 18

> *State –* A **state** is a behavior undergone by the system, a component, an actor or an operational entity, in some conditions imposed by the environment. A state can be broken down into substates. [Voirin] – Chapter 18

A mode machine (or respectively, state machine) is a set of modes (or, respectively, states) linked to one another by transitions. Modes and states cannot cohabit in the same machine: they are exclusive. At a given instant, a single mode or state (called "current") is active in each machine. One or more machines can be associated with the characterization of the system, a component, an actor or an operational entity. In this case, the current modes and states of each machine cohabit at a given instant. [37]

Figure 11-1 shows a very simple mode machine diagram.

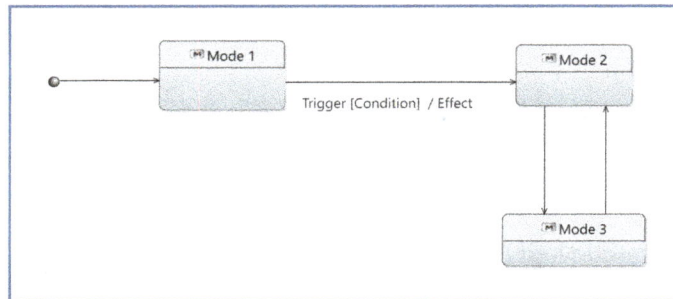

Figure 11-1 – Example of state and mode notation

Modes/States are shown as rectangles with rounded corners, with the name of the mode or state and a small symbol to distinguish modes from states. For modes, the small symbol contains the letter "M". For states, the small symbol contains the letter "S".

In Figure 11-1 two additional diagram elements are visible:

- The initial mode/state (solid filled circle) corresponds to the creation of the structural element containing the mode machine.

- The final mode/state (circle surrounding a small solid filled circle) corresponds to the destruction of the structural element containing the mode machine.

These concepts about creation and destruction of the element are actually object-oriented programming ideas taken from UML and SysML. However, they are useful even in a methodology such as *Arcadia*.

---

(37) This fine distinction between "modes" and "states" is unique to *Arcadia*. The rest of the worldwide systems engineering community uses "states" and "state machines" or "state charts". If you are talking to a systems engineer who is not a *Capella* user, you can expect a blank stare if you mention a "mode machine".

Basic States

We will continue to use the example of the simple Clock Radio. Once again, we will start at the system analysis level. The first step will be to associate a simple mode machine with the clock radio system.

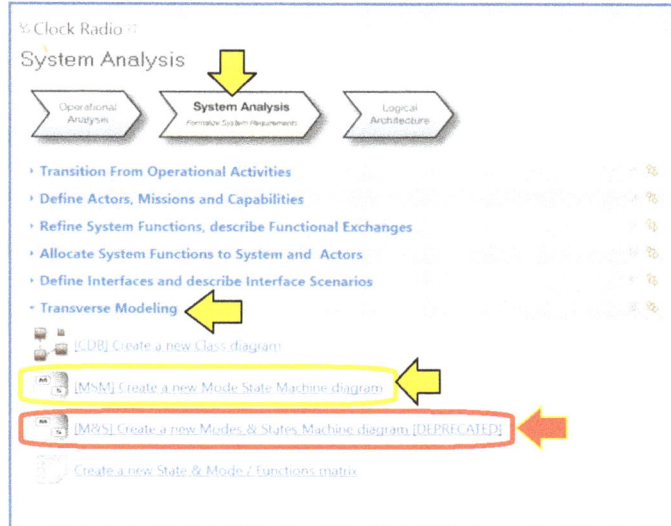

Figure 11-2 – Expand the Transverse Modeling topic

In the activity explorer, select the system analysis level and expand the topic "Transverse Modeling".

Two Versions of Mode/State Diagrams

As of *Capella* version 6.0, the tool still supports two versions of the Mode State Machine diagram. A long time ago, *Capella* version 1.1 added the possibility of creating concurrent regions (to represent parallelism between states), which was not possible in previous versions. Probably for compatibility reasons, the (very) old and (not so) new diagrams exist alongside each other. We strongly recommend the use of the [MSM] diagram type and not the "deprecated" [M&S].

Select "[MSM] Create a new a Mode State Machine Diagram".

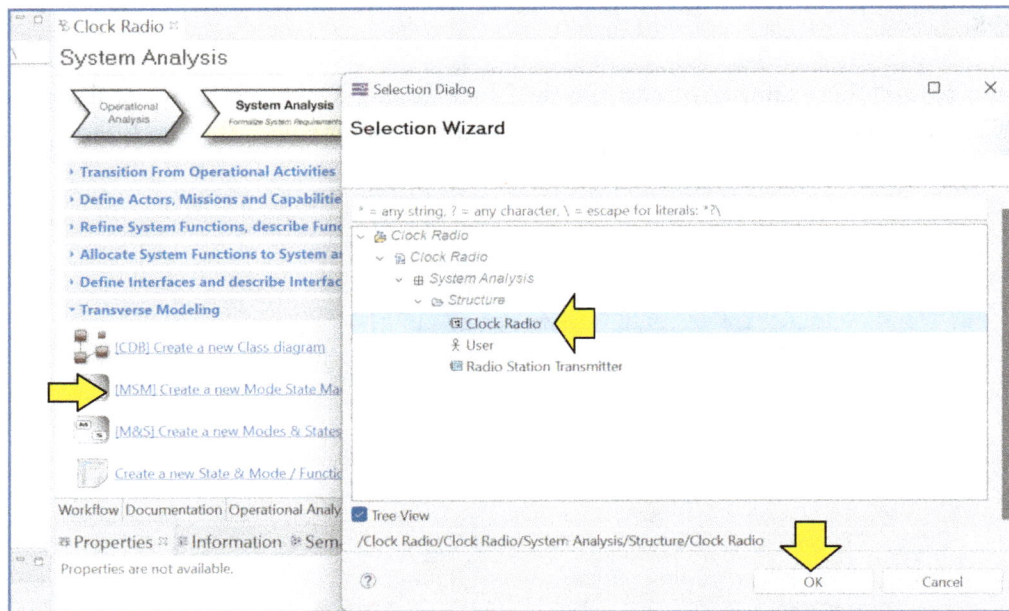

Figure 11-3 – Select the clock radio system

As explained previously, *Capella* opens a selection window, showing all the structural elements of the current *Arcadia* level. In our clock radio model, there are only three structural elements at this level: the system itself (that is, the clock radio) and two actors. Select the clock radio system to be the owner of the new diagram. [38] Click "OK".

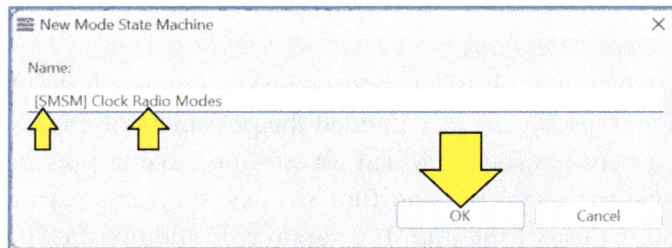

Figure 11-4 – Add S for the level and assign a sensible name

As usual, *Capella* proposes a default name for the diagram. In that case, that name is "[MSM] Default Region". Obviously, this is not a very intuitive or meaningful name. We will want to choose a better name. As usual, we will want to retain the acronym for the diagram prefaced with the letter "S" to indicate the system analysis level. Name the diagram: "[SMSM] Clock Radio Modes".

---

[38] Notice that we have not yet selected "mode" or "state". That selection will happen implicitly the first time we add one or the other to the diagram.

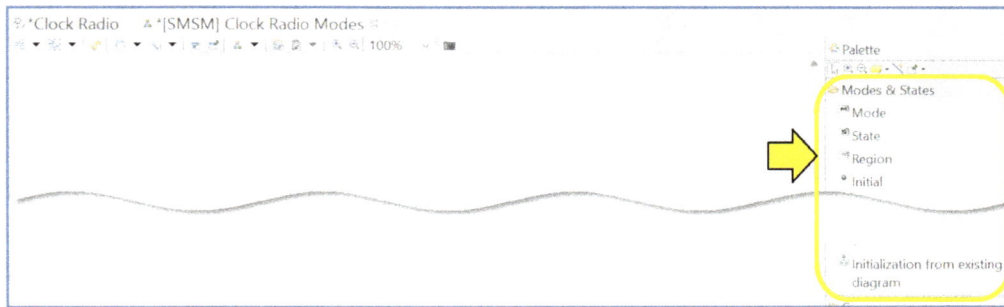

Figure 11-5 – The new diagram is empty

Capella creates an empty diagram (even if it is not formally a Blank) with a palette of specialized tools on the right.

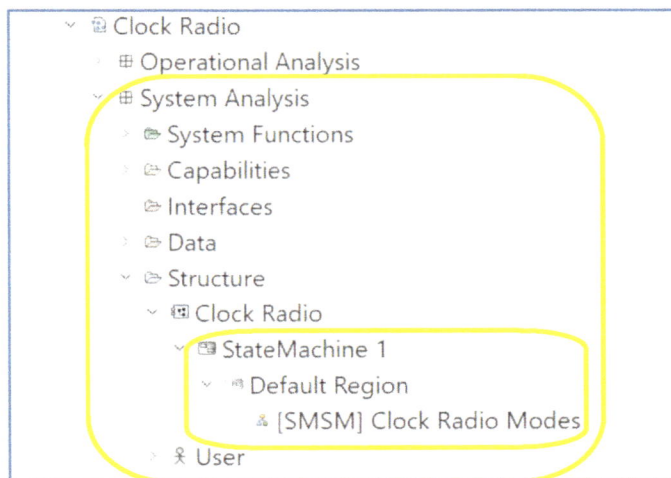

Figure 11-6 – The project explorer shows the newly created elements

To understand the default diagram name proposed by *Capella*, we have to go to the project explorer. *Capella* has automatically created a "StateMachine 1" under "Clock Radio". This state machine contains a "Default Region" (hence the diagram name) which in turn contains the diagram.

About Regions

What is this **region** concept? It comes from UML/SysML and enables the definition of parallel behaviors. A UML/SysML state machine comprises one or more regions, each region containing a graph of states connected by transitions.

Inside one region, one and only one state is active at a time, but each region denotes a behavior fragment that may execute concurrently with its orthogonal regions. Two or more regions are orthogonal to each other if they are either owned by the same state or, at the topmost level, by the same state machine.

By default, when you create a mode/state, *Capella* automatically creates one region inside so that you will be able to create substates, and/or concurrent regions.

In this example, the transitions will be caused by decisions of the "User" actor. As such, per the reasoning of *Arcadia* this will be a mode machine. Rename "StateMachine 1" to "Clock Radio Mode Machine".

The UML/SysML State Machine diagram is very complex and contains many subtle concepts. As *Capella* chose to mimic SysML on this topic, we will focus on the most important concepts here (Mode/State and transition) and give you some additional links if you wish dig deeper into the topic. Do not forget to adapt your level of modeling to the needs of your audience. [39]

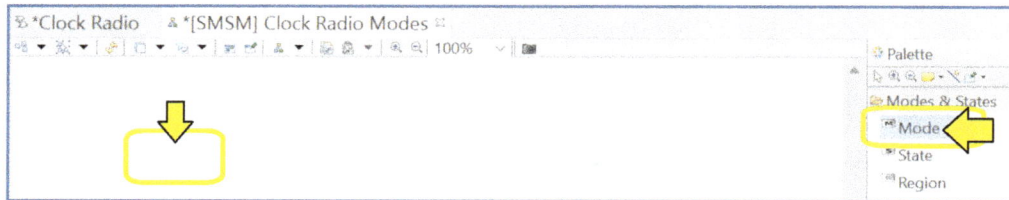

Figure 11-7 – Create a mode

Create a first mode by clicking on the "Mode" tool in the palette and then again in the background of the diagram.

---

[39] Many new modelers get very excited by the possibilities of describing modes and states and model far too much detail leaving their target audience utterly bewildered and unable to understand what the modeler was trying to express.

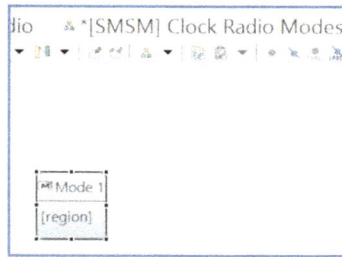

Figure 11-8 – Notation for the new mode

Capella creates a rectangle called "Mode 1", with an "M" symbol on top left, and an enclosed rectangle called [region]. AS we explained previously, as soon as you create a mode/state, *Capella* creates a default region so that will be able to create substates or concurrent regions…

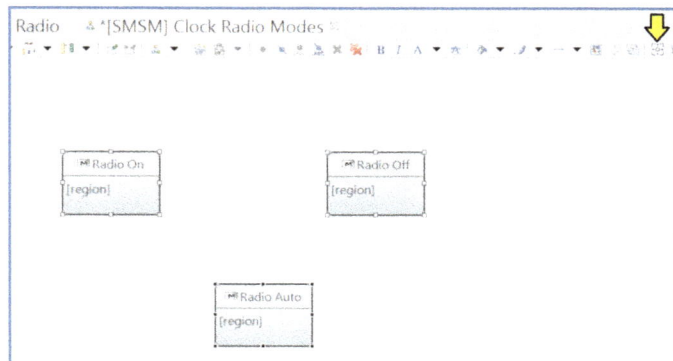

Figure 11-9 – Create additional modes

Rename this first mode "Radio On". Create two more modes "Radio Off" and "Radio Auto". Arrange, make them a little larger for better readability, and use the "make height and width the same size" tool to make them all a uniform size.

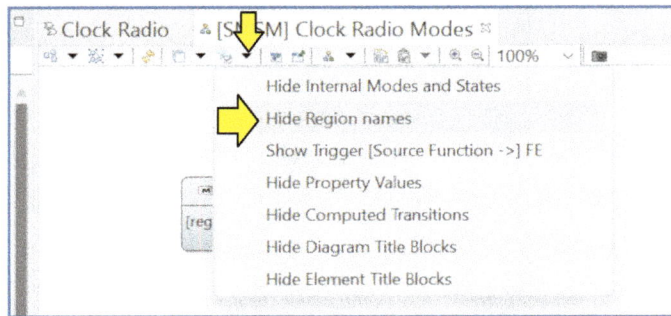

Figure 11-10 – Hide region names

Inside each mode box on the diagram, *Capella* shows the text "[Region]" which is the default name for the default region inside that element. These names do not add any value to our diagram; we will hide them using the "Hide Region names" filter from the "Filters" menu.

Figure 11-11 – Refresh to complete

Initially, nothing in the diagram will change! We need to click on the "Refresh diagram" tool at the top of the diagram.

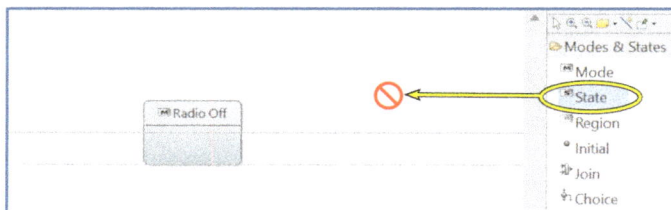

Figure 11-12 – Modes and states cannot be mixed

If we try to create now a third "state" instead of "mode", *Capella* forbids it. Remember, we cannot mix states and modes in the same Mode/State machine. Although the state tool does not disappear from the palette, it cannot be used.

We can add an initial **pseudostate** to the diagram.

Pseudostate – A **pseudostate** is an abstraction that encompasses different types of transient vertices in the state machine (mode machine) graph. A state machine (mode machine) instance never comes to rest in a pseudostate, instead, it will exit and enter the pseudostate within a single run-to-completion step.

Arcadia and *Capella* take state machine diagrams from UML and with them the different types of pseudostates. The two most common pseudostates are:

- **initial** – represents a starting point for a region; that is, it is the point from which execution of its contained behavior commences when the region is entered via default activation. It is the source for at most one transition, which may have an associated effect behavior, but not an associated trigger or guard.

- **terminate** – implies that the execution of the state machine is terminated immediately.

[UML2.5.1] – 14.2.3.7 Pseudostate and PseudostateKind

We will assume that the clock radio comes into existence in the "Off" state.

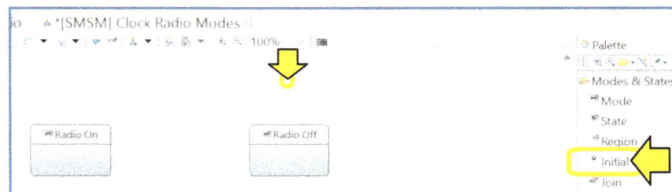

Figure 11-13 – Add an initial pseudostate

Click on "Initial" in the palette and then click again above the first mode.

Capella creates a black filled circle named "Initial4".

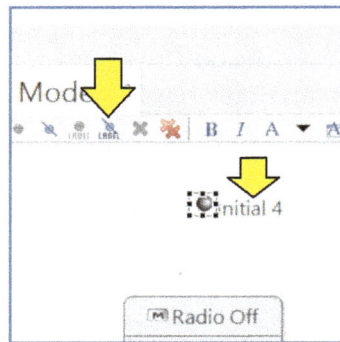

Figure 11-14 – Hide the label

The name "Initial4" is not particularly helpful. [40] Click on the "Hide label" button in the diagram toolbar to hide it.

Now, we are ready to create our first transition. We need a transition from the initial pseudostate to the "Radio Off" mode.

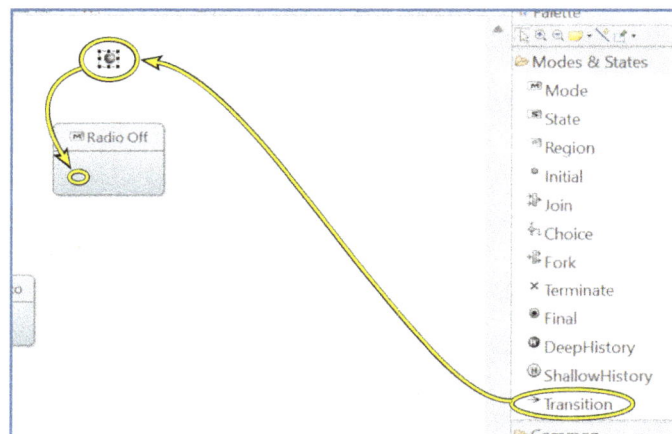

Figure 11-15 – Add a transition

Click on the "Transition" tool in the palette, then click on the initial pseudostate, and then move the mouse over the "Radio Off" mode and release the mouse button.

---

[40] Display of a name for the initial pseudostate is also not standard notation for UML/SysML.

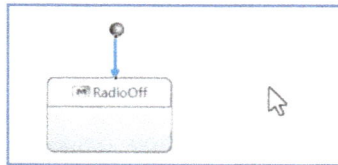

Figure 11-16 – Transition is displayed

The result is an arrow from the initial pseudostate to the "Radio Off" mode.

Figure 11-17 – Elements visible in project explorer

If we look at the project explorer, we can see that all mode machine elements – including the modes, the initial state, the transition to the "Radio Off" mode, and the diagram itself! – are located under the "Default Region" of the "Clock Radio Mode Machine".

If we press and hold the "CTRL" key before we click on the "Transition" tool in the palette, we can create several transitions without having to go back to the palette.

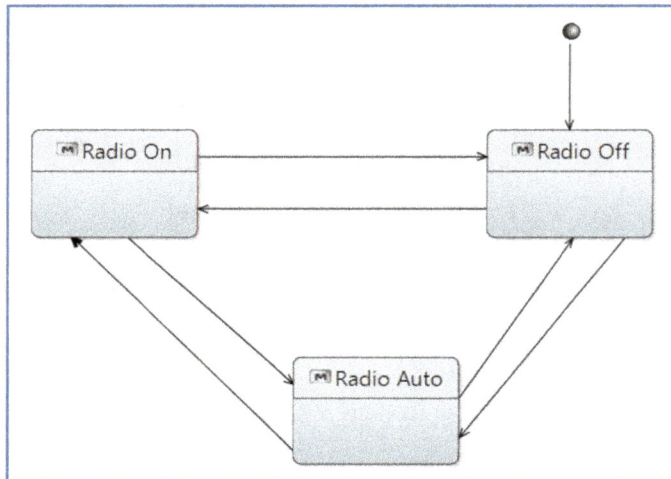

Figure 11-18 – Six transitions created

In just a few seconds we have created six transitions. Do not forget to click on the "ESC" key to interrupt the creation process.

By default, all lines are "oblique" in *Capella* (except in the breakdown diagrams, see Chapter 9). Many people prefer a rectilinear diagram style which *Capella* supports as well.

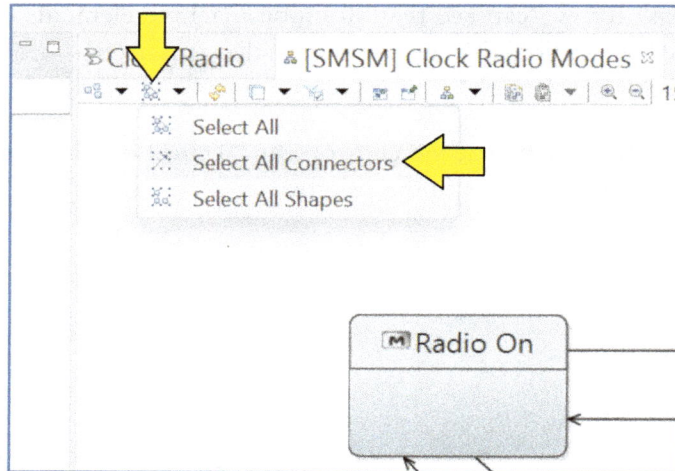

Figure 11-19 – Select all connectors

Select all lines by clicking on the "Select All Connectors" button of the diagram toolbar. All of the transitions will turn blue.

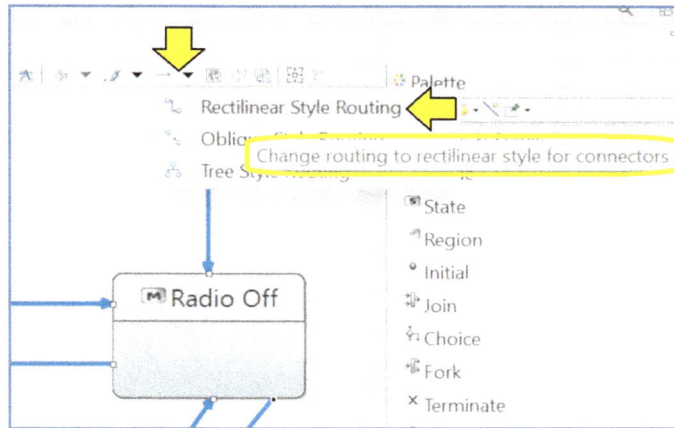

Figure 11-20 – Select rectilinear routing

Then, click on the "Rectilinear Style Routing" button of the same toolbar.

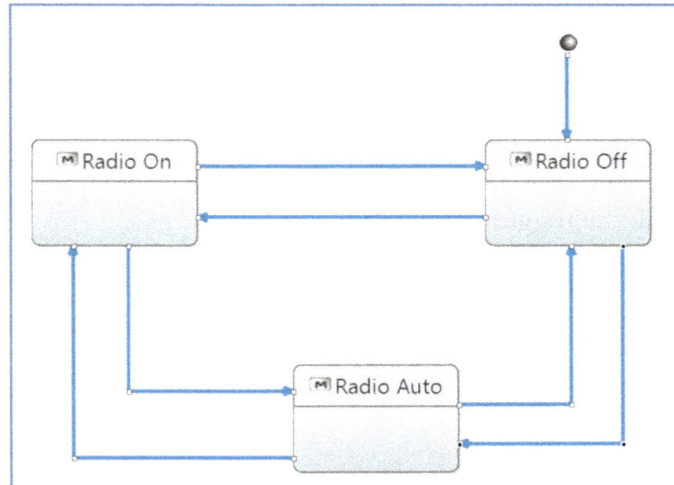

Figure 11-21 – Transitions are now rectilinear

All of the lines are now rectilinear, which is often more practical if we want to move them.

Next, we will take a closer look at the properties of a transition. Again, the three important concepts are: trigger, guard, and effect. The graphical notation for these concepts comes directly from UML/SysML and is the string: *trigger [guard] / effect* along the line of the transition.

Unfortunately, the properties sheet shown by *Capella* when we double-click on a transition is not laid out to match the sequence of elements that will be shown in the diagram. The properties sheet shows the "Guard", followed by the "Effects", and the lastly "Triggers".

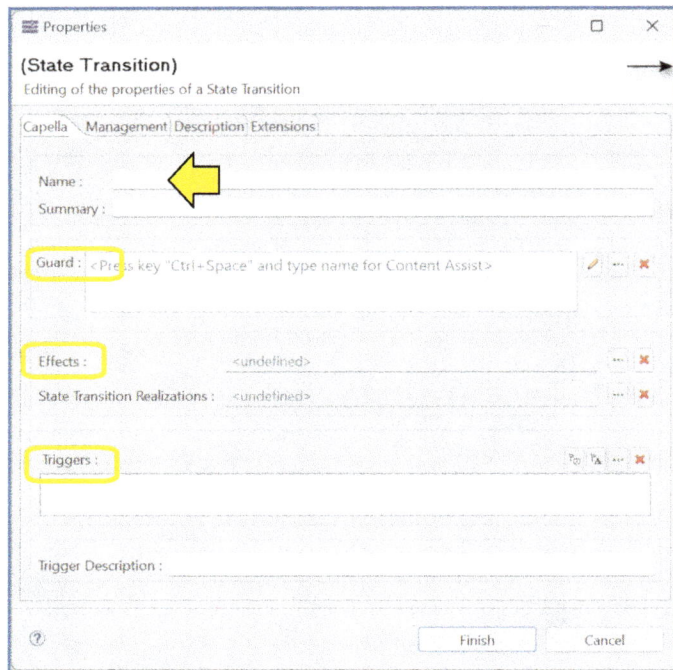

Figure 11-22 – Name is not used, and guard, effects, and triggers are out of order

The name of the transition is not used very much, as a transition is completely defined by its source and target modes/states, along with the trigger and guard.

On the second horizontal transition, between "Radio Off" and "Radio On", we would like to define an existing functional exchange as the trigger.

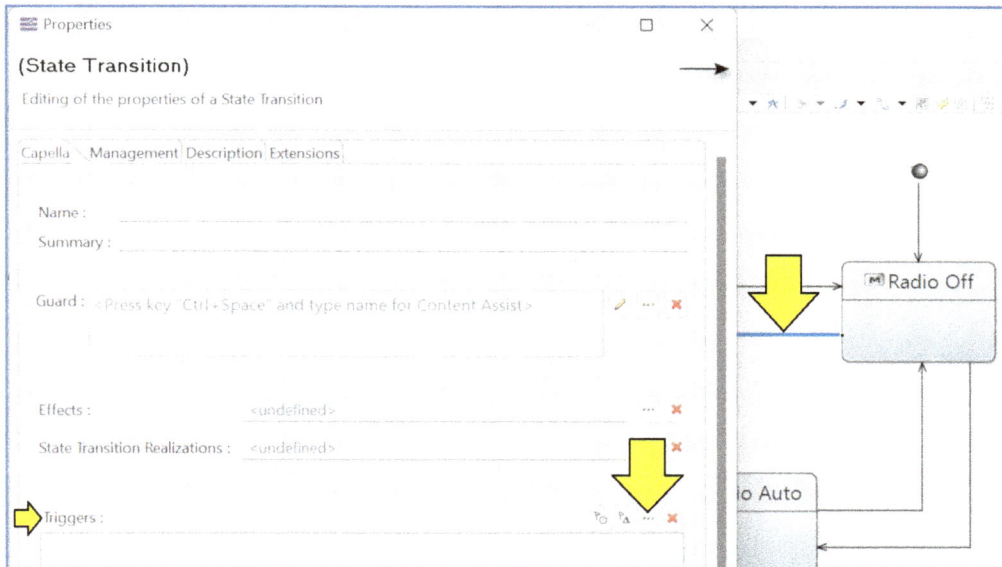

Figure 11-23 – Click on three dots to open triggers panel

Double-click on the transition to open the properties panel for the transition. Click on the "…" (three dots) button at the upper right corner of the "Triggers" field.

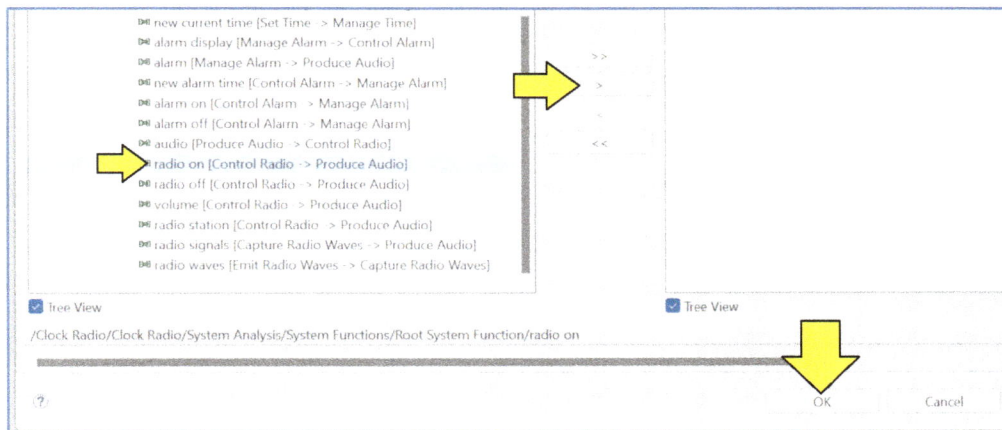

Figure 11-24 – Select the radio on functional exchange to be a trigger

Select the "radio on" functional exchange, click on the single arrow (>), and click "OK".

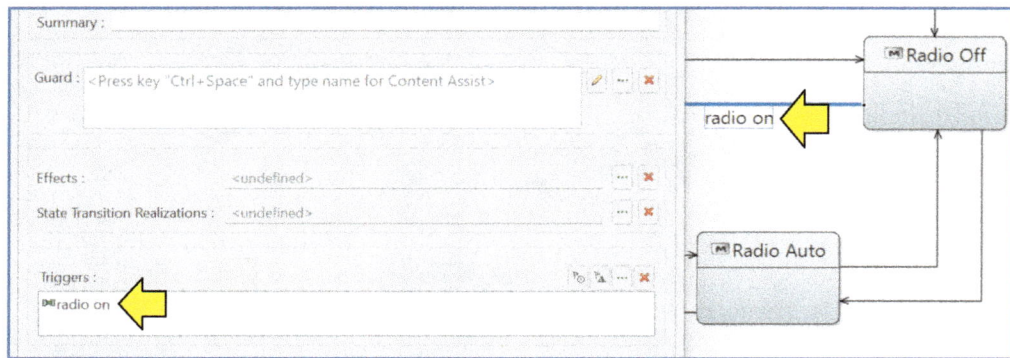

Figure 11-25 – Functional exchange is now present as a trigger

The functional exchange now appears in the "Triggers" field.

Click on "Finish" to close the panel.

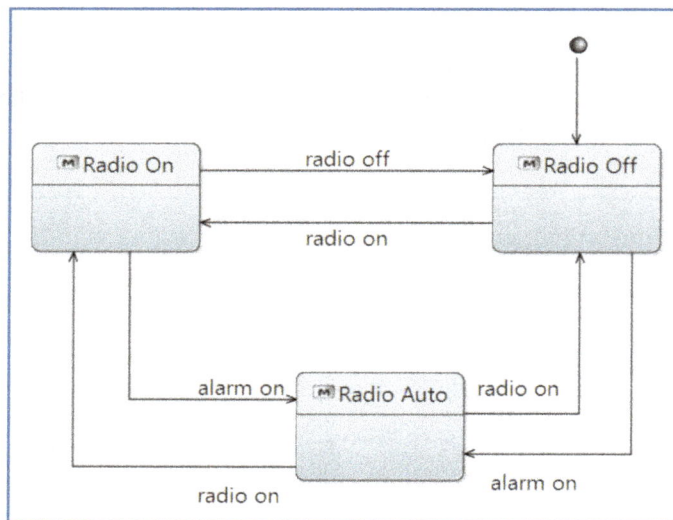

Figure 11-26 – Assign functional exchanges to all of the triggers

We can do the same on the other transitions, selecting the relevant functional exchange each time.

When the radio is on, it produces audio continuously. That is something we can express very well with *Capella*, by linking the "Radio On" mode with the "Produce Audio" function.

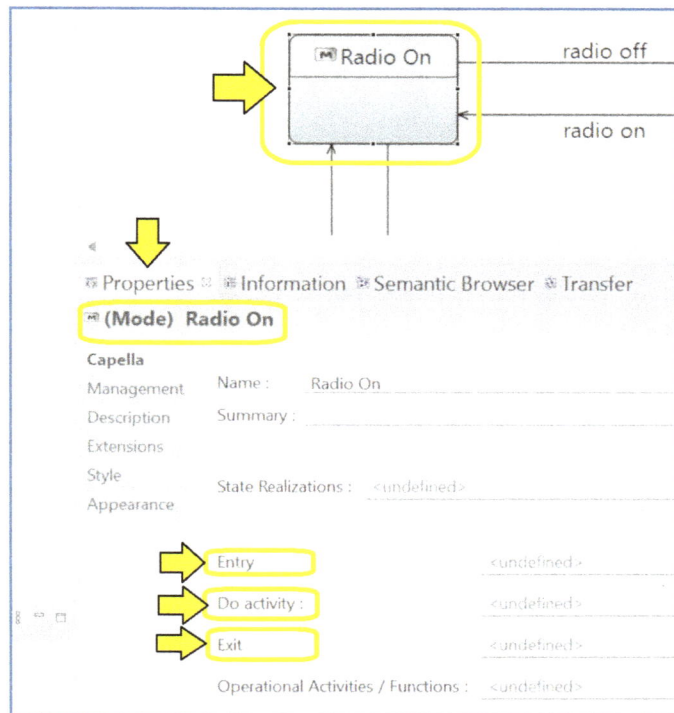

Figure 11-27 – Open the properties of the mode/state

Open the properties of the "Radio On" mode/state.

Internal Behavior – An **internal behavior** is associated with a state (mode) rather than with a transition. There are three types of internal behavior:

- **entry** – This label identifies a behavior, specified by the corresponding expression, which is performed upon entry to the state (mode).

- **exit** – This label identifies a behavior, specified by the corresponding expression, which is performed on exit from the state (mode).

- **do** – This label identifies a behavior, specified by the corresponding expression, which is performed as long as the modeled element is in the state (mode) or until the computation specified by the expression is completed.

[UML2.5.1] – 14.2.4.4 State

In this case, the "Do activity" concept is exactly what we need to express the fact that the function "Produce Audio" is performed as long as the clock radio is in the "Radio On" mode.

We just have to click on the green "..." of the "Do activity" field. *Capella* guides us by proposing only functions allocated to the current model element.

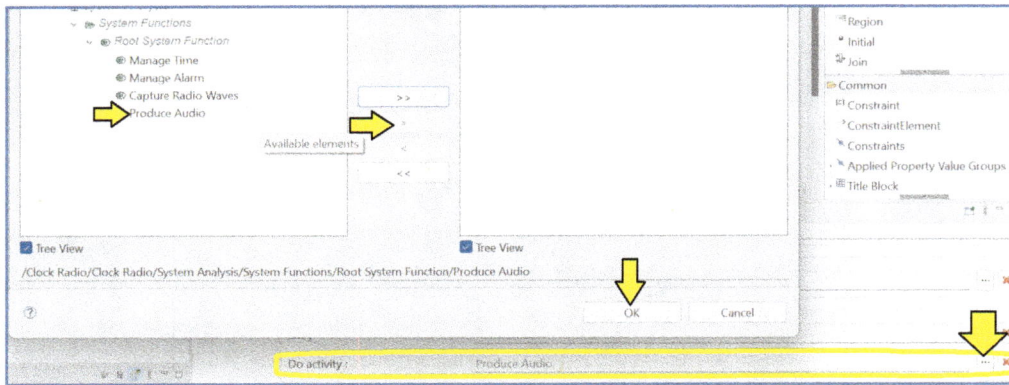

Figure 11-28 – Select the produce audio function

We can choose "Produce Audio" from the list of functions allocated to the clock radio.

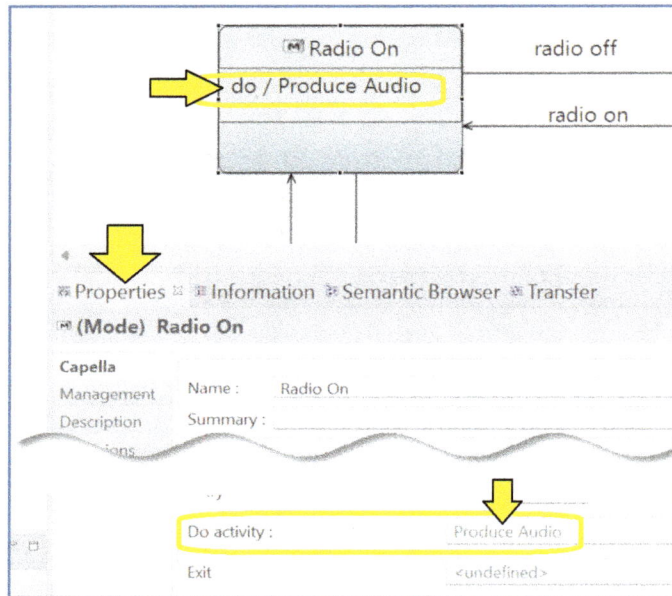

Figure 11-29 – A new compartment has been added for the behavior

Capella updates the diagram by adding a new compartment in the Mode box, with "do / Produce Audio" string inside.

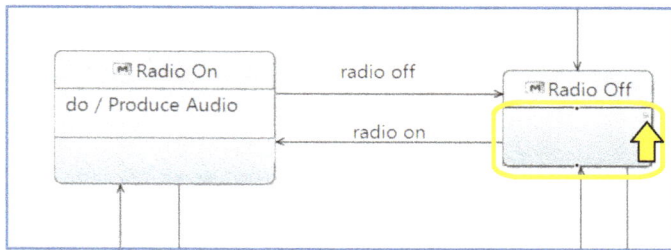

Figure 11-30 – Shrink the extra compartment

The region compartment looks empty. We can shrink it somewhat (although we can't eliminate it completely) by selecting the empty compartment and clicking the small minus sign that appears in the upper right corner.

Of course, shrinking that compartment causes the transition lines to move around. You may need to fix these up manually.

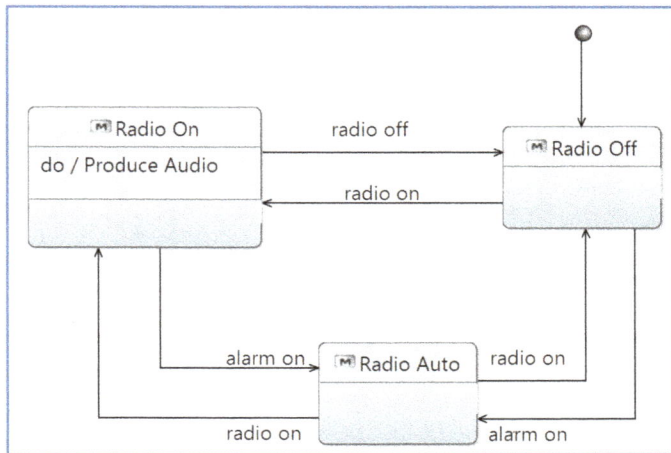

Figure 11-31 – Nice diagram, but something is still missing

The diagram looks nice. However, we have not yet expressed the important fact that when the alarm is on, the radio should produce audio when the time reaches the alarm setting.

Substates

In fact, when the clock radio is in the "Radio Auto" mode, it is either silent, waiting for the alarm time to arrive, or producing audio for a predefined duration. This set of possibilities can be modeled by two submodes (substates) inside "Radio Auto".

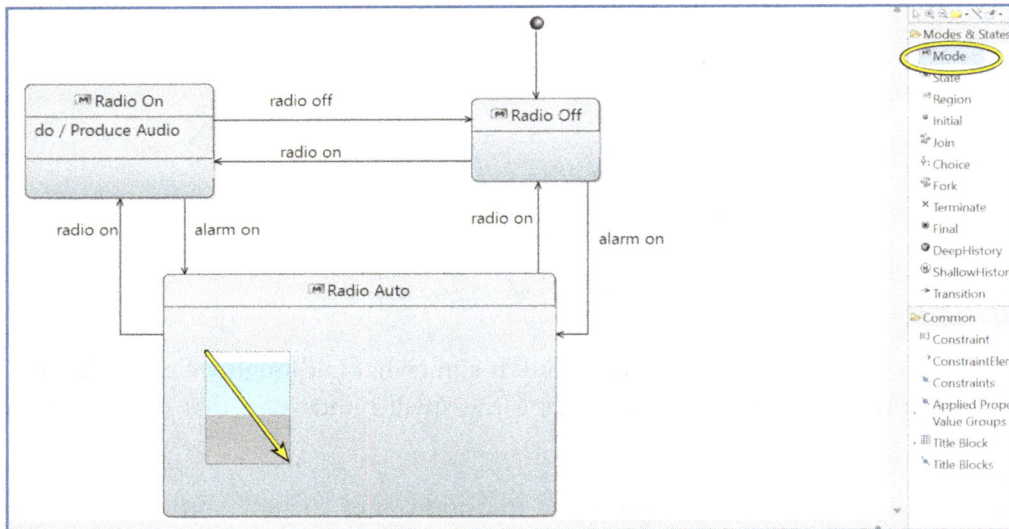

Figure 11-32 – Add a submode

First, resize "Radio Auto" in order to make space inside of it for two new modes. When you resize the mode, the transitions and their labels may jump around a bit. Be sure to arrange them nicely again before continuing. In the palette, select the "Mode" tool, and drag a new mode shape inside of "Radio Auto".

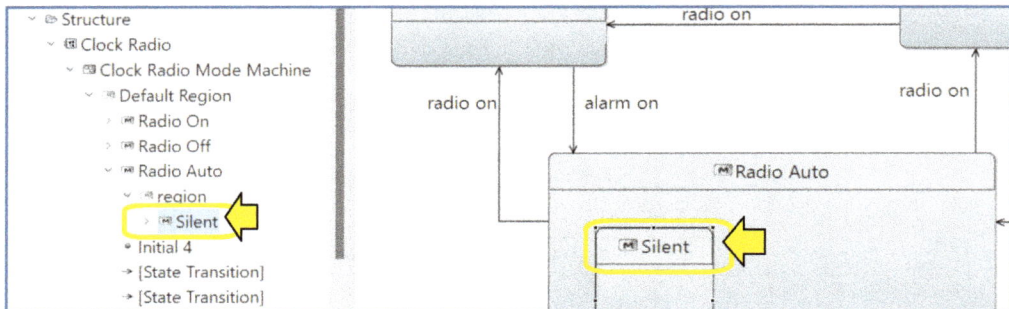

Figure 11-33 – Name the submode

Rename this new mode "Silent".

Notice that in the project explorer the new node is inside a region that is itself inside the "Radio Auto" mode. The three first-level modes are in the default region of the modemachine, and the submode is in turn in the (default) region of its container mode. This hierarchy is created automatically by *Capella*.

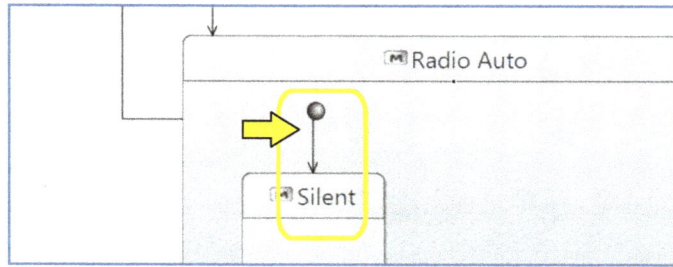

Figure 11-34 – Define the entry/default submode

Create an initial mode/state inside the "Radio Auto" mode that transitions immediately to the "Silent" mode. This is a functionally important part of the model and indicates that when the user places the clock radio into auto mode, it will always start off in the "Silent" submode.

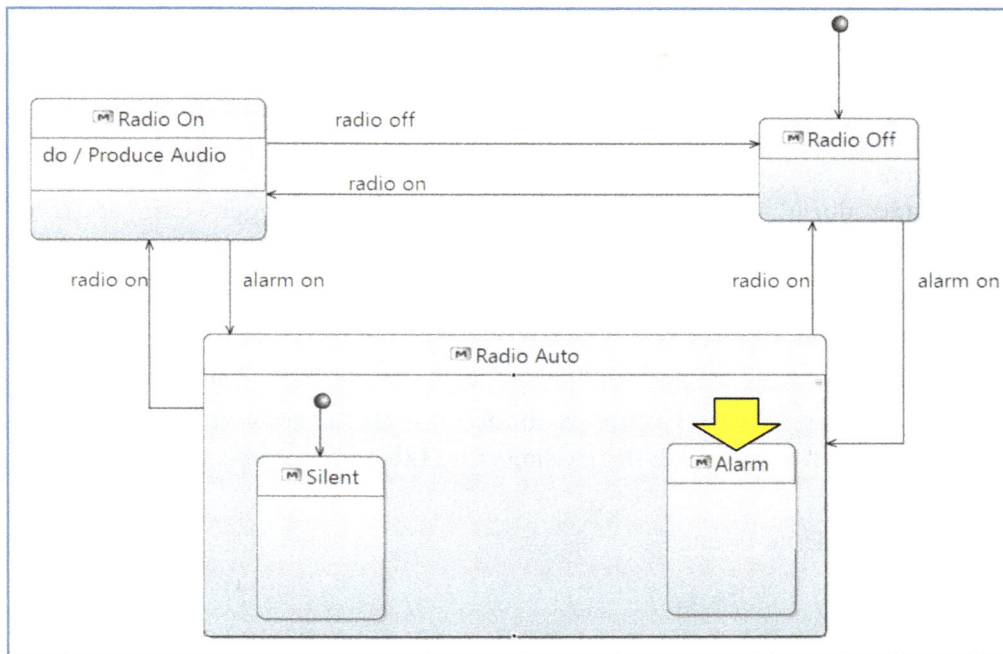

Figure 11-35 – Create second submode

Create a second "Alarm" submode which will contain the "Produce Audio" function. Since we are also going to add some detailed triggering information between "Silent" and "Alarm", make some more space between the two.

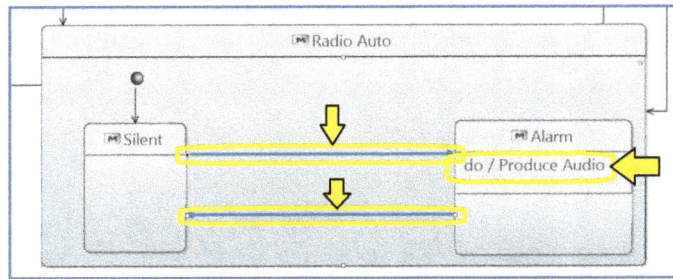

Figure 11-36 – Add do activity and triggers

Add the do activity inside "Alarm", and create two transitions between the submodes, as explained previous-ly.

On the first transition (from "Silent" to "Alarm"), we want to use a special trigger, indicating that the current time has just reached the alarm time. There is a nice concept from UML/SysML for this purpose called a **change event**.

Events as Triggers

Previously, we introduced the use of functional exchanges as triggers. Triggers can also be events. There are two types of events:

- **Time Event** – A Time Event is modeled using either the keyword *"after"*, followed by an expression that represents duration, counted from the entrance into the current state, or the keyword *"at"*, followed by an expression that represents an absolute time.

- **Change Event** – Change Events are modeled using the keyword *"when"*, followed by a Boolean expression, in which the passing from **false** to **true** sets off the transition.

[UML2.5.1] – 13.3.3.4 Time Events

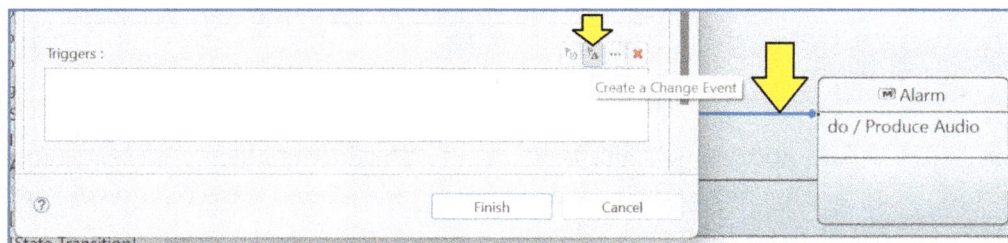

Figure 11-37 – Open the transition's property sheet

Double-click on the transition to open its property field. To the right of the triggers field, find the "Create a Change Event" button and click on it.

When we click on the "Create a Change Event" button, *Capella* opens a new property sheet where we can specify the name of the event and the Boolean expression.

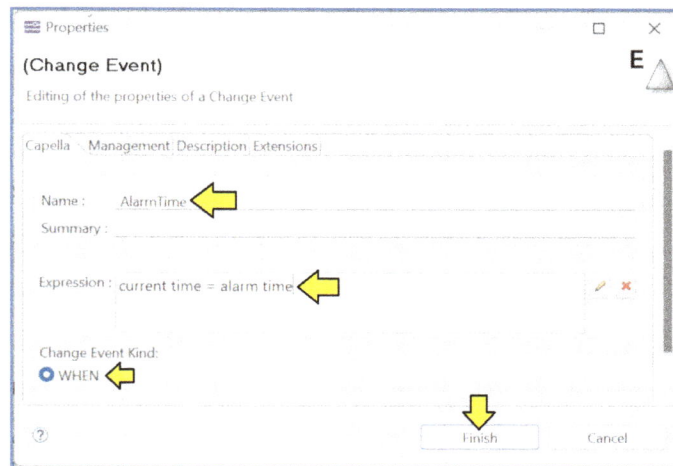

Figure 11-38 – Define the event

1) Name the event: "AlarmTime".

2) Set the expression to: "current time = alarm time".

3) Notice: "Change Event Kind: WHEN" – (Nothing to change here).

4) Click "Finish".

5) Notice that the triggers field in the transition's properties now includes an expression for the change event.

6) Click "Finish" again.

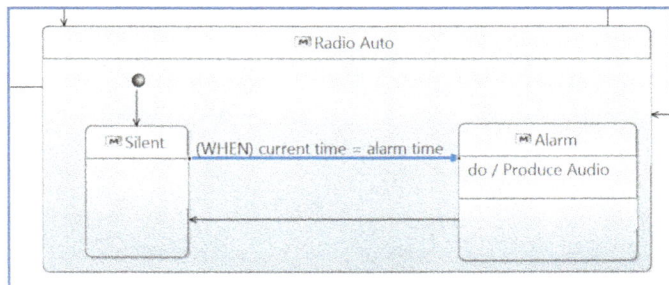

Figure 11-39 – Diagram is updated

Capella has updated the diagram.

To finish, we want to model a timeout for the clock radio. That is, the clock radio will produce audio for a certain period of time, and then stop automatically, without any external intervention. We can use here the concept of *time event*, with the keyword *after*.

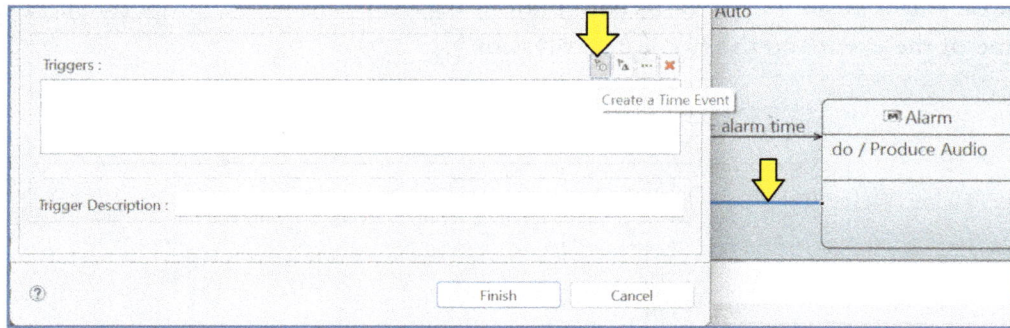

Figure 11-40 – Create a time event

Double-click the transition from "Alarm" to "Silent". In the transition's property sheet, click on the "Create a Time Event" button.

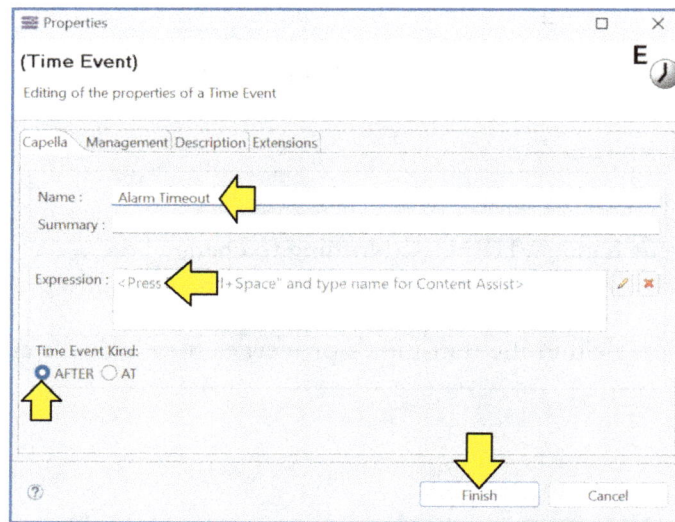

Figure 11-41 – Fill in the details of the time event

1) Name the event: "Alarm Timeout".

2) Leave the expression blank for the moment.

3) Set "Time Event Kind:" to "AFTER".

4) Click "Finish".

5) Notice that the triggers field in the transition's properties now includes an expression for the time event: "[AFTER] Alarm Timeout".

6) Click "Finish" again.

By leaving the expression blank, we have modeled the idea that there will be an alarm timeout, but that we don't know what the number will be yet. We could have also modeled an expression such as "5 minutes" if we already had a fixed requirement for the timeout.

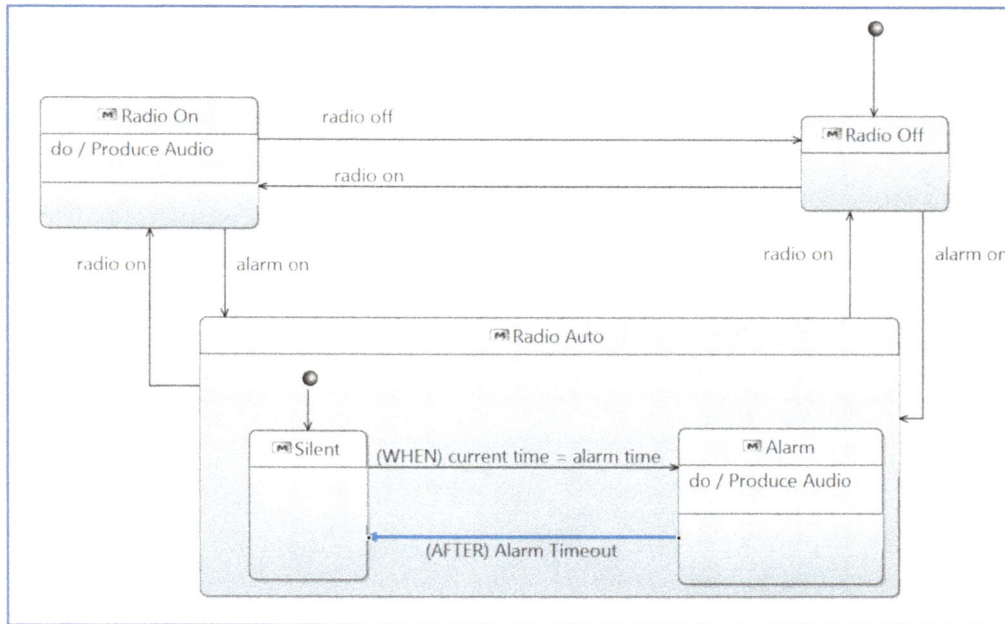

Figure 11-42 – The mode machine is complete

We now have a completed mode machine diagram. Of course, this is a fairly simple example. There are more details we could add. However, this is sufficient for the moment. [41]

States Within Scenarios

We will introduce one last nice feature of *Capella*: you can show modes/states of structural elements in scenario diagrams! Using states this way can be very helpful in clarifying specify "pre-conditions" or "post-conditions" when the scenario diagram (or part of a scenario diagram) represents a use case.

For instance, if we go back to the simple scenario that we created in the previous chapter, we may wish to clarify that before the user sends the "alarm on" message to the clock radio, the clock radio should be in the "Radio Off" mode.

---

[41] And when any model becomes "sufficient" or even "just barely good enough" you should stop modeling immediately!! The worst thing you can do for yourself is to continue modeling, and modeling, and modeling – adding extra detail that simply makes your model difficult for the users to understand and more difficult for you to maintain.

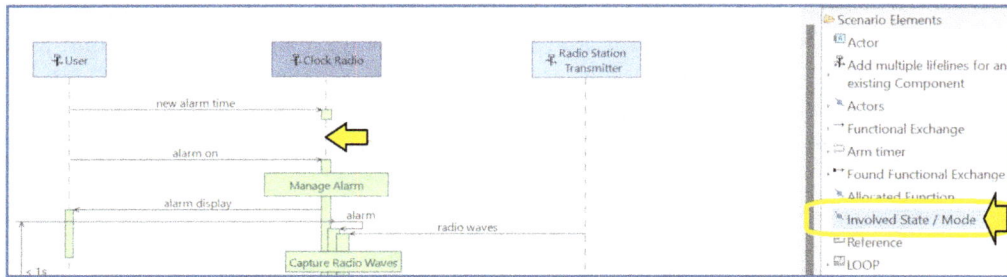

Figure 11-43 – Add a mode or state to a lifeline

Click on the "Involved State / Mode" tool in the palette and then click on the lifeline just before the reception of "alarm on". *Capella* opens a selection box with all modes/states of the structural element represented by the vertical lifeline, here the clock radio.

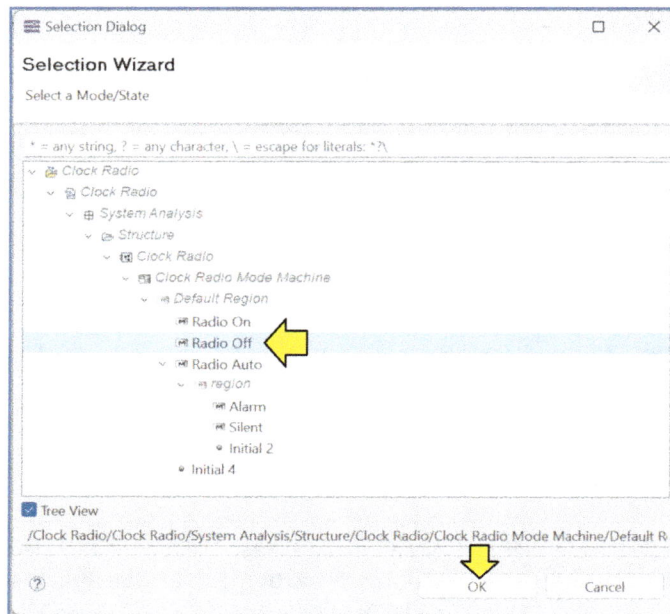

Figure 11-44 – Add the Radio Off state

Select "Radio Off" and click "OK".

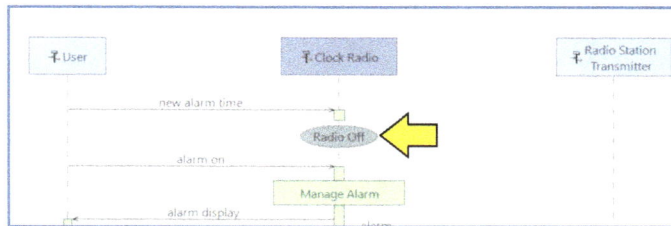

Figure 11-45 – The state is shown on the lifeline

Capella creates a grey oval with the name of the mode on the lifeline.

Further Study

Here is a list of further topics for more in-depth study:

- **Model Execution and System Simulation in Capella** – In this advanced video, you will find an in-depth explanation and illustration of Model Execution and System Simulation in *Capella* with a dedicated commercial add-on. It can even use MATLAB code embedded in functions.

  - https://url4ap.net/CH-11-MSM-Execution

- **Simulation with Python and MATLAB** – In this advanced video, you will find an in-depth explanation and illustration of System Simulation with Python and MATLAB in *Capella* with a dedicated commercial add-on, called DESS.

  - https://url4ap.net/CH-11-MSM-Phython

- **Model Simulation and Analysis** – In this advanced video, you will find an in-depth explanation and illustration of Model Simulation and Analysis in *Capella* with another dedicated commercial add-on, called FLE.

  - https://url4ap.net/CH-11-MSM-Sim1

- **What if You Could Simulate Your Model?** – In this advanced video, you will find an in-depth explanation and illustration of Model Simulation in *Capella* by bridging it with Simulink.

  - https://url4ap.net/CH-11-MSM-Sim2

- **Modeling States and Modes** – In this advanced video, you will find an in-depth explanation and illustration of additional concepts that could be interesting to model states and modes in a more powerful way.

  - https://url4ap.net/CH-11-MSM-States

Chapter 12 – Class Diagram Blank Diagrams (CDB)

In this chapter, we will take a close look at the data modeling concepts supported by *Arcadia*. In *Capella*, data modeling is implemented by the so-called "Class Diagram Blank" or [CDB], which, as the name suggests, is strongly inspired by the UML class diagram. However, the CBD has some unique features that we will introduce. We will also introduce the important notion of a *Capella* **library** which supports reuse of data type definitions throughout projects.

Introduction to Data Modeling

Data Modeling is a key systems engineering activity. The external interfaces of the system with its "actors", as well as the internal interfaces between logical or physical components are not fully specified until we have described the data flowing across those interfaces.

The word "data" should be understood here in a very general sense. A piece of data can represent many different things: a signal or an image, pure information (numeric values, strings, and similar), but also a physical quantity (temperature, pressure, speed, torque, and similar) or even fluids and matter.

In the preceding chapters we have started to describe interfaces between the system and its actors, or between internal components, by means of component exchanges, themselves being derived from functional exchanges between the corresponding allocated functions.

However, component or functional exchanges have little more than a name and perhaps an informal description. In order to unambiguously describe these exchanges, we need to model a more formal description of exactly what is flowing across the exchanges.

Another key engineering task is to avoid multiple definitions for a single piece of data in different areas of the system. We need to be able to declare that several exchanges carry the same type of data, without having to redefine the shared data for each exchange. In order to support this sort of globally unique data definition, *Arcadia* provides advanced mechanisms for modeling data structures with the desired level of precision. *Arcadia* also defines mechanisms to link data structures to functional or component exchanges, as well as function or component ports.

Considering the above more carefully, we need to introduce a few definitions.

Basic Type – **Basic types** are the basic elements of data modeling. *Arcadia* defines four kinds of simple type: **Enumeration**, **NumericType**, **StringType**, and **PhysicalQuantity**.

- *Arcadia* basic types are closely related to **primitive types** in UML. However, they are not identical. The UML distinction between "Integer" and "Real" is implemented as a setting within the *Capella* NumericType. The UML "UnlimitedNatural" type does not seem to be supported at all.

- [Roques] – section 4.9, page 155

- [UML2.5.1] – section 21.1

We can consider the basic types to be the "atoms" of *Arcadia* data modeling.

Class – A **class** in *Arcadia* is simply a list of data items which can either be basic types or other classes.

1) *Arcadia* classes can be linked together with the UML relationships of aggregation, composition, association, or generalization.

2) Unlike UML classes, *Arcadia* classes do not have methods or operations.

3) As such, an *Arcadia* class can be thought of as being similar to a UML **DataType**.

[Roques] – section 4.9, page 156

With classes defined, we now have the "molecules" of the data model.

Exchange Item – An **exchange item** is an ordered set of references to elements routed together, during an interaction or exchange between functions, components and actors. An exchange item is defined by a name, the list of referenced elements, and an optional communication mechanism (event, operation, flow, shared data). Each element referenced in an exchange item has a name and a type (basic type or class). [Roques] – section 4.9, page 151

Exchange items are the data definitions for functional exchanges (which can be bundled into component exchanges).

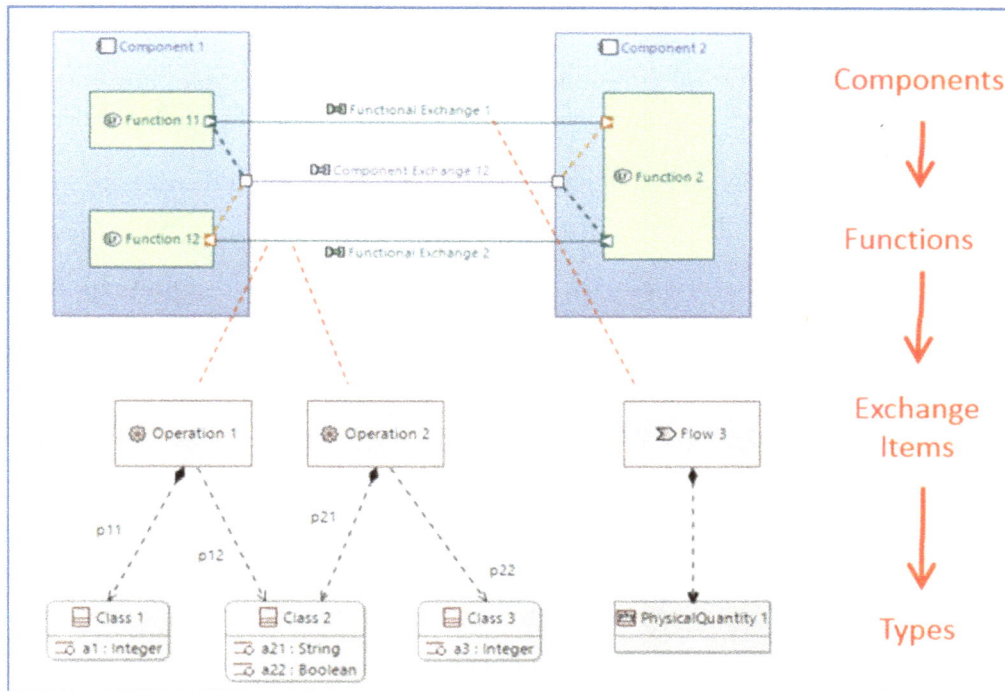

Figure 12-1 – Overview of the data model

Exchange items, classes, and types are modeled in *Arcadia* diagrams called **Class Diagram Blanks**. These **CDB** are available at each *Arcadia* level and are placed under the theme of "Transverse Modeling" (just like Mode and State diagrams) in the activity explorer.

Arcadia Levels and Data Modeling Concepts

These data modeling concepts are the only model elements in *Capella* which can be used at more than one *Arcadia* level.

A type, class, or exchange item defined at the system analysis level can be used at all of the lower levels, all of the way down to the physical architecture level. There is no need to "transition" definitions of types, as is done for functions or actors.

The opposite is not true: a type, class, or exchange item defined at a given *Arcadia* level cannot be used at higher levels.

Library Models

Introduction

A library is a *Capella* model intended to be shared between several projects.

The basic idea is to allow, within a given project, the separation of reusable model elements, so that they can be referenced later in other projects. These reusable elements are mainly definitions of types and classes of the domain, as well as common physical components. This list is not restrictive, however: a *Capella* library can contain the same model elements and the same *Arcadia* levels as classical *Capella* projects.

One *Capella* project may reference several libraries. A library can also reference in turn more general libraries. But a plain *Capella* project cannot be referenced by other projects or libraries.

A project can reference a library in either READ (default) or READ/WRITE mode. In the latter case, this means the library contents can be modified from the project itself, without having to specifically open the library.

Creating a Library of Basic Types and Classes

Let's return to the clock radio example and perform some data modeling on it.

We will need very general concepts such as a timestamp. These concepts are highly reusable. It makes sense to create a library which could be helpful for our future projects in the same domain.

A *Capella* library is just a specific type of project. We need to define the project as a library at creation time. A project cannot be turned into a library later.

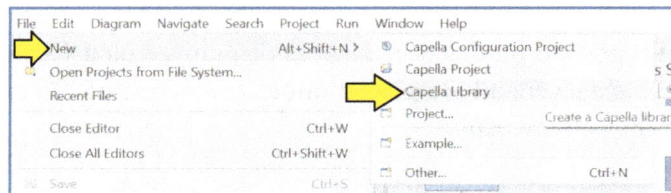

Figure 12-2 – Create a new library

From the *Capella* toolbar, select "File", then "New" and then "Capella Library".

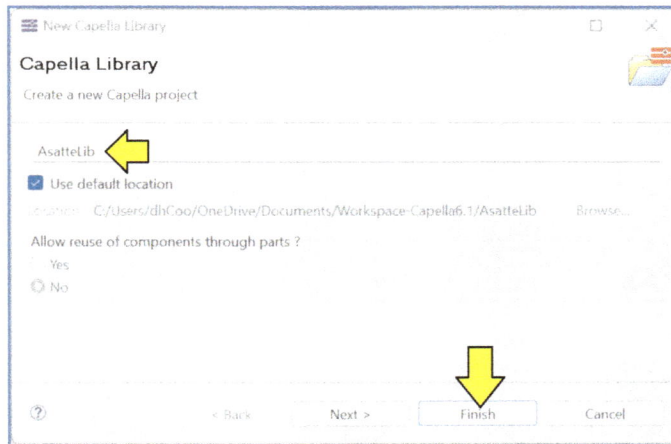

Figure 12-3 – Name the new library

Capella opens a definition panel for the new library. This pop-up window is similar to the one we obtain when we create a plain *Capella* project. Name this new library "AsatteLib" and click on "Finish".

Capella creates a new project in the same workspace. Since this new project is a special "library" project, the icons in the project explorer are slightly different. By default, we get the five usual *Arcadia* levels and the same contents as a plain *Capella* project.

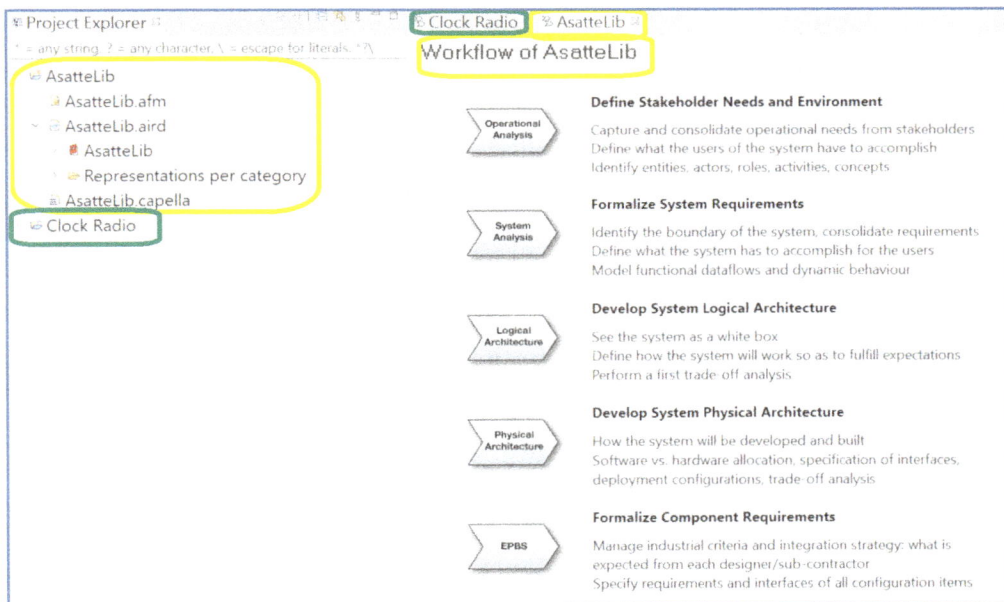

Figure 12-4 – New activity explorer window

Capella opens also a new activity explorer along with the existing diagrams of our clock radio project.

As previously mentioned, data model elements in *Arcadia* can be used at the level where they are defined as well as levels below. In most cases, we will simply want to define data items at the operational analysis level so they can be freely used anywhere.

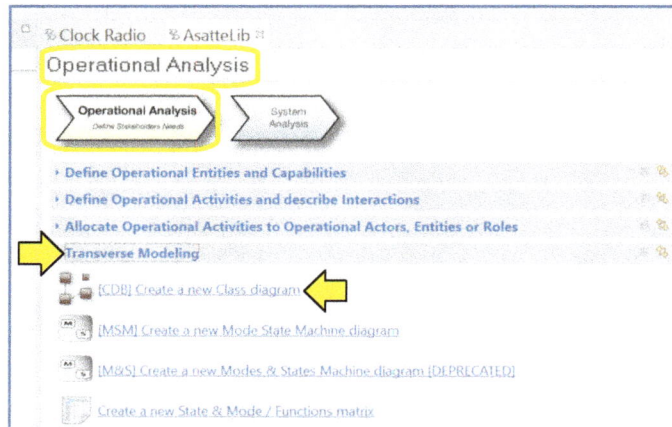

Figure 12-5 – Create class diagram at the operational analysis level

Click on "Operational Analysis" in the activity explorer, and then open the "Transverse Modeling" group to select the "[CDB] Create a new Class diagram" command.

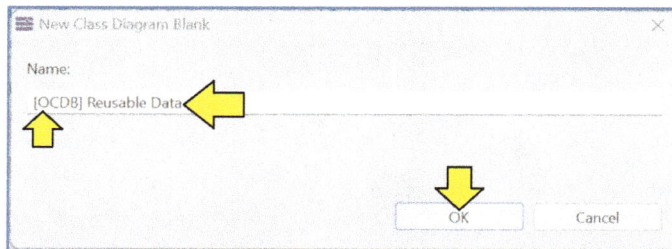

Figure 12-6 – Name the diagram and add the level manually

As usual, we just need to give the diagram a name, for instance: Reusable Data. Again, *Capella* omits the first letter for the *Arcadia* level in the acronym (as was the case for scenario diagrams). Add the "O" manually to make the full diagram name "[OCBD] Reusable Data". Click "OK".

The Class Diagram Blank created behaves like a normal blank diagram. However, palette divided into three groups:

- **Classes** – for data type definitions (basic types, classes, properties, and similar).

- **Common** – for common diagram elements such as title blocks (not unique to the CBD).

- **Communication** – for exchange items.

Note: the CDB palette is very complex, and you most likely will not need to use all of the concepts in your projects. Moreover, the property sheets of these model elements are also very extensive, allowing those who wish to need to do so to develop their formalizations in great depth! In particular, we will not use the "Interface" concept that you find in the Communication group. This concept is really similar to the UML interface concept, and usually only well understood and used by object-oriented software experts.

In our library, we will only create type definitions (which are very common and reusable). We will create the exchange items in the clock radio project itself. That is, the exchange items are much more likely to be specific to the clock radio and not easily reusable in other projects.

Figure 12-7 – The CDB palette has three sections

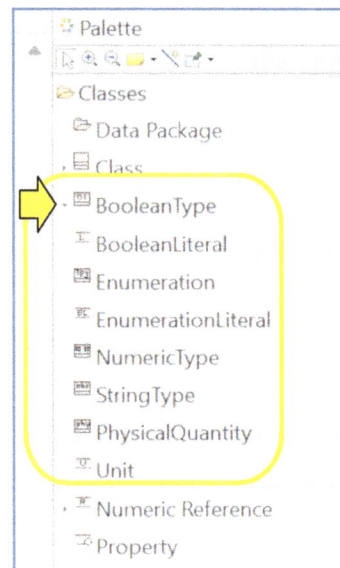

With this strategy in mind, we will focus on the "Classes" group of the palette.

All the basic types are grouped under "BooleanType". Click on the small arrowhead in front of "BooleanType" to open the list of basic types.

The basic types predefined by *Capella* are:

- BooleanType

- Enumeration

- NumericType

- StringType

- PhysicalQuantity

"BooleanLiteral" and "EnumerationLiteral" are not really types in and of themselves but rather helper tools for completing the definition of "BooleanType" and "Enumeration-Type".

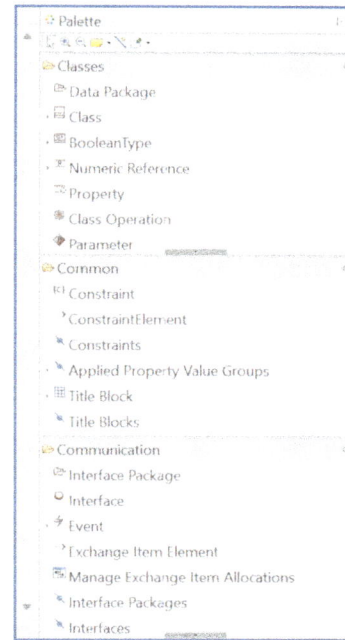

Figure 12-8 – Basic types are grouped under BooleanType

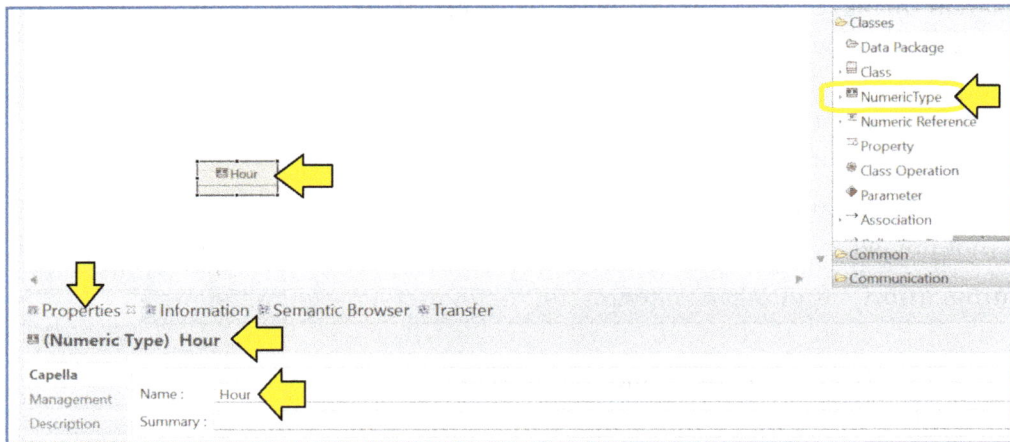

Figure 12-9 – Create a numeric type

Create a numeric type called "Hour". [42]

As mentioned previously, the property sheet is quite complex. For the moment, we will just specify here minimum and maximum values: 0 to 23.

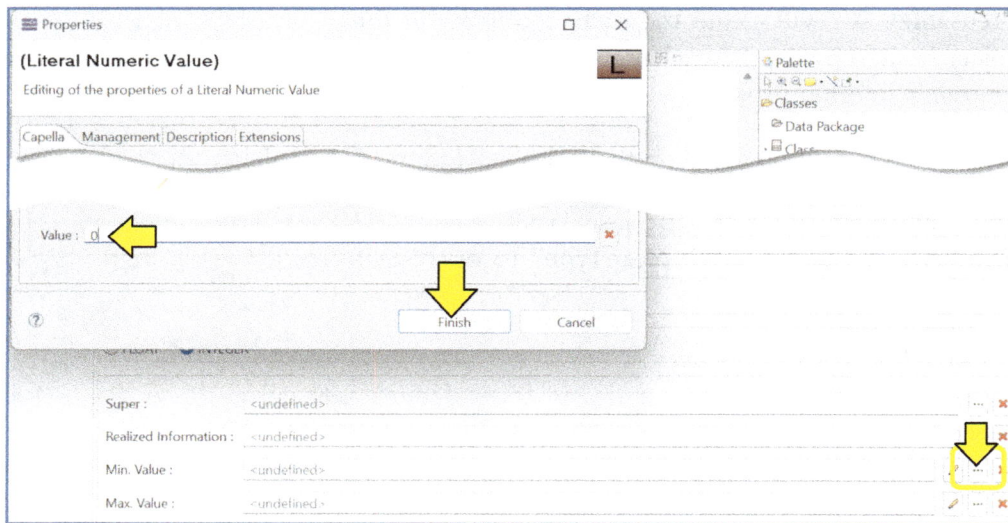

Figure 12-10 – Set the minimum value to 0

Click on the pencil to the right of the "Min. Value:" field to open an additional property sheet, called "Literal Numeric Value". Enter "0" for the value and click "Finish". Set the max value to 23 using the same technique.

---

[42] Notice that the *Capella* palette for the CDB is a bit tricky. It may have listed all the basic types under "BooleanType" previously, but if you select "NumericType" the tool will decide that you are more interested in that choice and make that the top (container) tool.

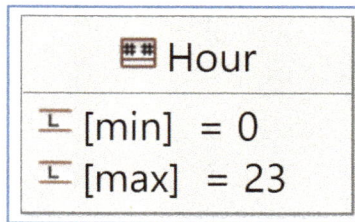

Figure 12-11 – The hour rectangle is updated

The Hour rectangle is updated in the diagram with both literal values.

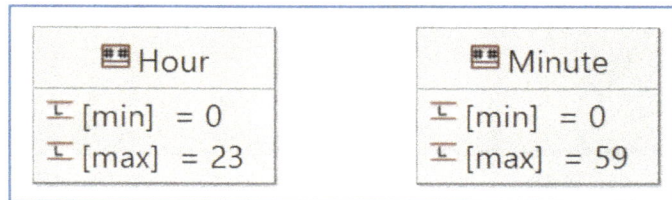

Figure 12-12 – Create another numeric type for minute

Create a second numeric type called "Minute" using the same method.

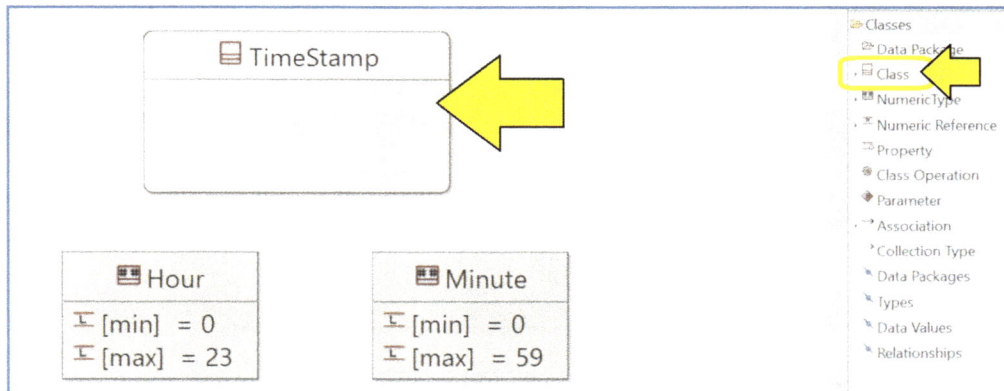

Figure 12-13 – Create the TimeStamp class

Now if we want to define a data type with two properties, one value of hour and one value of minute, we need to create a class in *Capella*. Name the new class "TimeStamp".

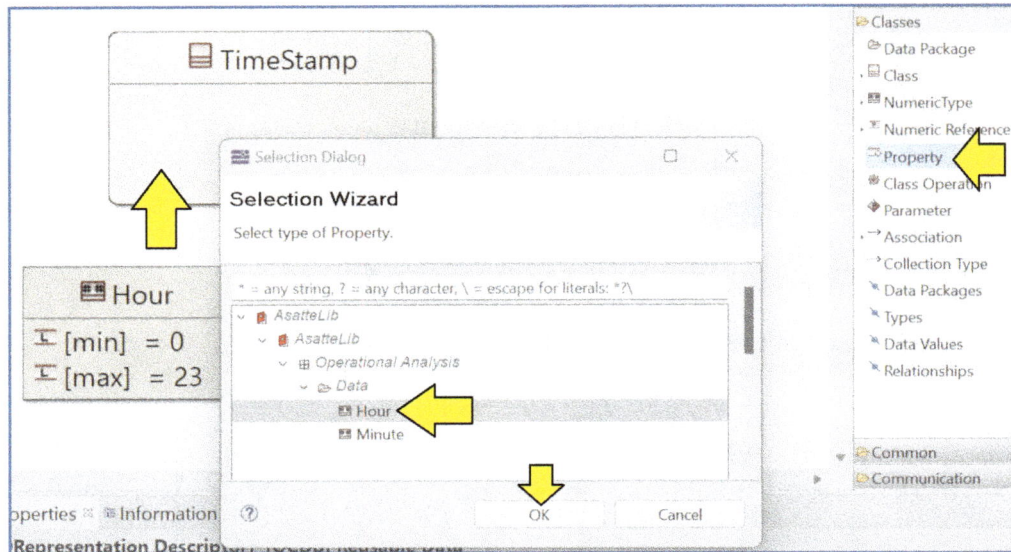

Figure 12-14 — Select the types for the properties

In the palette select the "Property" tool. Click inside the "TimeStamp" class. A selection dialog will appear. Select "Hour" and click "OK".

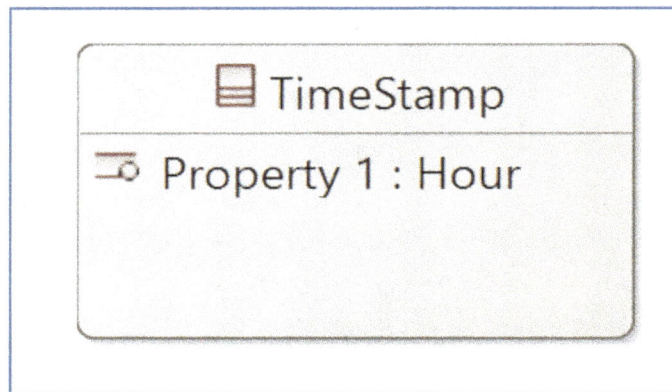

Figure 12-15 — The property appears inside the class

The property appears inside the class, albeit with a default name of "Property 1".

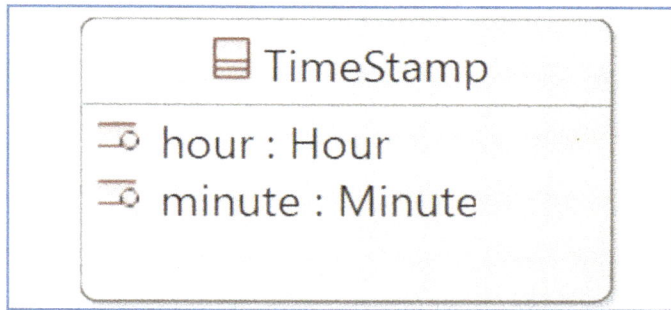

Figure 12-16 – Rename first property and create another

Rename it "hour" and create a second property "minute" of type "Minute".

If we open the project explorer, we can see the model elements created in the "Data" package of level "Operational Analysis" of the library.

Note: if we want to be allowed to use TimeStamp as a type in turn, for properties of another Class, we need to declare it as primitive (tick the checkbox "Is Primitive").

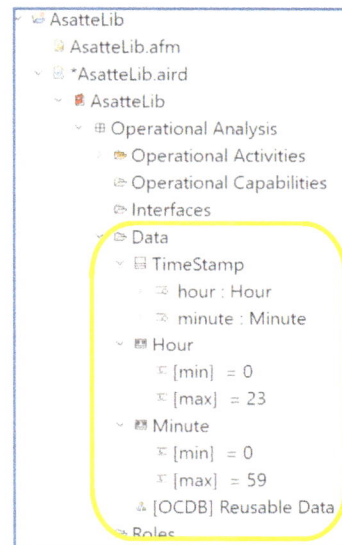

Figure 12-17 – Elements are visible in the project explorer

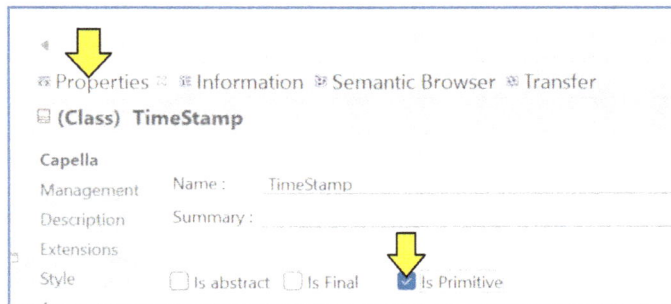

Figure 12-18 – Make TimeStamp Primitive

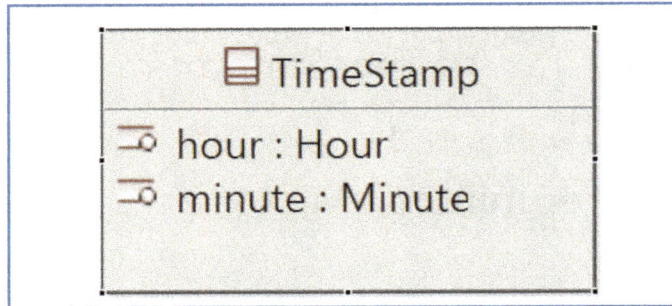

Figure 12-19 – Shape and color change slightly

The shape and color of the box change slightly after setting the class to be primitive.

With that, we have enough content in our library for the moment. Save the project. [43] Next, we can turn our attention back to the clock radio project.

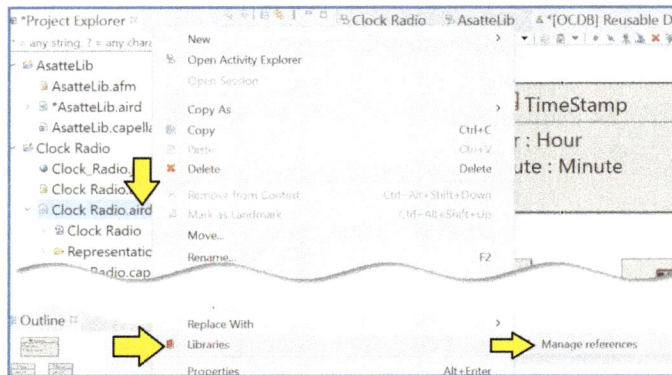

Figure 12-20 – Add library reference in clock radio project

From the project explorer, right-click on the .aird file, click "Libraries" and then "Manage references".

---

[43] The library and the clock radio are separate projects. You will need to save each of them separately.

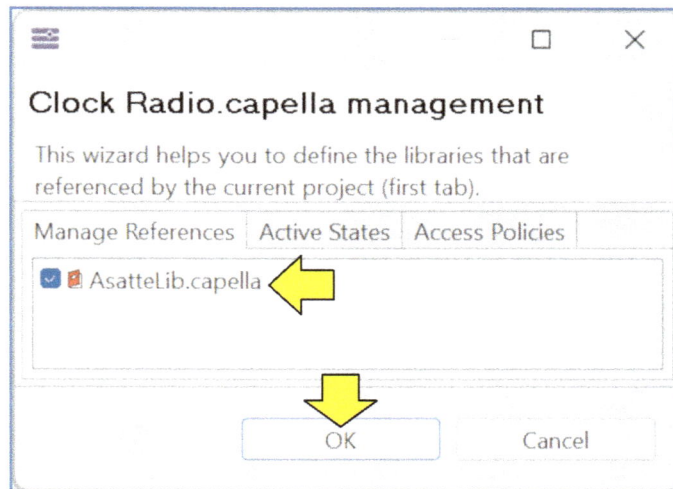

Figure 12-21 – Select the AsatteLib

Capella opens a wizard asking us to select the libraries we want to reference. Select the AsatteLib and click "OK".

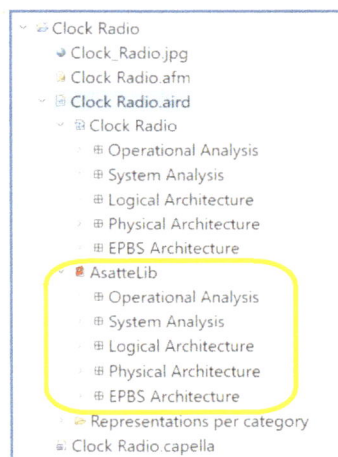

Figure 12-22 – Library visible inside Clock Radio project

We can now see the "AsatteLib" library contents in the project explorer inside the clock radio project.

Exchange Items

We will now return to the system analysis level of the clock radio project. Open the "[SAB] Functional View" diagram. We will use this diagram to enhance the modeling precision for some of the functional exchanges.

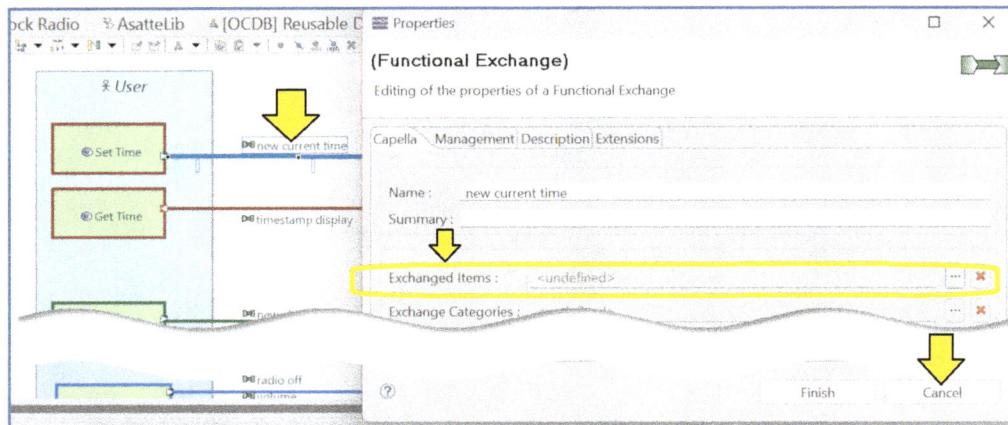

Figure 12-23 – No exchange items have been defined yet

Double-click on the "new current time" functional exchange to open its properties panel. Checking the exchange items field, we see that no exchange items have been defined. Click "Cancel" to exit.

We cannot directly link functional exchanges to types or classes. First, we will need to create some exchange items. We will create several of them in another [CDB], at the system analysis level of our project.

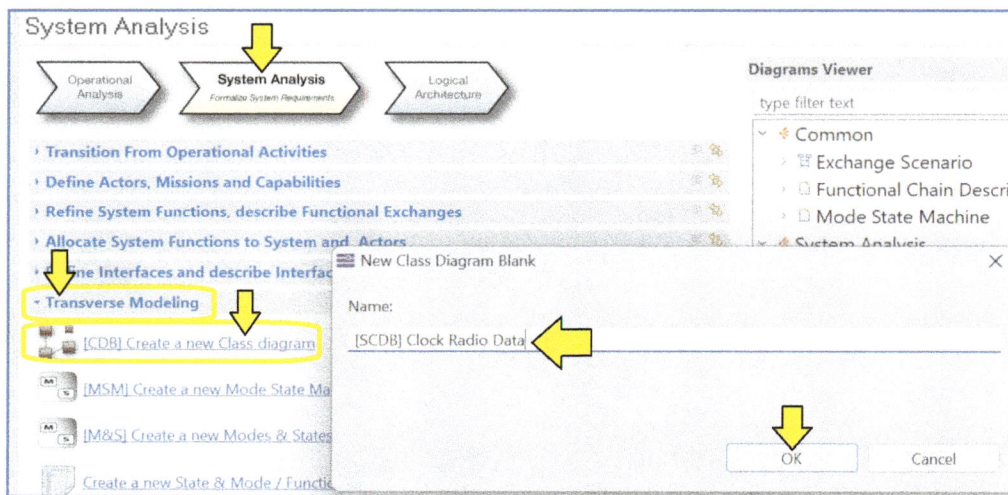

Figure 12-24 – Create a system class diagram blank for the exchange items

Return to the activity explorer, at the system analysis level and open the "Transverse Modeling" group. Click on "[CDB] Create a new Class diagram" and rename the new diagram "[SCDB] Clock Radio Data".

In the palette we will find the exchange item tools in the "Communication" group hidden under the "Event" command.

We will be creating the following exchange items:

- CurrentTime, with the "Operation" mechanism,

- AlarmTime, with the "Operation" mechanism,

- TimeDisplay, with the "Flow" mechanism.

In fact, at system analysis level, the communication mechanism applied to a given exchange item might not yet be known. In this case, the undefined "joker" communication mechanism can be used. The choice of communication mechanism can be refined later using the property sheet of the exchange item.

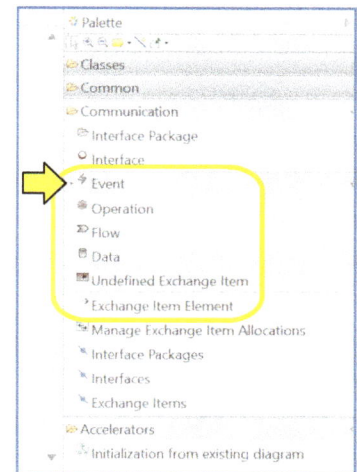

Figure 12-25 – Exchange items

Communication Mechanisms for Exchange Items

As we said, an exchange item has a name and a communication mechanism. There are four predefined communication mechanisms in *Arcadia*:

- **Event** – asynchronous mechanism where an event is sent by an element and received by one or several others.

- **Flow** – flow of matter, energy, or data, information, or similar.

- **Operation** – process carried invoked by one element and carried out by another.

- **Shared Data** – data modified by an element and read by others.

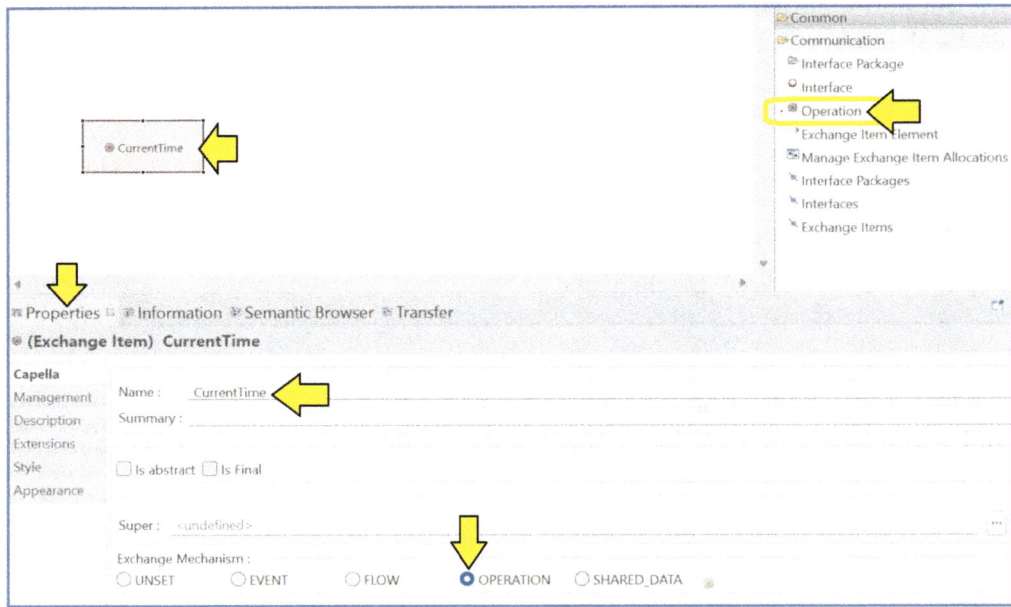

Figure 12-26 – Click in diagram to create operation exchange item

Start by clicking on "Operation" in the "Communication" group of the palette. Next click inside the diagram somewhere. *Capella* creates an exchange item with a specific icon, that we can rename "CurrentTime". Create "AlarmTime" the same way.

Next, we will want to link these two exchange items with the "TimeStamp" class which was defined in the library.

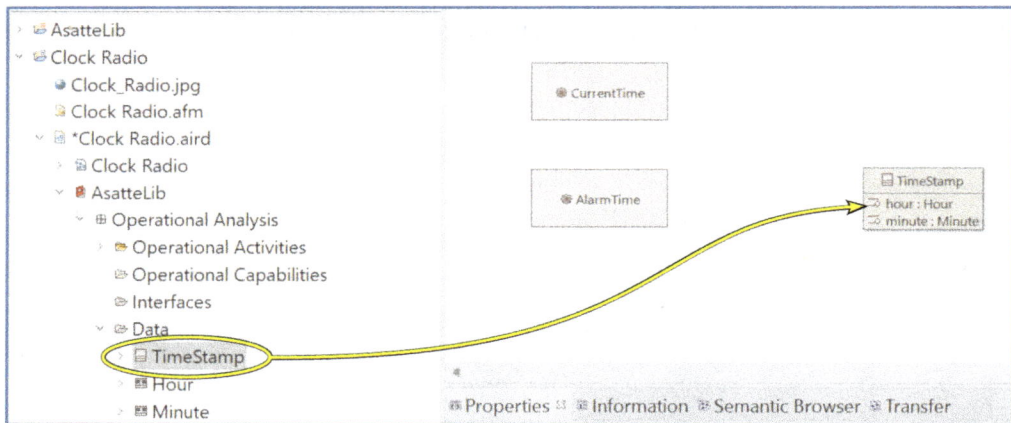

Figure 12-27 – Drag the TimeStamp from the library to the diagram

We can just drag and drop the "TimeStamp" class from the project explorer to the diagram.

Note: be sure to drag from the image of "AsatteLib" inside the clock radio project rather than from the original "AsatteLib" project.

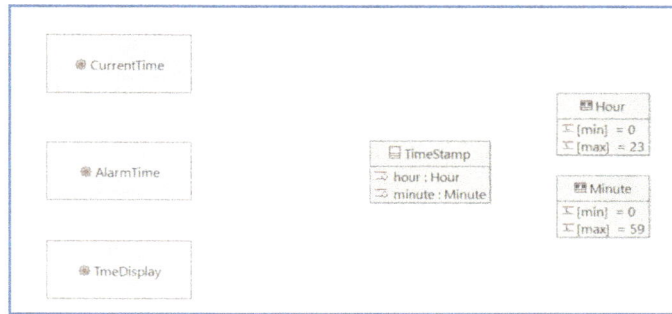

Figure 12-28 – Add hour, minute, and another operation for time display

We can also drag and drop the "Hour" and "Minute" numeric types. Adding them to the diagram does not really change the structure of the model. However, having them on the diagram adds context for the "TimeStamp" class. Create a third exchange item called "TimeDisplay" to complete the diagram.

Finally, we need to link the exchange items to the types.

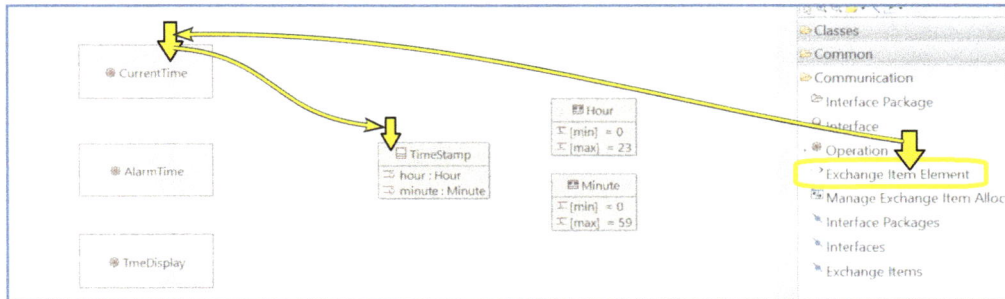

Figure 12-29 – Drag from CurrentTime to TimeStamp

Click on the "Exchange Item Element" command in the palette. Select first an exchange item, and then a type.

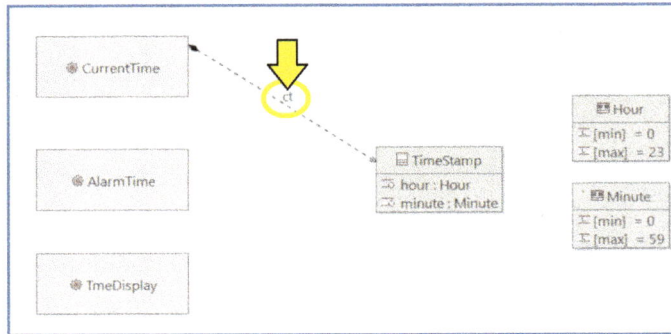

Figure 12-30 – Give short name to exchange item element or hide it completely

The name of the "Exchange Item Element" is not very meaningful, except for operations, where it can describe the different parameters. One common practice is to put the initials of the type in the label. It is also possible to hide the label on the diagram with the diagram top toolbar.

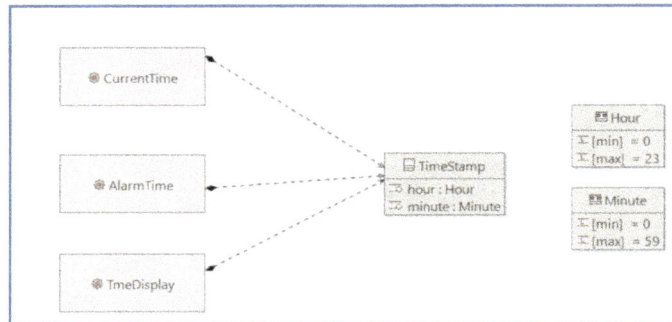

Figure 12-31 – Finished diagram

Link all three exchange items to the same TimeStamp class. With that, the diagram is complete.

Functional Exchanges

Now that we have created several exchange items, we can start to link them with the functional exchanges in our clock radio model.

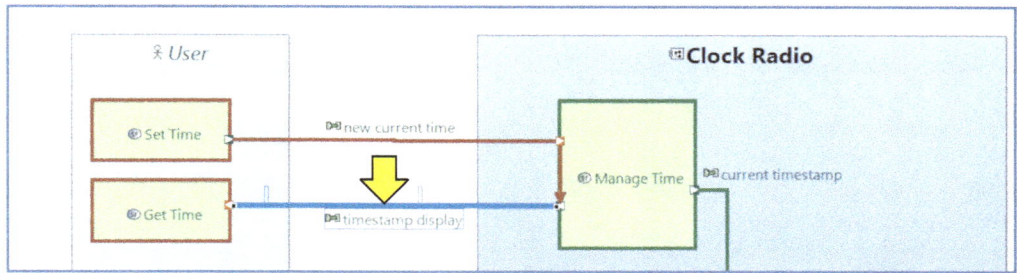

Figure 12-32 – Select timestamp display functional exchange

Reopen the "[SAB] Functional View" diagram that we created in Chapter 6. Select the functional exchange "timestamp display" that connects the function "Manage Time" to the function "Get Time".

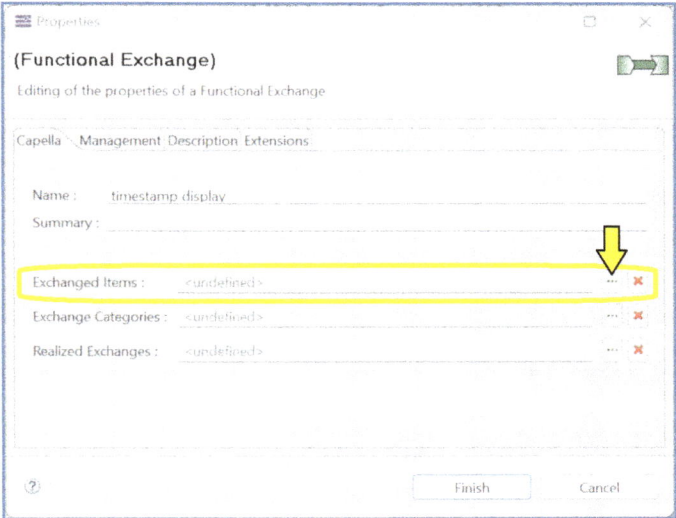

Figure 12-33 – Open the property sheet of the functional exchange

Each functional exchange can carry one or more exchange items. To assign exchange items open the property sheet of the functional exchange and go to the "Exchanged Items" field. Click on the three green dots. *Capella* will open a transfer dialog enabling us to select existing exchange items.

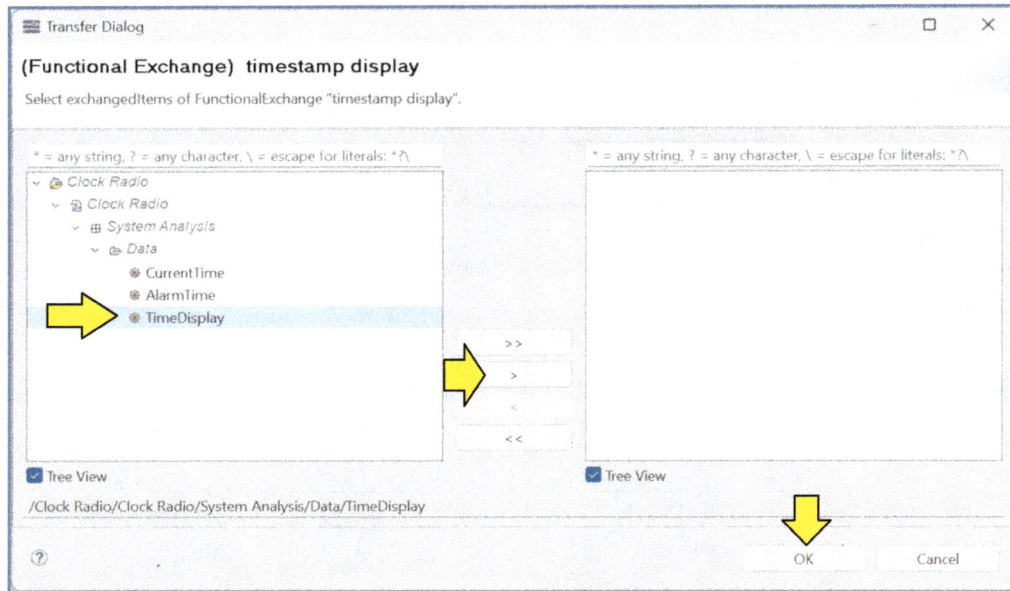

Figure 12-34 – Link the TimeDisplay exchange item

Select "TimeDisplay", click single arrow (>) to assign the exchange item to the functional exchange, and click on "OK".

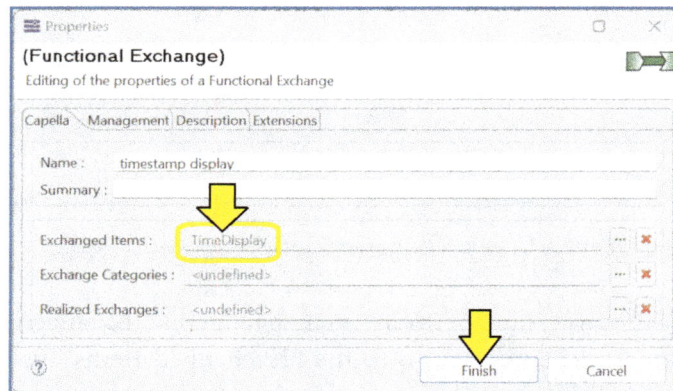

Figure 12-35 – The exchange item is now visible

The exchange item appears in the "Exchanged Items" field. Click on "Finish".

Figure 12-36 – Link the CurrentTime exchange item

Use the same process to assign the "CurrentTime" exchange item to the "new current time" functional exchange that connects the function "Set Time" to the function "Manage Time".

Figure 12-37 – Link the AlarmTime exchange item

Finally, use the same process to assign the "AlarmTime" exchange item to the "new alarm time" functional exchange that connects the function "Control Alarm" to the function "Manage Alarm".

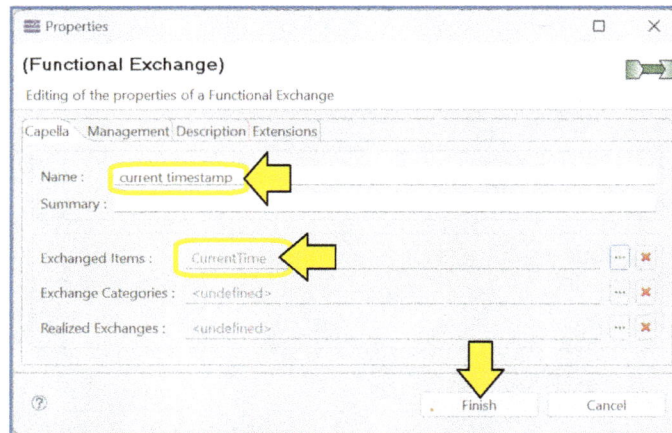

Figure 12-38 – Associate exchange item CurrentTime with a second functional exchange

An exchange item can be carried by several functional exchanges. In order to demonstrate this point, we can associate "CurrentTime" with a second functional exchange "current timestamp".

Again, this one-to-many relationship can work both ways:

- A functional exchange can carry multiple exchange items.

- An exchange item can be carried by multiple functional exchanges.

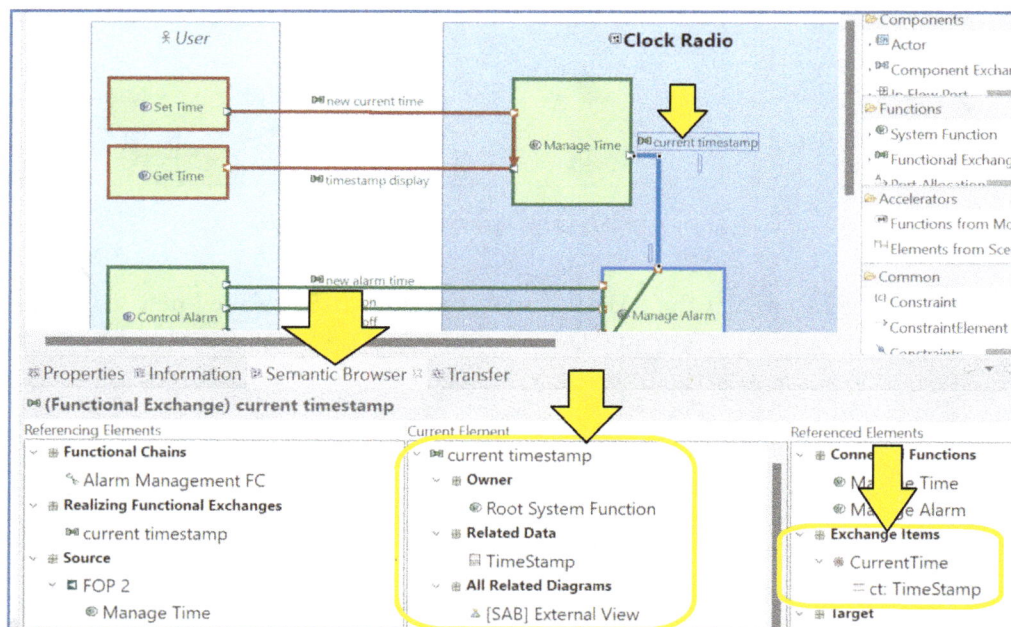

Figure 12-39 – Exchange items are visible in the semantic browser

This new relationship does not show automatically in the diagrams, so how can we visualize it efficiently? As often is often the case with *Capella*, the semantic browser provides the information we are looking for.

If we select the "current timestamp" functional exchange, and select the semantic browser, we can see the "CurrentTime" exchange item in the right column, underneath "Referenced Elements".

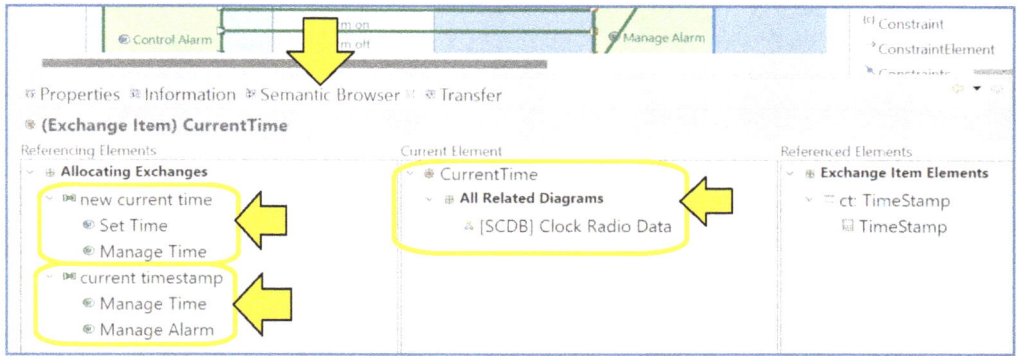

Figure 12-40 – Cross referencing from exchange item to functional exchanges

Next, if we select the "CurrentTime" exchange item and click on "F9" to make it the current element of the semantic browser, we can easily see the two functional exchanges it is referencing.

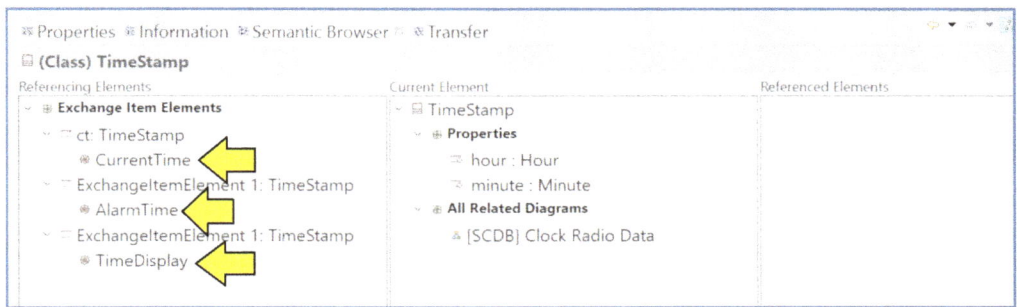

Figure 12-41 – Cross referencing from class to exchange item

And now when we select the "TimeStamp" class and click on "F9", we get all three exchange items referencing it.

A more efficient way to link exchange items with functional exchanges is to use the **Mass Editing View**, a recently added feature of *Capella*.

In order to demonstrate the mass editing view, create two additional exchange items (of type "Event"), "AlarmOn" and "AlarmOff", and link them to the functional exchanges "alarm on" and "alarm off".

Next, display the functional exchanges in the project explorer, select any functional exchange in the diagram, right-click "Show in Project Explorer". You may need to agree to disable some filter that is hiding the functional exchanges.

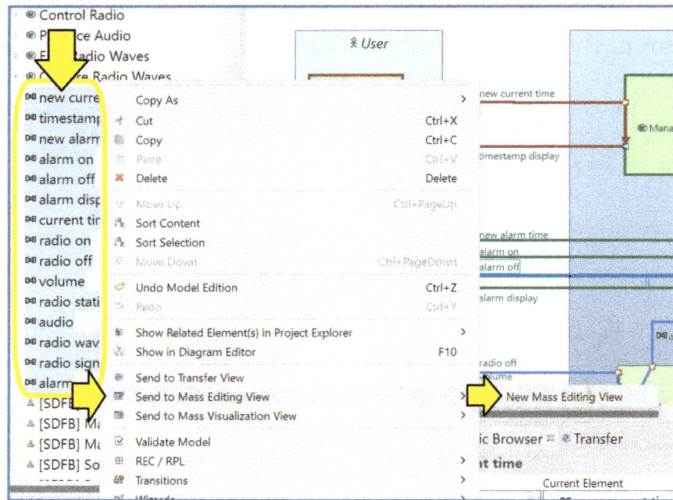

Figure 12-42 – Send to mass editing view

Select all of the functional exchanges in the project explorer, right-click, select "Send to Mass Editing View" and then "New Mass Editing View".

Figure 12-43 – Mass editing window appears

Capella creates a new window near the semantic browser, called "Mass Editing 1", displaying a lot of default columns corresponding to properties of the elements selected, in our case functional exchanges.

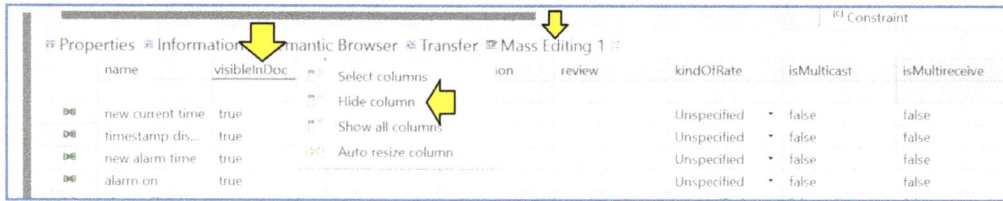

Figure 12-44 – Hide columns that we don't currently need

A lot of these columns are not really interesting for what we want to do so we will hide them in order to focus on exchange items. We can right-click on a specific column and Click "Hide Column" to hide it. Go ahead and hide most of the columns so we can focus on the exchange items.

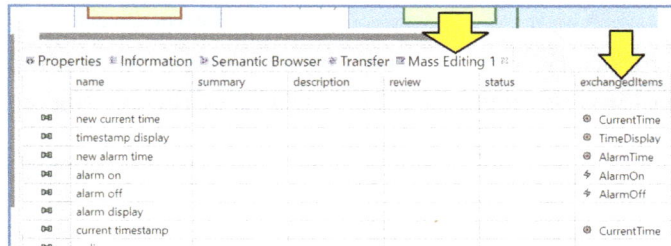

Figure 12-45 – Simplified view

We can see all the functional exchanges to which we already have associated exchanged items.

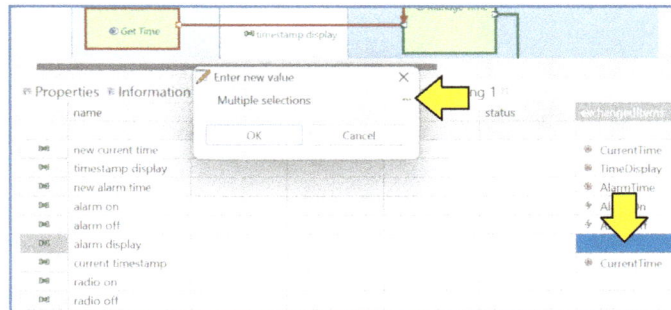

Figure 12-46 – Click to assign exchange items directly

We can also continue the process directly in this view. Click in an empty cell of the "exchangedItems". A pop-up will appear. Click on the three dots. This procedure will open the same transfer dialog as used previously and you can continue to assign exchange items in this manner.

Impact on Diagrams

As mentioned previously, exchange items do not appear automatically on previously defined diagrams, even those showing functional exchanges. We need to apply specific filters, depending on the type of diagram, to display the exchange items.

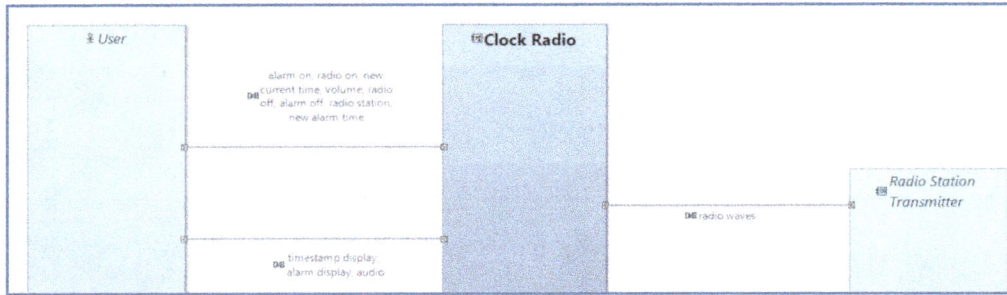

Figure 12-47 – Previous state: Only component exchanges shown, but with functional exchange names

For example, if we go return to the "[SAB] External View", another diagram that we created in Chapter 6, we can add the exchange items to the view.

In this diagram, we chose to display only component exchanges, but with the names of the allocated functional exchanges. The goal was to provide a good synthesis of the external interface of the system.

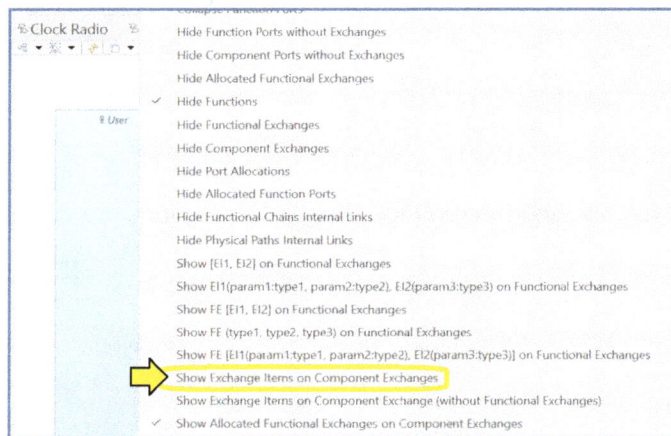

Figure 12-48 – Show exchange items on component exchanges

First, select the interface between the user and the clock radio. Then pull down the filter menu and select "Show Exchange Items on Component Exchanges". Remember that we have to "refresh" the diagram after any "Show" command.

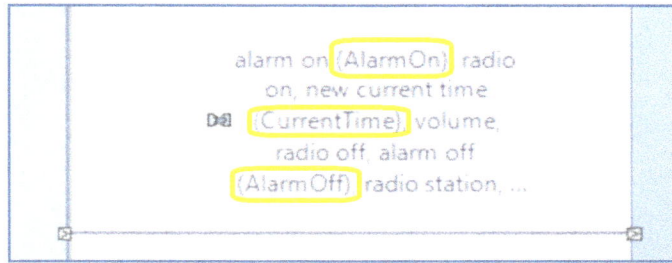

Figure 12-49 – Exchange items appear in parenthesis

The names of the exchange items appear in parenthesis after their related functional exchange.

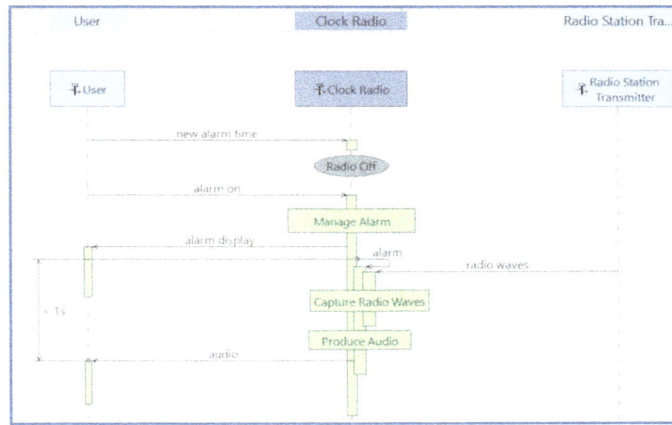

Figure 12-50 – Revisiting the scenario diagram

Next, we will take a look at the scenario diagram. Open "[SES] Clock Radio Main Success Scenario" from Chapter 10.

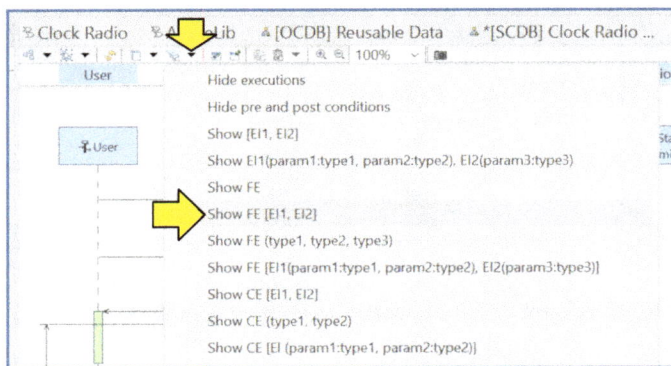

Figure 12-51 – Select the filter to show exchange items

We just need to find the appropriate filter to add the names of the exchange items. This time, we need to know our *Arcadia* acronyms, as the filter is: "Show FE [EI1, EI2]". FE means functional exchange, and EI means exchange item, of course! Once again, we need to refresh the diagram to show the changes.

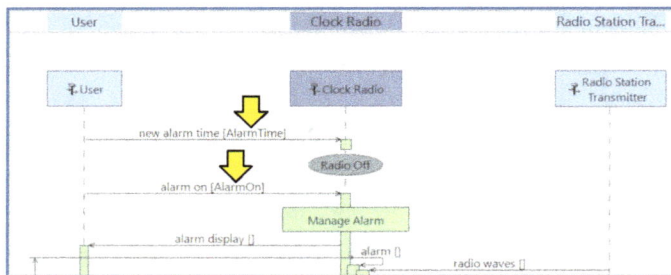

Figure 12-52 – Exchange item names appear in square brackets

After a "refresh" diagram, as usual, we now see the names of the exchange items appear between brackets (and not parenthesis this time) after their related sequence message (referencing a functional exchange). Once again, if nothing appears between the brackets, it means that we have not finished assigning exchange items to the functional exchanges.

Further Study

The design of *Arcadia* class diagram blanks is heavily influenced by UML. Here are some additional references you can look at to get some additional perspective.

- **Pascal's UML book** – Pascal released a convenient, compact book on UML in 2004.

    - [Roques(UML)] – Chapters 3, 4, and 7.

- **Data Modeling in Capella – Basic Constructs** – These web pages explain in detail the basic constructs available in the CDB diagram of Capella.

    - https://url4ap.net/CH-12-CDB-Basic

- **Data Modeling in Capella – Advanced Constructs** – These web pages explain in detail the advanced constructs available in the CDB diagram of Capella.

    - https://url4ap.net/CH-12-CDB-Advanced

- **Data Modeling in Capella – Methodology** – These web pages give detailed methodological recommendations for powerful data modeling in the CDB diagram of Capella.

    - https://url4ap.net/CH-12-CDB-Method

Chapter 13 – Challenges Ahead

Congratulations! If you have made it this far, you are now able to make all of the basic *Arcadia* diagrams using *Capella*. You should be well-positioned to practice and continue to extend and deepen your modeling skill.

So far, so good. However, unless you are a millionaire recluse with some very complicated hobbies, your *Arcadia* diagrams will not be useful unless you can use them to interact with other humans. Usually, those other humans will be coworkers in a large and complicated organization – an organization working on projects of sufficient complexity to warrant the use of *Arcadia*. This is the point at which things get tricky. You are going to encounter some surprising objections and misconceptions. As you go deeper, you will also encounter some surprises in your own use of the technology. The manner in which you end up using *Arcadia* may end up being rather different from what you imagined when you started.

A fully exhaustive discussion of these factors is far beyond the scope of this book. In fact, an exhaustive treatment of the subject would not end up being a book about engineering, but rather a book about cognitive psychology, sociology, or political science. However, we can at least give you a brief introduction to a few of the things you are likely to encounter early in your journey. The following sections cover some of the more common challenges.

Common Objections

As you get excited about the potential of *Arcadia* to help your organization and start trying to spread the vision, you are going to run into some objections. Actually, you are going to run into a *lot* of objections. In the next few sections, we will cover a few of the most common objections and give you some ideas about how to consider and discuss them in a nuanced manner.

First – Is MBSE Right for Your Project?

Before we delve into the most common objections for introducing *Arcadia* and/or MBSE, we have to consider the fact that the people objecting may be right: *Arcadia* and MBSE might not be a good fit for your project. In fact, the percentage of projects that will benefit from a determined deployment of MBSE techniques is actually fairly small.

Figure 13-1 – The cat-eating-cheeseburger application

For example, consider a team that is working on the cutting-edge mobile social media application shown in Figure 13-1. Would this team benefit from a careful MBSE workup of their application? Probably not. This team would face several challenges that would make *Arcadia* or MBSE adoption difficult:

1) **Requirements –** The team has no idea what the end-user requirements are. In fact, the end users have no idea what the end-user requirements are either.

2) **Concept –** They don't really have one. Their goal is simply to get the cat-eating-cheeseburger application on the market as fast as possible and see if anyone is interested.

3) **Quality –** No one cares if the cat fails to eat a cheeseburger every now and then. In fact, the failure to eat a cheeseburger might end up being entertaining in its own right. [44]

4) **Schedule –** The application needs to ship yesterday and to be updated every few hours after that until customers lose interest in it.

Introducing a very slow and meticulous architectural planning process to the cat-eating-cheeseburger team isn't likely to be helpful. In fact, it is likely to contribute to the failure of the team if it slows down their application deployment and they get beaten to market by the competing Shiba Inu application.

So, what kinds of projects might benefit? Generally, the two ingredients that a project needs to be a good candidate for *Arcadia* and MBSE are:

1) **Overwhelming Complexity –** The project needs to be so complicated that it is impossible for the team members to keep track of all the requirements in their heads or even with spreadsheets.

---

[44] In the 1980s when David Hetherington was developing software for IBM, these sorts of behaviors were often referred to as "Feechurs", as in: "That's not a bug, it's a *Feechur*".

2) **Severe Consequences for Mistakes** – Mistakes generally need to have the potential to instantly end executive careers, cause unimaginable financial losses, or cause horrific injuries.

However, even when a project has plenty of both of these ingredients and painful failures are already happening regularly, many project leaders can be amazingly resistant to any sort of behavior change that might help get the problems under control. Don't be surprised if you meet a lot of resistance, even if the failures are already starting to happen.

Needless to say, if your project resembles the cat-eating-cheeseburger application, you should consider learning *Arcadia* purely as a personal hobby. You can save yourself a lot of frustration and disappointment by not even trying to convince the cat-eating-cheeseburger team to change their development approach and start modeling things carefully before implementing them.

Objection – Why Can't We Just Use PowerPoint?

Figure 13-2 – PowerPoint is excellent for compelling presentation of high-level concepts

PowerPoint is an outstanding presentation tool with some strong drawing features. However, PowerPoint is not really an engineering tool and has a number of limitations in terms of Systems Engineering:

- **No Standard Syntax** – Everyone's diagrams look different. PowerPoint's drawing features are great for high-level concept presentations, but PowerPoint struggles as you try to draw with any level of consistency and precision.

- **No Underlying Data Model** – If you make a 100-page presentation full of something called "fluid" and you find later that you need to change that to read "liquid" you will need to go back and manually find all the "fluid" boxes in the presentation and change them one-by-one.

- **No Tool Integration** – Although it is out of the scope for this book, MBSE tools like *Capella* are easy to integrate with other engineering management tools such as requirements database systems. PowerPoint does not really integrate with anything.

Make no mistake: PowerPoint is an excellent tool. However, it is not a substitute for a modeling tool like *Capella*.

Objection – Why Can't We Just Use Visio?

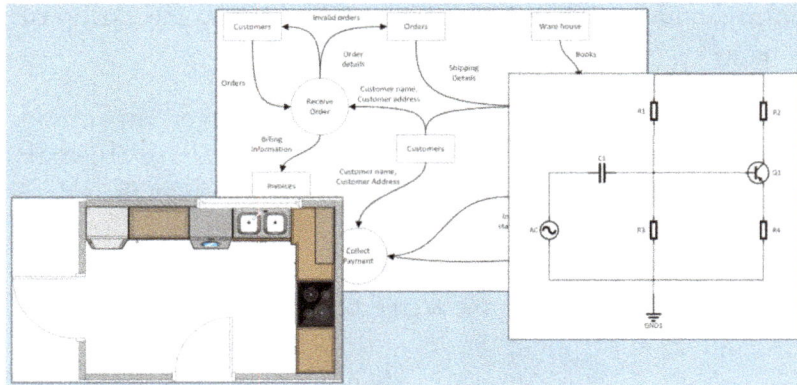

Figure 13-3 – Visio sample diagrams

Visio is a powerful diagram drawing tool produced by Microsoft. Visio can produce a wide variety of technical and business drawings including organization charts, flow charts, floor plans, and circuit diagrams. That having been said, Visio has no underlying model or tool integration.

Objection – Why Can't We Just Use Spreadsheets?

Spreadsheets are also very powerful tools. However, *Arcadia* is designed to manage relationships between the elements of your model. Spreadsheets are excellent for managing row/column style relationships, but the relationships in a complex system are more of a web or a network and are difficult to represent in a spreadsheet.

Objection – Why Can't We Just Use Text Documents?

Just writing things out in text is an intuitive approach that does not scale well. Again, text is a sort of two-dimensional medium. As you start adding lots of links and cross-references to the text to make it represent the more graph/node interlinked nature of complex requirements, the text gets harder and harder to understand.

If you need a concrete example of the problem with complicated interlinked text requirements, spend a pleasant afternoon reading through your country's tax regulations.

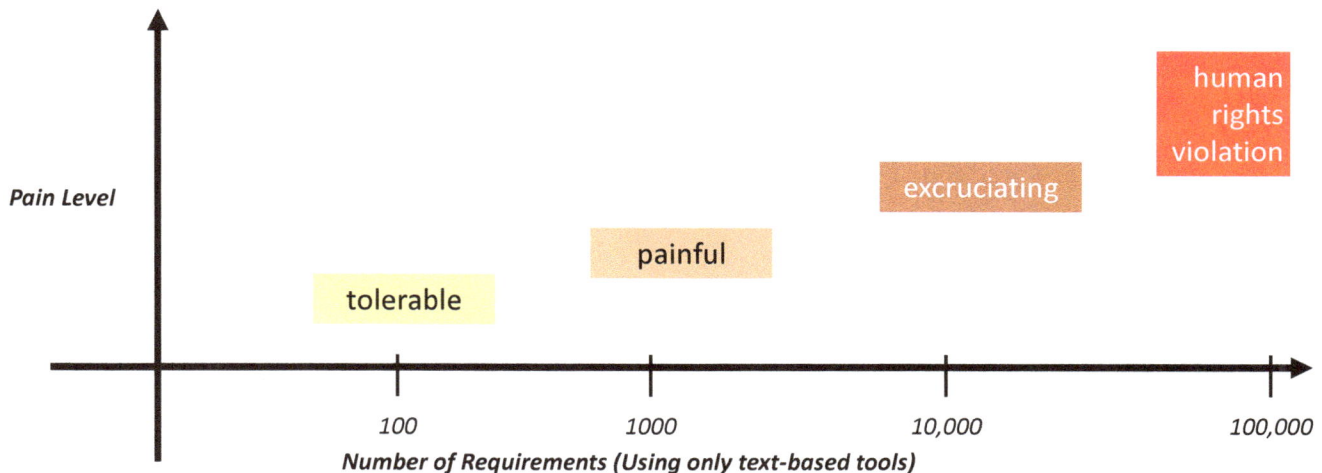

Figure 13-4 – The increasing level of pain in text-only requirement systems

As shown in Figure 13-4, managing requirements (and hence specifications) purely and only with text is only manageable when the number of requirements is small.

- **~100 Requirements** – The situation is manageable. You can write the text-based requirements in a document or perhaps put them in a spreadsheet.

- **~1,000 Requirements** – The problem has become painful. Very likely by this point you have given up on the spreadsheet and switched over to a purpose-built requirements management tool such as IBM DOORS, Siemens Polarion, Visure, Jama, or another similar product.

- **~10,000 Requirements** – The problem has become excruciating, even with the purpose-built requirements tool.

- **~100,000 Requirements** – You start to get visits from human rights protection organizations.

Common Misconceptions

In this section we introduce some common misconceptions about MBSE that you will encounter as you start introducing MBSE to your organization. These are not really objections, as much as assumptions commonly made by people who do not yet have enough information about the subject.

Misconception – The PowerPoint Compiler

Neither *Capella* nor *Arcadia* itself are "PowerPoint Compiler" technologies. *Arcadia* and *Capella* exist to enable a common understanding between a diverse set of stakeholders. These stakeholders each have specialized tools for their domain and these tools do the compiling of the final product.

Misconception – Project Speedup

One more subtle misconception is that all of the improved clarity provided by *Arcadia* and MBSE techniques will lead to a dramatic project speedup.

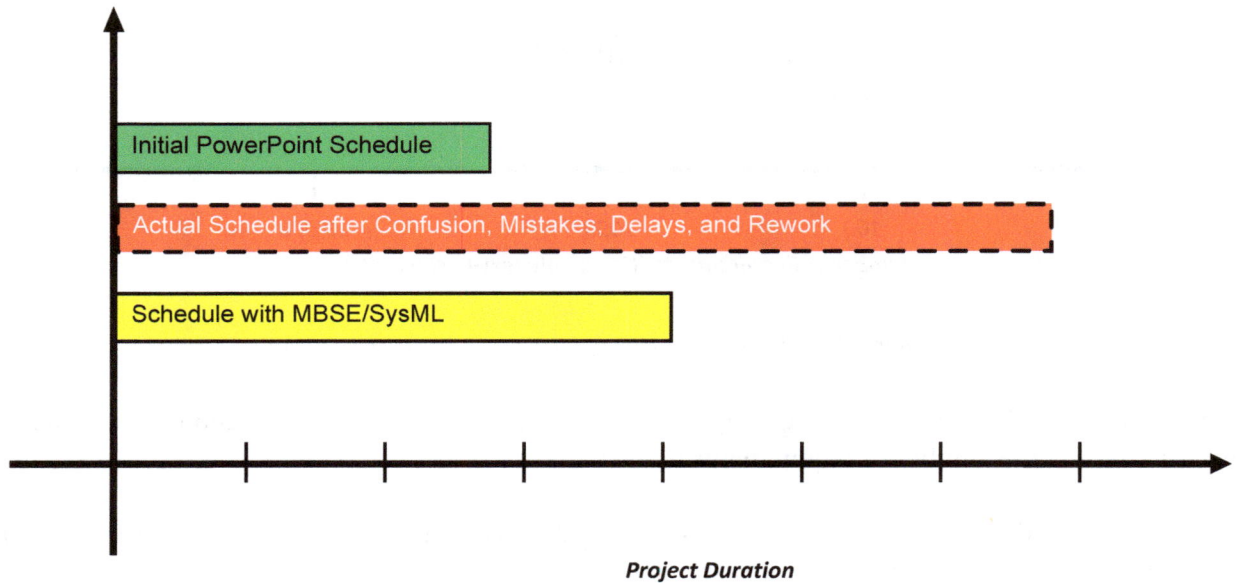

Figure 13-5 – Do MBSE and Arcadia speed up the project?

Adding *Arcadia* (or SysML or any other MBSE technique) to a project, does not create the impression of a dramatic project speedup. Quite the opposite, creating high-quality *Arcadia* diagrams takes a significant amount of time, especially when the team is still just learning to use the tools.

What about the overall project schedule? Disciplined use of *Arcadia* by a team will probably bring in the **actual project schedule** considerably, especially if *Arcadia* is used in a disciplined manner, project-after-project. The challenge, however, is that using *Arcadia* carefully in a project will tend to create the impression of a "bureaucratic slowdown". The team will tend to compare the progress of the project using *Arcadia* to the imagined *PowerPoint Project Schedule* and grumble that things are moving painfully slowly.

Subtle Problems

Over Modeling

MBSE and *Arcadia* have a subtle danger: modeling too much.

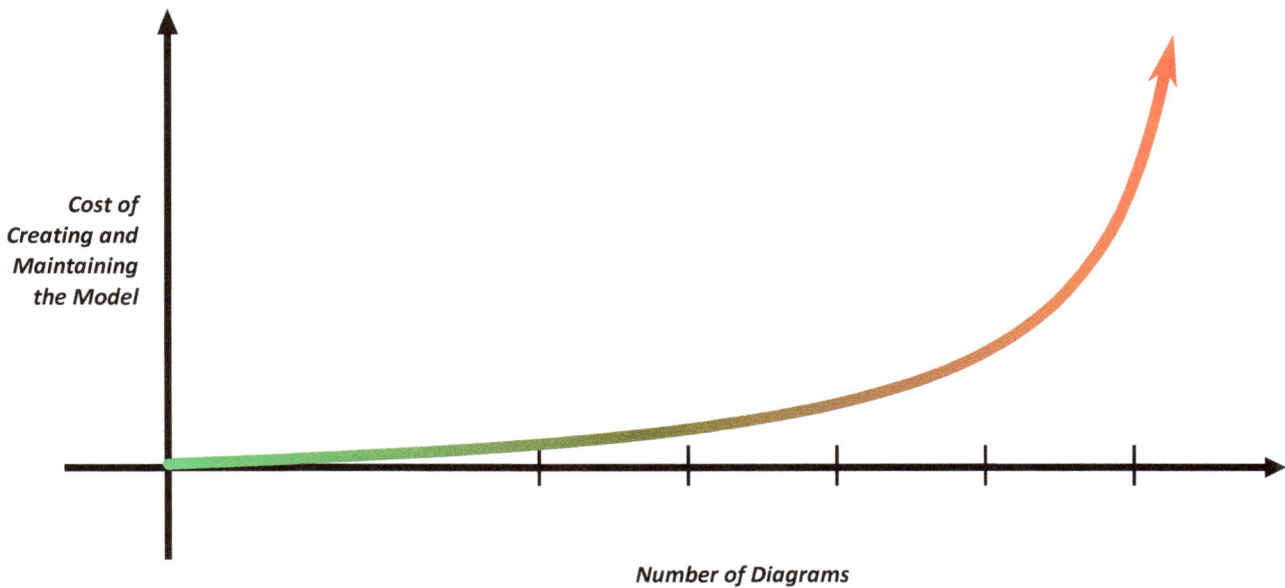

Figure 13-6 – *The more detailed the model, the more expensive*

The problem is that model development has its own costs. As the scope of your model expands, these costs start to multiply with each other. You have to think carefully about the lifecycle of the model. If stakeholders are going to continue to reference the model, then the model must be maintained. If the stakeholders start encountering inaccuracies in the model or portions of the model that no longer reflect realty, the model's value as the *Shared Vision for the System* will deteriorate rapidly. On the other hand, the more you model, the more costly it becomes to maintain the model. As you start using the modeling power to link all sorts of different aspects of the system concept together, you also create a maintenance burden in keeping all those links up-to-date and accurate. [45] [46]

---

[45] One of the early leaders in the UML arena is said to have taken the position that models should be produced on whiteboards and that the whiteboards should be erased at the end of each discussion.

[46] See Scott Ambler's excellent discussion of "Just Barely Good Enough" models: http://agilemodeling.com/essays/barelyGoodEnough.html

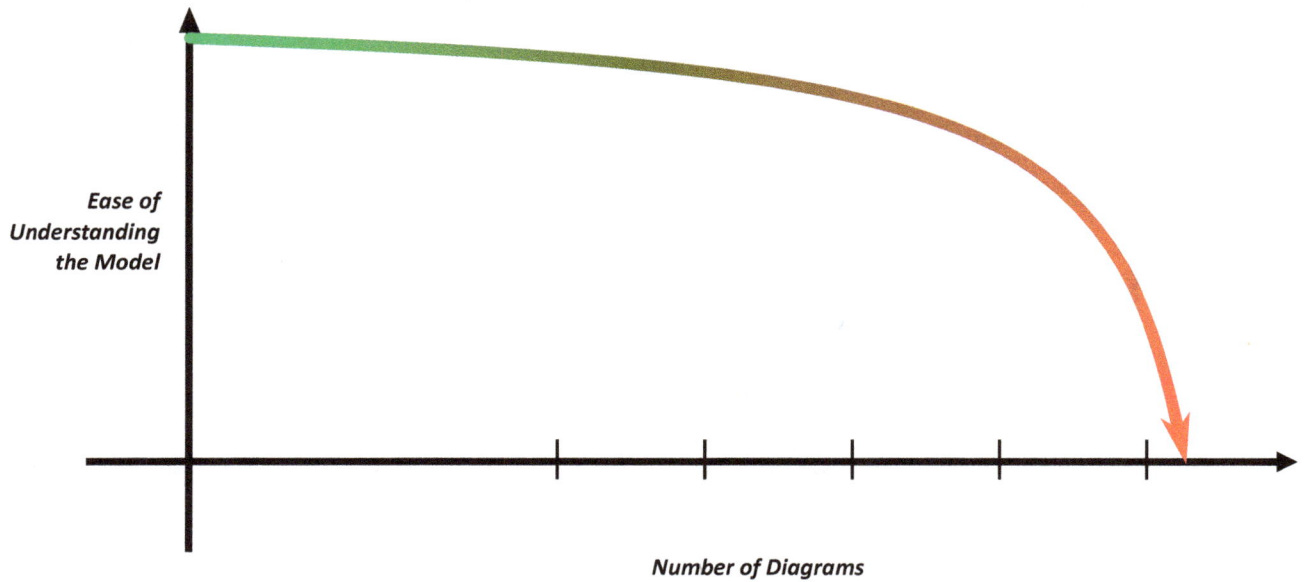

Figure 13-7 – Complex models are hard to understand

The second aspect of this problem is the ease of finding things for anyone other than the person who first entered the model in the tool. As the model increases in complexity, it quickly becomes difficult for anyone other than the person who did the initial modeling to find anything. In fact, if the person who did the initial modeling goes away and works on another project for a year, even that original modeler can find it surprisingly difficult to find things in a complicated model.

Appendix A – Planning, Ordering, Installing, Getting Help

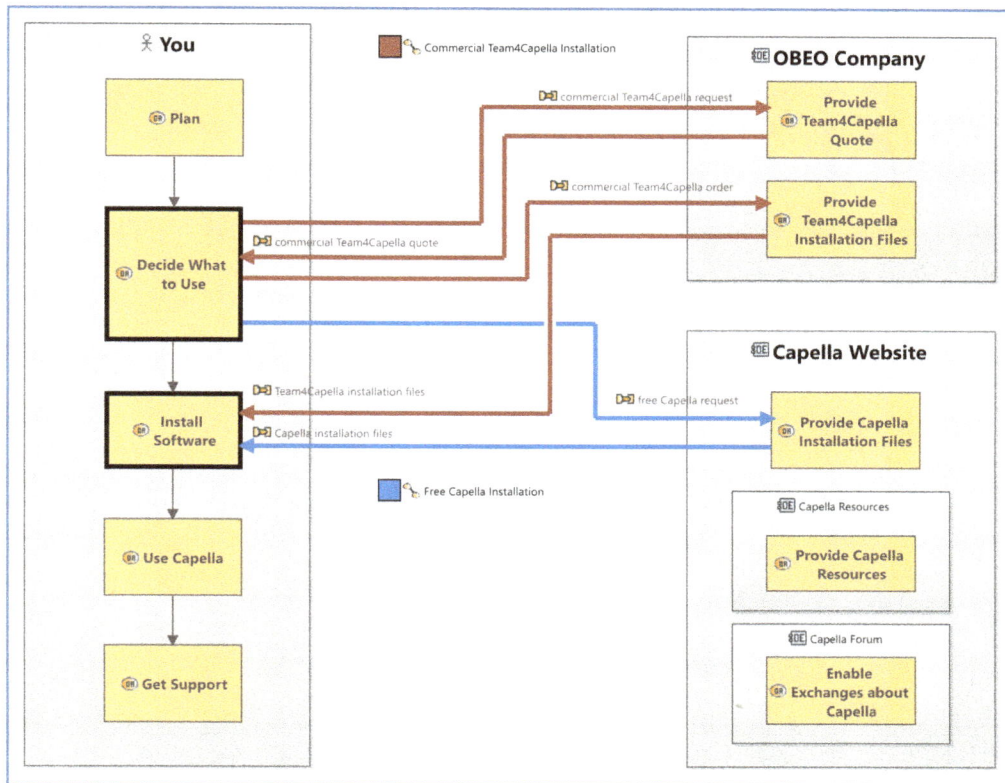

Figure A-1 – Overview: planning, ordering, installing, getting help

Figure A-1 shows the general flow of topics which are covered in the following sections.

1) **Planning** – *Capella* is an engineering tool, not a social media application. Installing and learning to use *Capella* will involve a significant commitment of time and effort. You will want to do some planning before you get started.

2) **Free Version** – All things considered, the most practical approach to getting started will be to download and install the free open source version from the *Capella* website.

3) **Commercial Version** – Once you have completed your first pilot project and feel the need for a multiuser version with several people modeling simultaneously on a bigger project, the most straightforward approach will be to contact the Obeo company.

4) **After Installation** – We have also included some information about routine needs after installation such as finding online help and asking questions to the community.

Note that the screen captures in the following sections are current for *Capella* version 6.1.0 and *Capella* and Obeo websites as of August 2021. Both the tool and websites are subject to change.

Planning

What to Buy

As mentioned previously, we recommend that you start by downloading and installing the free open source version from the *Capella* website:

https://www.eclipse.org/capella/download.html

All the screen captures in this book were made using the free open source version.

If you need information on the multiuser commercial version, Obeo provides an overview of the *Team for Capella* tool here:

https://www.obeosoft.com/en/team-for-capella

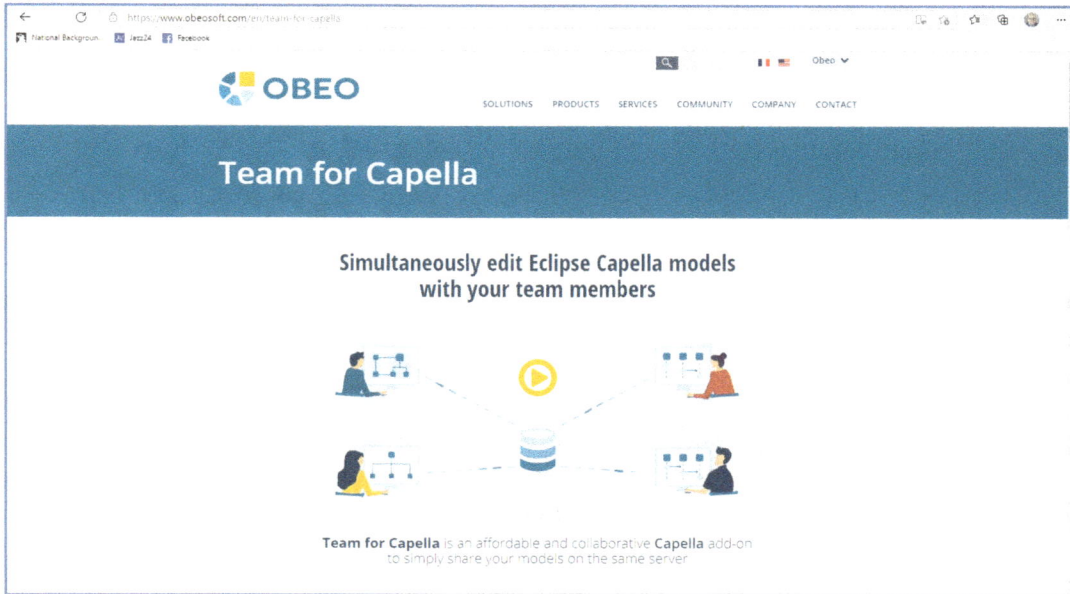

Figure A-2 – Team for Capella

Either version will work fine for learning Arcadia. Beginners may appreciate the cost-free open source version and its ease of installation.

Getting a Price

Obeo does not directly publish a full price list on its website. The most straightforward approach to investigate pricing is to go to *Team4Capella* product overview website, scroll down, and click on the "Request a Quote" link.

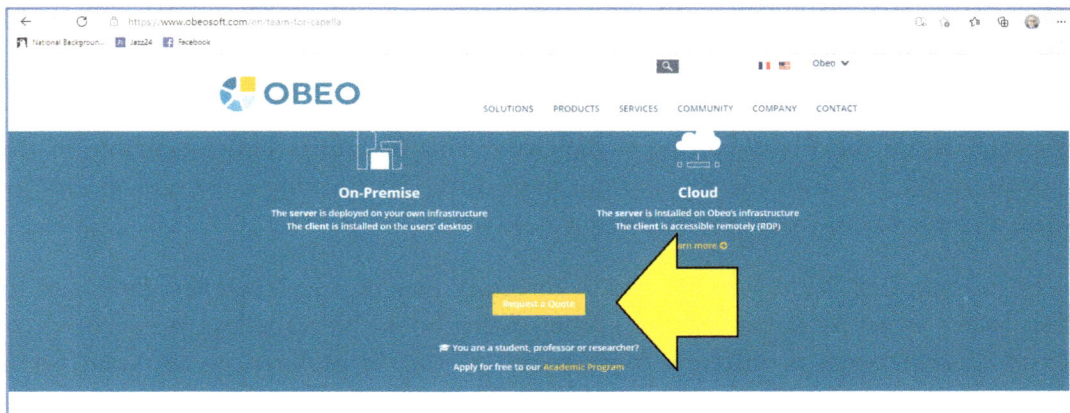

Figure A-3 – Scroll down and click the orange box to request a quote

Windows Environment

The installation prerequisites for *Capella* can be found on the *Capella* GitHub repository here:

https://url4ap.net/Capella-Prereqs

Choosing an Installation Directory

Capella does not come with a standard installer. You will need to decide in advance where you want the program files to be stored on your hard drive.

Warning: *Capella* is based on the *Eclipse* platform. The folder hierarchy and namespace of the *Capella* and *Eclipse* plugins are quite long. Because of Windows folder path length restriction, avoid entering a long installation location. If the path is too long, *Capella* may fail to open some files. Also, the *Eclipse* platform has historically had some issues with path names containing spaces. Choosing an installation directory such as:

"C:\Program Files\Capella"

may cause unpredictable errors such as problems with the Description Editor. We recommend: *"C:\Capella"* or even: *"C:\CP"*

Deciding Where to Put Your Project Files

Capella is based on the *Eclipse* platform. As such, when you start the tool, you will be prompted to select your **workspace**. The workspace is the *Eclipse* central hub for user's files. Different types of projects such as *Capella* projects and Java projects can share the same workspace. Workspaces are used for builds, version management, sharing, and resource organization. Like folders, projects map to directories in the file system. A location in the file system is specified during the project creation.

Note that *Eclipse* platform and some plug-ins use the workspace to store global settings and preferences in a hidden sub-folder ".metadata".

The default workspace location that *Capella* will propose when you first start the tool will be a subdirectory called "workspace" inside the *Capella* program directory. This default is understandable in that the tool knows nothing about the layout of the rest of your hard drive. That having been said, storing your project data in the same directory as the tool is usually not a good idea.

For many users, the standard Windows *"Documents"* directory may be a good choice. However, the management of a disk folder structure is beyond the scope of this book. Nevertheless, we recommend that you store different *Capella* projects in different workspaces, as *Capella* preferences are valid for a complete workspace, not for one project. For this reason, we will use a different directory as workspace for each chapter of this book.

Configuration Management

If you are working alone on a single personal computer, you won't have to think much about configuration management beyond any normal backups you may routinely create. However, if you are going to work in a team and you want to manage your projects with a source code control system such as Git, you will find that there is a certain amount of complexity involved. You will need to think about two configuration management problems:

1) **Managing the Tool** – Are you going to allow each developer to set up his own configuration of add-ons? Are you going to force the entire team to use a common setup? For a discussion of the ins and outs of this topic see:
   https://url4ap.net/Capella-CM-Tool

2) **Managing the Workspace** – As mentioned above, the *Eclipse* platform implicitly stores an assortment of information in metadata files. Sharing these is tricky. For a discussion of the ins and outs of this topic see:
   https://url4ap.net/Capella-CM-Workspace

Installing

Now we are ready to install *Capella*. The first stop is the main download page:

https://www.eclipse.org/capella/download.html

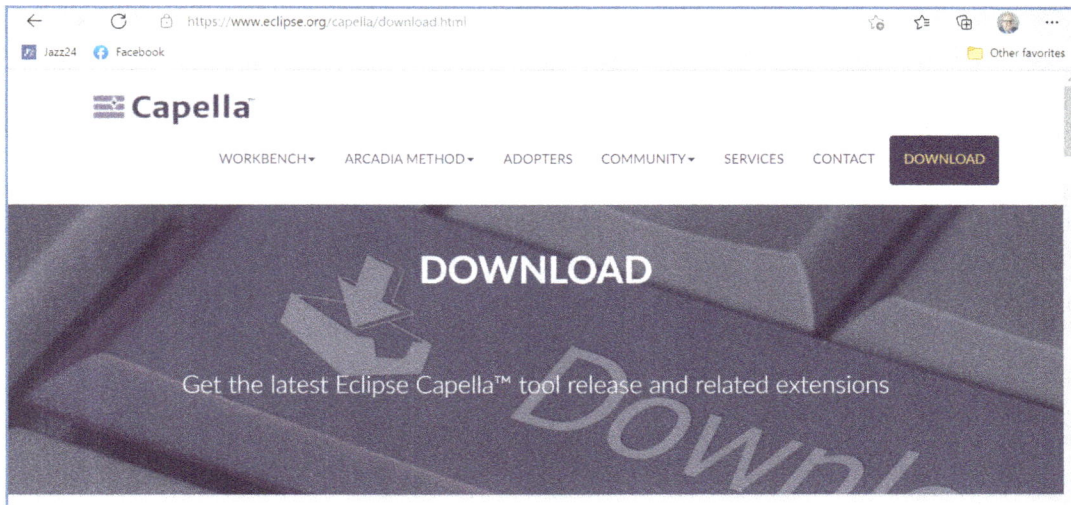

Figure A-4 – Install Capella download page

Download Software

To download *Capella*, scroll down until you find an orange "GET CAPELLA..." box for the latest version of the tool. Click on this box to begin the download.

The installation file is fairly large (about 0.7GB) and may take a while to download, depending on the speed of your local internet service. [47]

Install Software

Before installing the downloaded software, you may want to review the supplementary information on the main download page. Simply scroll down to the next selection below the download link.

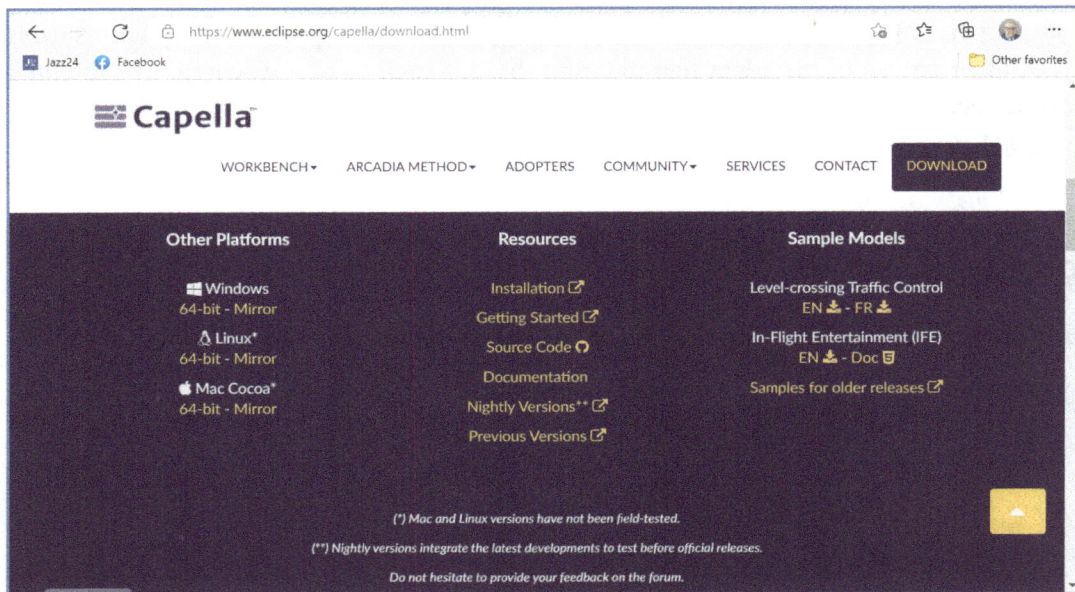

Figure A-5 – Capella additional installation information

In addition to the download link, this section of the main download page contains a lot of useful information. For instance, you can choose to download Linux or Mac versions. You can also download previous versions. If you are adventuresome, you can download the advanced nightly versions. You even have access to the source code. Additional sample models are available as well.

The "Installation" link leads you to a GitHub page with prerequisites mentioned above. You will also find information about installing previous versions there.

---

[47] The speed of the hosting server does not seem to be a concern. David Hetherington has fiber-to-the-home service in Austin, Texas and was able to download the file in about 30 seconds.

Create the installation directory that you planned for in in *Choosing an Installation Directory* on page 356 and unzip the contents of the downloaded file to that directory.

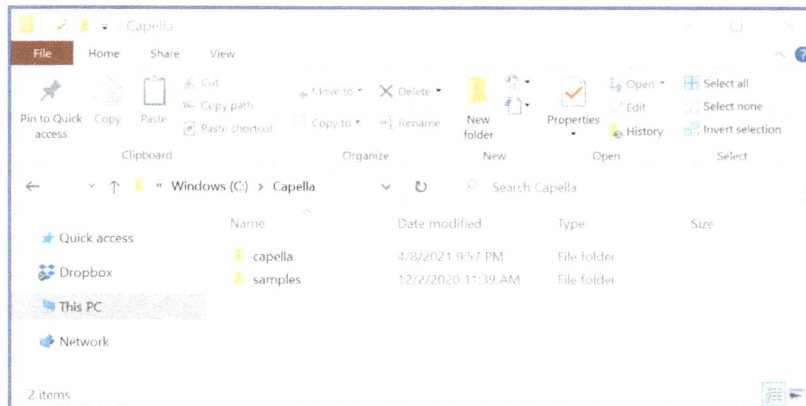

Figure A-6 – Capella installed

Since there is no installation program, if you would like a shortcut on your desktop, you will need to create it manually.

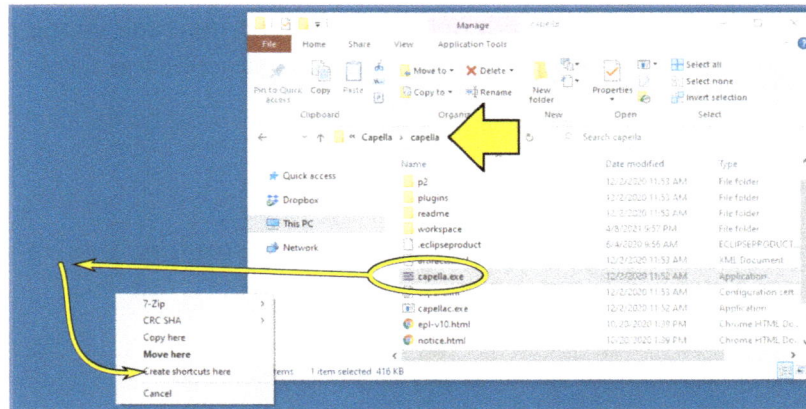

Figure A-7 – Right-click and drag capella.exe to the desktop to create a shortcut

In the *Capella* installation directory, you will find a subdirectory "capella". In this subdirectory, find the file "capella.exe". Right-click on this file and drag it to the desktop. When you release the right mouse button, a context menu will appear that will allow you to create a shortcut.

Installing Add-Ons

A number of free and commercial add-ons are available for *Capella*. The main site for such add-ons is:

https://www.eclipse.org/capella/addons.html

There are two general approaches for installing add-ons:

- *Eclipse* update sites
- "dropins"

We recommend dropins as they are simpler to install and configure.

The configuration of add-ons can be somewhat complicated if multiple users are involved. If you are interested in making use of add-ons, we recommend that you read the detailed instructions and guidance provided here:

https://url4ap.net/Capella-Add-On-Configuration

After Installation

Routine Software Updates

Capella is not updated automatically. Unlike some other Windows applications or applications that might run on your smartphone, there is no program update function in *Capella*. When a new version of the tool is released, you simply download the new version and unzip it. Note that you can maintain multiple parallel versions on your system simply by installing them in different directories.

Generally, there is a major release each year with one minor release sometime during the year. The best way to stay abreast of updates is to participate in the community as described in *Contact Capella Community* on page 369.

Configuring for Unicode

If you need to include East Asian text (Japanese, Chinese, or Korean) in your model, you will need to perform a few configuration steps on *Capella* first.

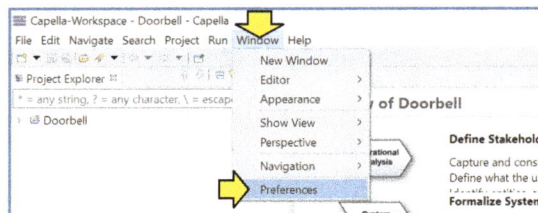

Figure A-8 – Open preferences

From the top menu bar select "Window" and then "Preferences"

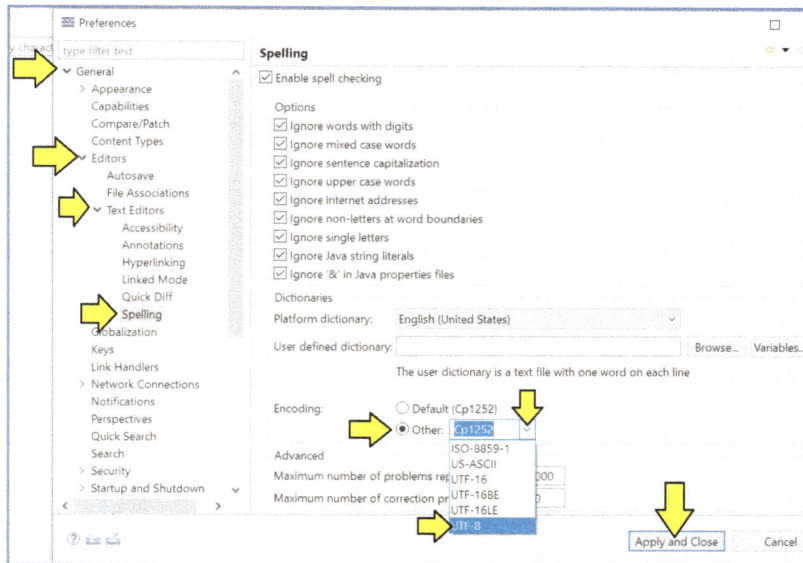

Figure A-9 – Open preferences

1) Expand "General".

2) Expand "Editors".

3) Expand "Text Editors".

4) Select "Spelling".

5) *Capella* defaults to the Windows Latin-1 codepage 1252. [48]

6) Deselect "Cp1252" and select "Other:"

7) Pull down the menu and select "UTF-8". [49]

8) Click "Apply and Close".

---

[48] CP 1252 was last updated by Microsoft for Windows 98 to include the Euro symbol. This codepage only covers English, French, Spanish, and German. This codepage has not been in common use by Windows itself or web pages on the internet for 20 years or so. You really probably don't want to be using CP 1252 for English, French, Spanish, or German anymore either.

[49] UTF-8 is a method of encoding that preserves 7-bit ASCII compatibility for plain English and punctuation at the cost of using two bytes per special European character, three bytes per East Asian character, and four bytes for some special supplementary characters. In this manner, UTF-8 can encode the full range of human text languages in the Unicode standard while maintaining (at least English) backward compatibility with the decades of existing English language text data and software. UTF-8 is now the dominant codepage for the worldwide web.

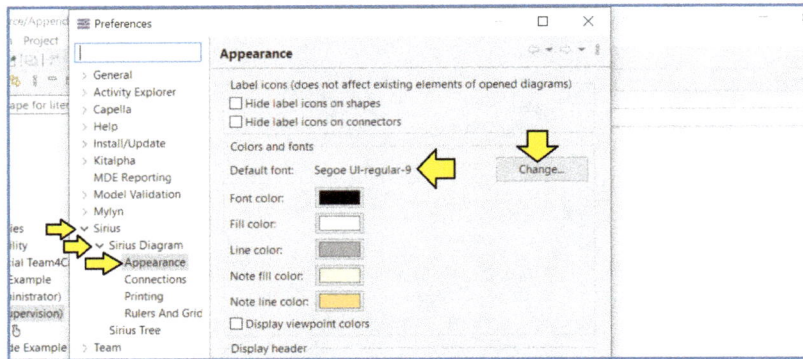

Figure A-10 – Set font if desired

Capella comes with the user interface font set to "Segoe UI". This font seems to cover at least Japanese and European languages as well as common symbols and emojis. However, if you would prefer to use another font (perhaps MS Mincho for Japanese) the UI font can be set in the "Appearance" panel in the "Sirius" section of the preferences menu.

In the examples files directory *Examples\Appx_A_Ordering_Installing* you will find a small example model containing a sample diagram using Unicode characters.

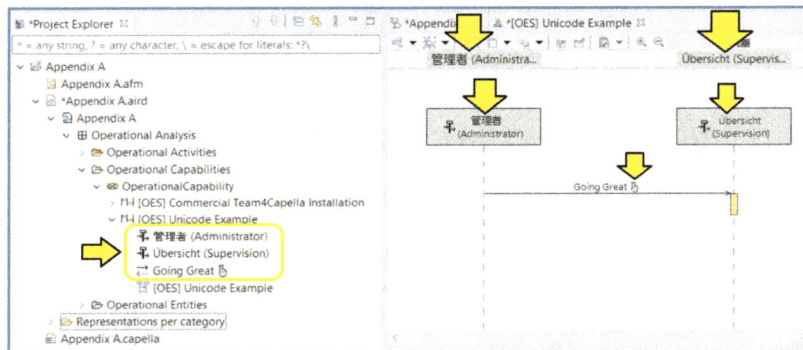

Figure A-11 – Example of model with Unicode characters

In the example model, we have:

- a Japanese administrator "管理者" ("kanrisha");

- who is sending a status message: "Going Great";

- to the somewhat nebulous "Übersicht" ("supervision") in Germany.

If we look carefully at the status message, however, we see that our administrator is less than entirely confident of the status and hence the message includes the emoji "fingers crossed". [50]

Figure A-12 – Emoji

Archiving and Importing Capella Projects

Saving a project that you are going to stop working on for a while is slightly more complicated in *Capella* than some other tools because of the underlying *Eclipse* platform. It is not quite as simple as saving and restoring a simple file. The entire file structure as well as the *Eclipse* metadata need to be saved together. For this purpose, we will need to create an **archive file** of the project workspace.

Exporting to an Archive File

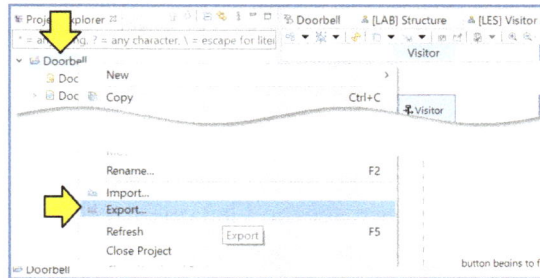

Figure A-13 – Export the project

Select the project you would like to archive in the project explorer, right-click, and select: "Export…"

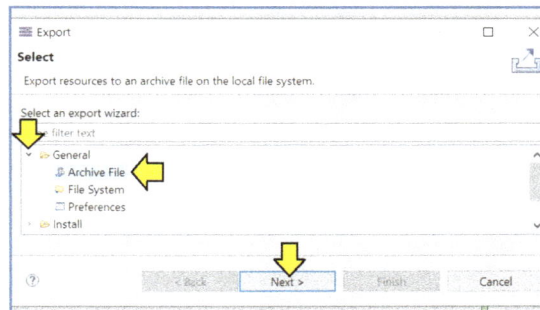

Figure A-14 – Select Archive File

In the next panel, expand "General" and select: "Archive File". Click: "Next".

---

(50) The emoji "fingers crossed" is Unicode Ü. See: https://emojipedia.org/hand-with-index-and-middle-fingers-crossed/

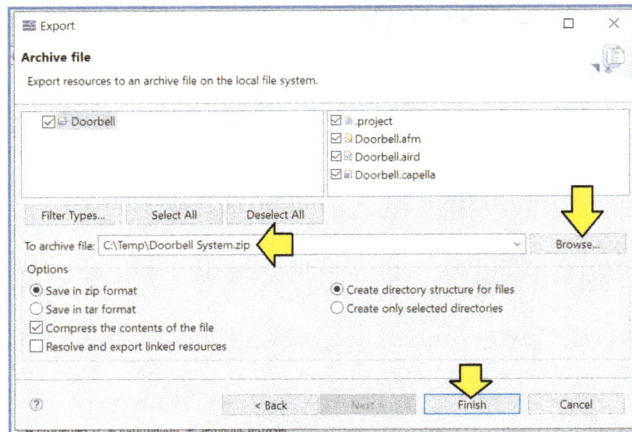

Figure A-15 – Set filename and finish

In the next panel, browse to the target directory, set the filename for the archive file, and click: "Finish".

Deleting a Project from the Workspace

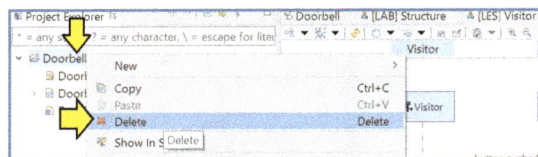

Figure A-16 – Delete project from workspace

In order to remove a project from your workspace, select the project you would like to remove in the project explorer, right-click, and select: "Delete".

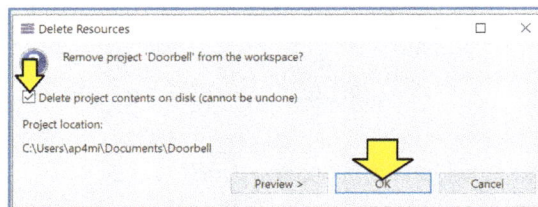

Figure A-17 – Decide whether to delete files on the disk

In the next panel, you will be presented with an option to delete the working files on the hard disk. In most cases, doing so is probably a better idea. Create an archive file. Save it to a secure location. Don't leave old working files on your disk. After you have decided whether or not to delete the working files, click: "OK".

Importing from an Archive File

Importing an archived file (such as one of the example *.zip files for this book) can be a bit tricky. *Capella* offers a somewhat confusing selection of different paths that all look similar, but do not all produce the result you are expecting, namely that *Capella* will be back to where it was when you archived and deleted the project previously. Follow these steps carefully and you should end up where you want to go.

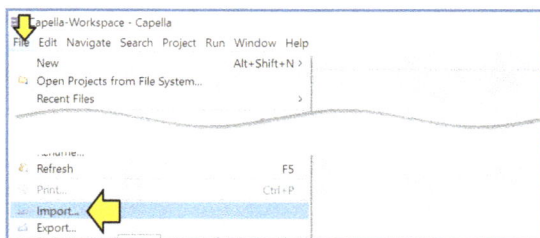

Figure A-18 – Select import project

The first step is fairly straightforward. Select "File" and then "Import..."

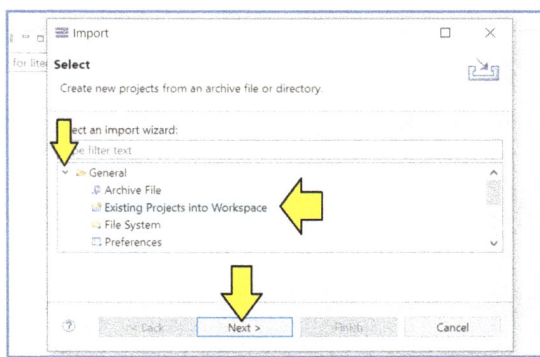

Figure A-19 – Select existing projects

Expand the "General" category and select "Existing Projects into Workspace". Click "Next".

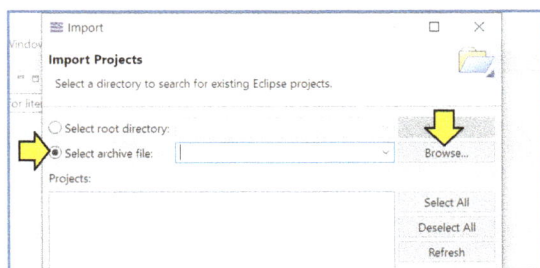

Figure A-20 – Click the button for select archive file

Click the radio button for "Select archive file:" and then "Browse..."

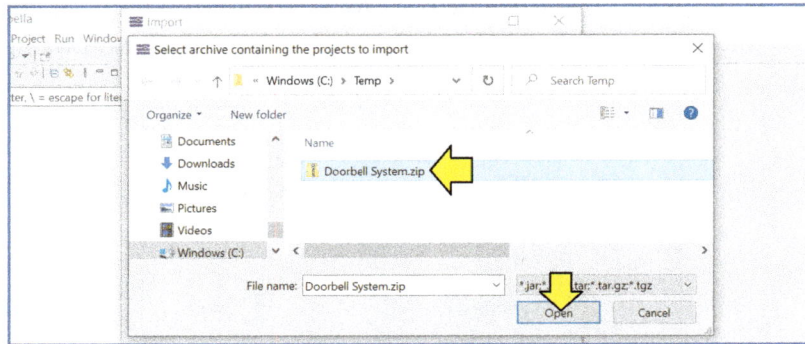

Figure A-21 – Select file and open

Navigate to the archive file, select it, and click "Open".

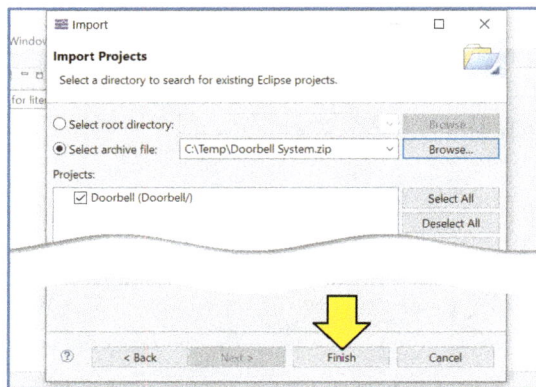

Figure A-22 – Click Finish

Click "Finish".

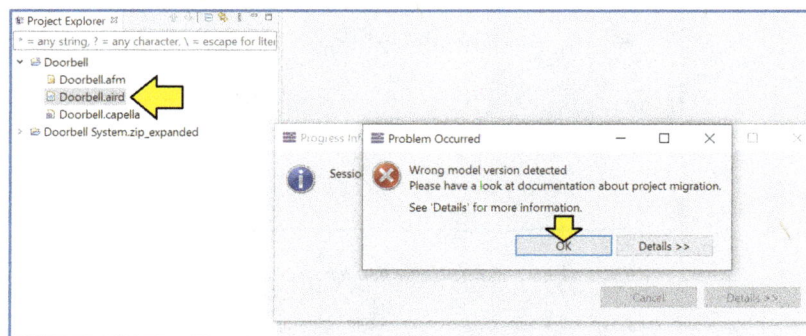

Figure A-23 – Wrong model version warning

If the version of the model in the archive is the same as your current installation of *Capella* you can open the model by simply double-clicking on the *.aird file.

If the model was archived from a previous version of *Capella*, you will need to migrate the model to the current version first. Fortunately, the migration procedure is quite easy.

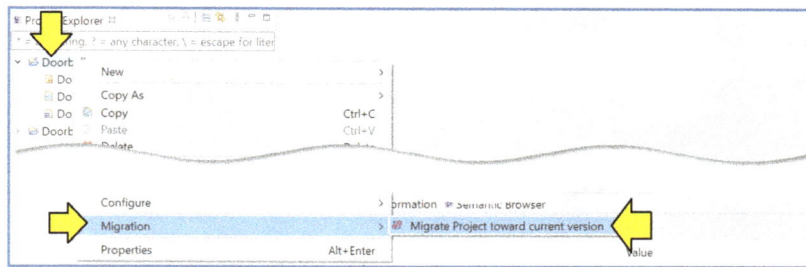

Figure A-24 – Migrate the model

Select the model, right-click and select "Migration" and then "Migrate project toward current version".

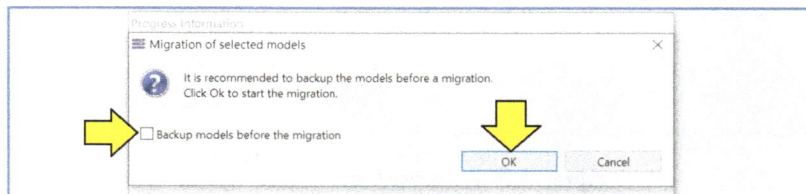

Figure A-25 – Be sure to deselect backup models

By default, in the next panel *Capella* will have an option selected to back up the models before migration. You don't need this option as you already have an archive file. If you leave this option selected, the migration procedure will clutter your model with unneeded backup copies. Clear this selection and click "OK".

After migration finishes, you can open your imported model by double-clicking on the *.aird file.

Capella Online Documentation

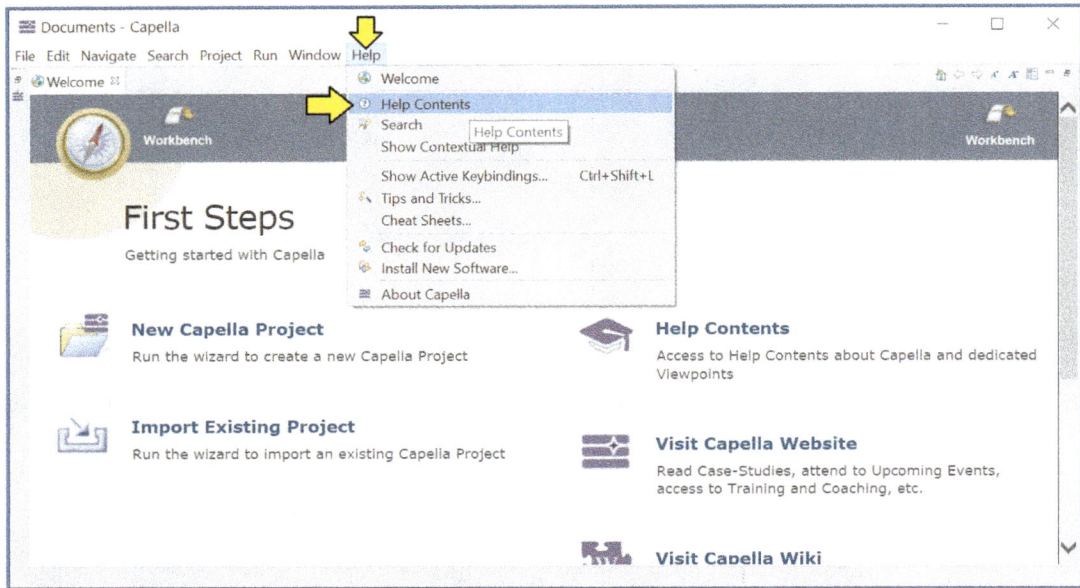

Figure A-26 – Opening Capella online help

Capella contains extensive online help which can be easily accessed by clicking on "Help" and then "Help Contents".

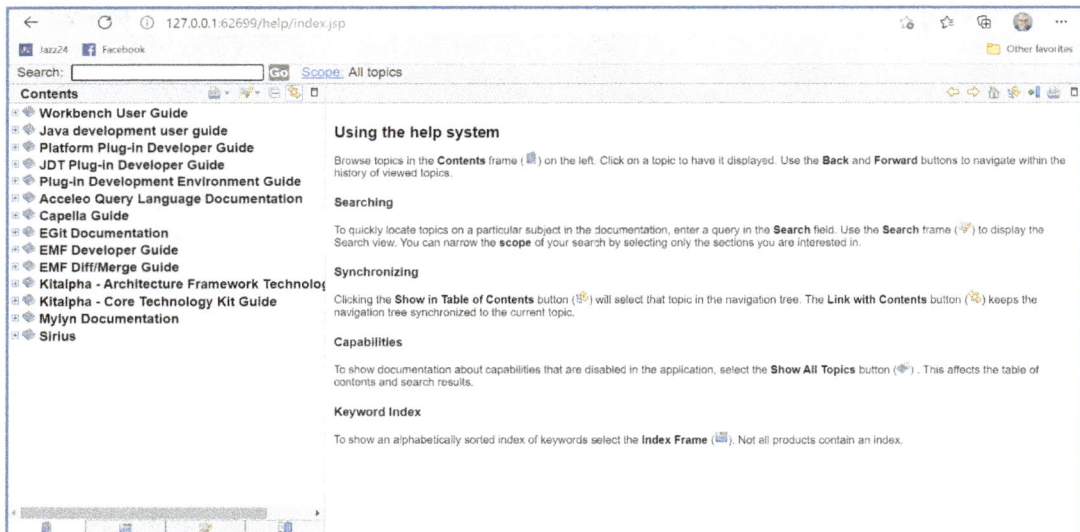

Figure A-27 – Capella help in the browser

The online Help will open in a web browser. The most informative part is the "User Manual".

Contact Capella Community

The best place to get in touch with other *Capella* users and experts is through the Capella Forum:

https://forum.mbse-capella.org/

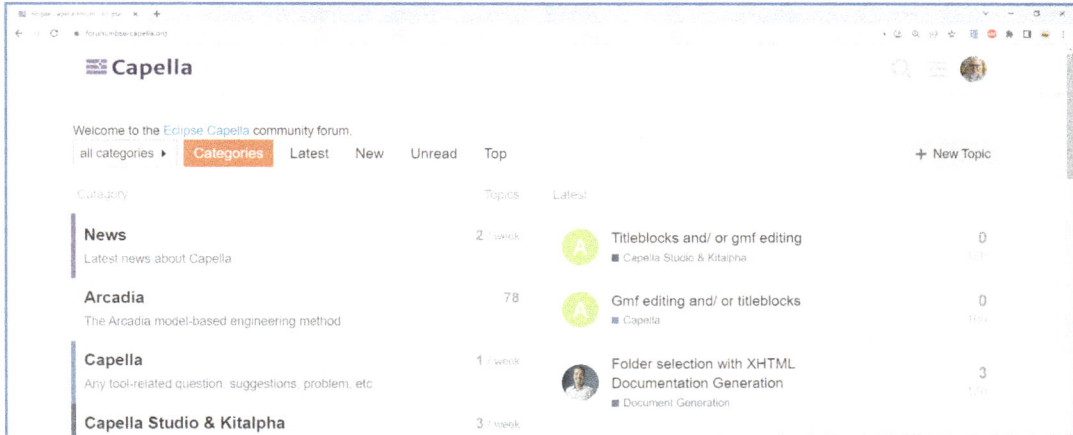

Figure A-28 – The Capella forum

Capella YouTube Channel

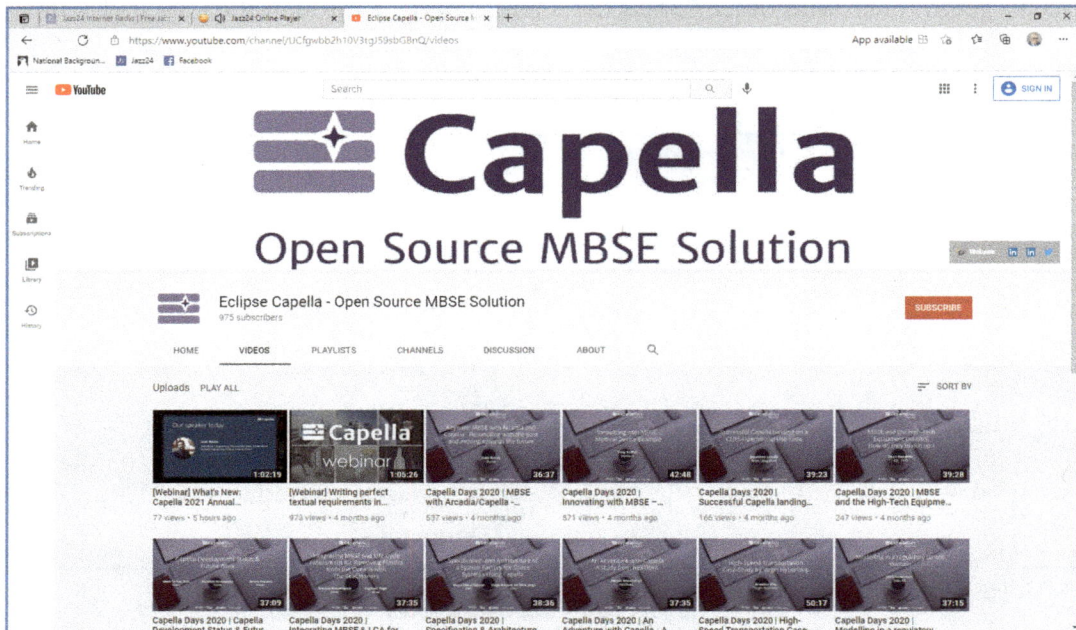

Figure A-29 – Capella YouTube channel

The "Eclipse Capella - Open Source MBSE Solution" YouTube Channel is a great training resource with more than 20 webinars:

https://www.youtube.com/c/EclipseCapella/videos

Pascal Roques (PRFC) Online Resources

Pascal also has a YouTube channel, with short videos, both in French and English:

https://url4ap.net/PRFC-Videos

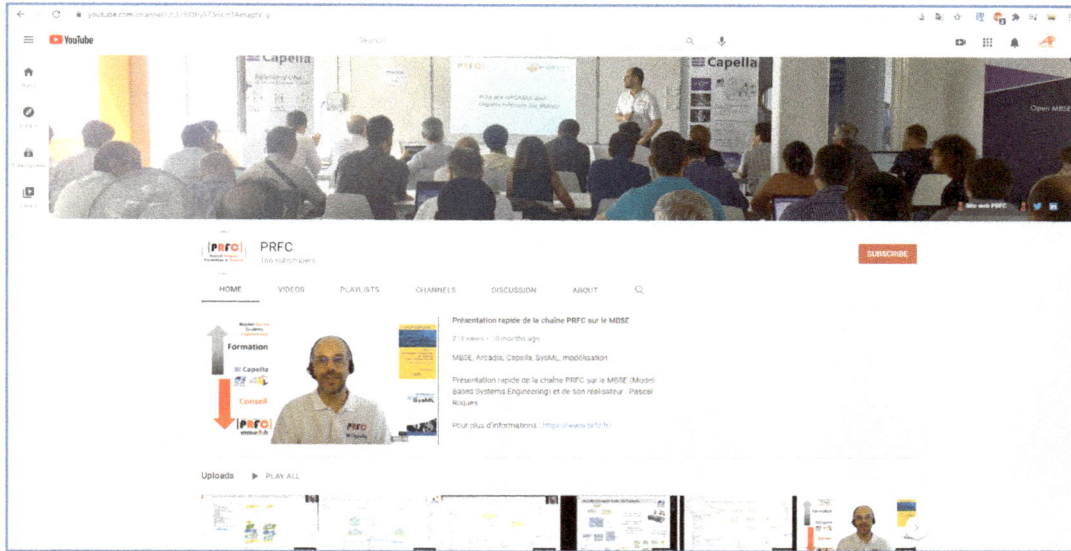

Figure A-30 – Pascal's YouTube channel

Most of the videos are short extracts from real Capella training sessions, focusing on a specific topic of interest. These videos cover:

- **Simple Subjects for Beginners –** such as "Blank Diagrams Basics in Capella", which explains the differences between a Graphical Object and a Model Element. Other topics include the difference between "Delete from Diagram" and "Delete from Model" commands, as well as the creation and modification of functional chains.

- **Advanced Features of the Tool –** such as "System Subsystem Transition Add-On for Capella", which demonstrates the use of the System/Subsystem Transition Add-On in the EOLE case study, or "REC/RPL et Physical Path avec Capella" (in French) explaining how to create a reusable component (REC) in a Library, and how to instantiate it in a project as a replica (RPL).

Appendix B – Further Reading and Information

As the authors of this book, we hope you have found the material useful and easy to understand. However, this book is merely a beginning. In the following sections, we have listed a number of good books that you can read for more in-depth coverage of Arcadia, Capella, and other subjects related to model-based systems engineering. We have also included some nformation on web-based seminars that might be useful.

Arcadia and Capella Books

There are really only two other books about Arcadia and Capella on the market at the moment.

- **Pascal Roques** – *Systems Architecture Modeling with the Arcadia Method: A Practical Guide to Capella (Implementation of Model Based System Engineering), 1st Edition.* ISTE Press - Elsevier, 2017

  Systems Architecture Modeling with the Arcadia Method is Pascal's first book on Arcadia and the Capella tool. The focus is on introducing the Arcadia method, using Capella.

- **Jean-Luc Voirin** – *Model-based System and Architecture Engineering with the Arcadia Method, 1st Edition.* ISTE Press - Elsevier, 2017

  This book describes the fundamentals of the Arcadia method and its contribution to engineering issues such as requirements management, product line, system supervision, and integration, verification and validation (IVV). It provides a reference for the modeling language defined by Arcadia.

SysML Books

In the model-based systems engineering space, SysML cannot be ignored. These books provide excellent coverage of SysML.

- **Delligatti, Lenny** – *SysML Distilled: A Brief Guide to the Systems Modeling Language.* Addison-Wesley Professional, 2013

  SysML Distilled is an excellent, compact, all-around text for engineers who want to develop a more thorough understanding of SysML. In fact, this book is my primary "go to" reference when I have questions about the more subtle/advanced details of SysML. Prior to working as a systems engineer, Lenny Delligatti had experience both as a naval officer and as a high school mathematics teacher. This leadership and instructional background comes through in his well-organized, and straightforward textbook.

- **Douglass, Bruce Powel** – *Agile Systems Engineering, 1st Edition.* Waltham, MA, USA:Morgan Kaufmann, 2015

Bruce Powel Douglass was involved in the development of specifications for both UML and SysML and is a prolific author, having written extensively about the use of UML for realtime applications.

The scope of this book is somewhat broader than the other two, including more guidance on methodology. The treatment of SysML is brief – one chapter, 80 pages. However, that 80 pages includes some important insights, such as his treatment of:

- role versus type (page 88)
- the relationship of block definition, internal block definition, and package diagrams (page 94)
- model organization for a large project (page 104)
- value properties in blocks, parts, and instances (page 108)

This book also goes into depth on the development and management of use cases and requirements.

- **Friedenthal, Sanford and Alan Moore, Rick Steiner** – *A Practical Guide to SysML, Third Edition: The Systems Modeling Language.* Morgan Kaufmann, 2014

  This book is the authoritative reference on SysML and its authors are key leaders in the ongoing development of the standard. Since the SysML standard itself includes limited explanatory content, this book is helpful as a sort of window into the intent of the leaders of the SysML standard development.

UML Books

Arcadia is not an implementation of UML. Nevertheless, many aspects of *Arcadia* have been heavily influenced by UML. It may be helpful to have a good UML book for reference.

- **Booch, Grady and James Rumbaugh, Ivar Jacobson** – *The Unified Modeling Language User Guide, Second Edition.* Addison-Wesley, 2005

  This is the authoratative book from the three consultants that created UML in the 1990s. While this book does not cover every last detail of UML, it is pretty thorough. It is very well written with high-quality graphics. Be sure to get the second edition as there were significant changes in UML 2.0.

- **Roques, Pascal** – *UML in Practice.* Wiley, 2004

  This is an earlier book by Pascal Roques (author of this book) introducing UML in a series of case studies. It is shorter and perhaps less comprehensive, but somewhat easier to understand as the material is presented as a series of worked examples.

Systems Engineering Books

While this book shows how to model very simple examples, your success in using *Capella* for larger projects will depend heavily on your mastery of some classical systems engineering skills. These books are good sources of information about classical systems engineering activities.

- **Cockburn, Alistair –** *Writing Effective Use Cases.* Addison-Wesley, 2001

  Use cases are used in both software development and systems engineering. Writing effective use cases is not easy. Where do you start? What do you include? How much is too much? This book is regarded by many to be best book ever written on this difficult topic.

- **INCOSE –** *INCOSE Systems Engineering Handbook: A Guide for System Life Cycle Processes and Activities, 4th Edition.* Wiley, 2015

  This is the handbook of the International Council on Systems Engineering. As such, it is a handy, compact compendium of standard systems engineering techniques – including, but not limited to model-based systems engineering.

- **Jenney, Joe with Mike Gangl, Rick Kwolek, David Melton, Nancy Ridenour, Martin Coe –** *Modern Methods of Systems Engineering: With an Introduction to Pattern and Model Based Methods, 4th Edition.* CreateSpace Independent Publishing Platform, 2011

  This book gives a solid overview of general systems engineering techniques with an overview of model-based and pattern methods as promised by the title. However, what I found most valuable was actually chapter 2 with its brief but fascinating accounts of:

  - The development of the highly innovative Cord 810 in 1935 by a 3-person design team in about six months – including the production of 100 prototype vehicles. This feat would be utterly unimaginable in today's automobile industry.
  - The development of World War II aircraft in which the designers had offices on balconies overlooking the line that was producing the first aircraft. This close proximity allowed them to step out and look at the aircraft in seconds, as well as walk down a stairway and touch it within a minute.

  This chapter does an excellent job of setting the correct frame of reference: the real problem is neither fancy tools nor abstract modeling grammars; the real problem is the complexity of the human communication. Even though chapter 2 is brief, for me this chapter alone is worth the purchase price of the book.

Model-Based Engineering Books

These books provide some alternate approaches to and a broader view of the model-based engineering discipline.

- **Beatty, Joy and Anthony Chen** – *Visual Models for Software Requirements*. Microsoft Press, 2012

  Joy Beatty and Anthony Chen present an excellent step-by-step method for developing requirements for typical IT end-user applications. The authors present a method for using Microsoft Visio to produce diagrams for clarifying the structure and behavior of the IT application to be developed. Visio is chosen because it is more widely available, less expensive, and easier to learn to use than most SysML tools.

- **Brambilla, Marco and Jordi Cabot, Manuel Wimmer** – *Model-Driven Software Engineering in Practice*. Morgan & Claypool Publishers, 2012

  One challenge in looking into model-driven and model-based engineering methods is that most of the key textbooks were written ten or more years ago. This concise book is an excellent up-to-date review of activity in the field. The book is particularly useful for navigating the blizzard of acronyms and related standards.

  I particularly like Chapter 5 "Integration of MDSE in Your Development Process". Since the initial surge of enthusiasm in the early 2000's, model-driven approaches have suffered a number of setbacks and there are quite a few disillusioned veterans around. [51] Chapter 5 covers some of the different sociological problems that model-driven development can introduce into an organization such as fears about job security. Chapter 5 is a must-read for anyone considering introducing model-driven or model-based engineering techniques to a larger product development organization.

- **Kelly, Steven and Juha-Pekka Tolvanen** – *Domain-Specific Modeling: Enabling Full Code Generation*. Wiley IEEE Computer Society Press, 2008

  One of the weaknesses of SysML is that its rich descriptive power comes with a significant learning curve. Certain aspects of SysML are inherited from UML and can be confusing to users without a background in computer science. The direction of arrows that show a relationship between two blocks can seem counterintuitive to a mechanical engineer. The ends of a relationship between an engine and a crankshaft may be marked as "target" and "client" by the tool. What does that mean!?

  Domain-Specific Modeling attacks this problem by deliberately sacrificing the idea of a universal, interoperable set of graphics. Instead, DSM tools help small organizations make simple, customized tools that draw the graphical diagrams using any set of icons, shapes, arrows, or other conventions that make sense to that particular organization. In an example mentioned often by the authors at conferences, the diagrams that are intuitive to a team that designs railroad stations might not look anything like the diagrams that are intuitive to a related team that designs the railroad network.

---

[51] Many of these disillusioned modeling experts were key early founders of the Agile movement and now resist documentation in any form, especially any sort of modeling.

This book lays out the Domain-Specific Modeling approach and introduces the MetaEdit+ tool for creating such focused end-user modeling tools.

- **Tim Weilkiens** – *SYSMOD - The Systems Modeling Toolbox: Pragmatic MBSE with SysML.* MBSE4U, 2020

   SysML is generally agnostic about methodology. Tim does a nice job in this book of laying out a step-by-step methodology that has some similarities to Arcadia but is based on SysML.

Requirements Engineering Books

Once you start working on any really large system, especially one with safety-critical components, there will be no way to avoid large sets of text-based requirements. These books introduce the specialized "requirements engineering" field.

- **Hull, Elizabeth and Ken Jackson, Jeremy Dick** – *Requirements Engineering, 3rd Edition.* Springer, 2010

   This is a compact and concise textbook suitable for a university-level introductory course on requirements engineering. For readers with no background in systematic requirements management, this book does a good job of covering basics like requirements elaboration. [52]

- **van Lamsweerde, Axel** – *Requirements Engineering, From Systems Goals to UML Models to Software Specifications.* Wiley, 2009

   This is a thoroughly researched book with an emphasis on formal methods. Professor van Lamsweerde is a proponent of formally structured specification languages – a restricted form of English that reads a little bit like a programming language. He is also a recognized authority on goal-oriented requirements engineering. After thoroughly covering the basics of requirements engineering, this book uses structured specification languages and goal-orientation to teach a method for developing the UML models for a software application.

- **Leffingwell, Dean** – *Agile Software Requirements: Lean Requirements Practices for Teams, Programs, and the Enterprise.* Pearson Education, Inc., 2011

   Dean Leffingwell's book actually contains little about requirements. On the other hand, his book is a superb distillation and discussion of the various organizational techniques that fall under the umbrella of "Agile" As he points out, project scope/cost, project schedule, and project quality are an iron triangle. You can only control two of these at a time. Any enlightened executive team that is responsible for the purchase and development of a safety-critical system will do well to think carefully about requirements prioritization and

---

[52] Requirements elaboration is a very basic topic that gets surprisingly little treatment in some of the other larger books on requirements.

the possible application of some modified Agile approaches to get the best possible scope coverage while not compromising on safety or schedule.

- **Pohl, Klaus** – *Requirements Engineering: Fundamentals, Principles, and Techniques*. Springer, 2010

  If you can only afford one formal reference text for requirements engineering, this is the one you should buy. Klaus Pohl's book is the most comprehensive requirements textbook that I have encountered so far. It is clear and well-structured. It very helpfully lays out the evolution of key ideas in the field of requirements engineering with meticulous attribution of the original sources.

  Klaus Pohl also addresses some of the messy reality that is glossed over in other more academic treatments of requirements engineering. I particularly like his treatment of the interaction between "What" and "How" in section 2.3.3 on page 27.

Appendix C – About the Authors

Pascal Roques

Pascal has 30 years of experience in consulting and training about modeling with Arcadia/Capella, SysML, and UML.

After his initial experience as a design engineer, Pascal Roques joined Verilog in Toulouse early 1986. He worked there as a consultant and trainer for almost ten years on modeling tools.

In mid-1995, Pascal Roques opened a new office in Toulouse for the Valtech Agency. Working from the new office, he continued to do consulting and training in modeling. In 2001, he joined Valtech Training where he took over responsibility for the modeling offerings in Valtech's training catalog. In this role, he led more than 150 UML and SysML training sessions in France and abroad.

Figure C-1 – Pascal Roques

Pascal Roques is now an independent consultant and trainer focusing on Model-Based Systems Engineering (MBSE). Since 2011, he has led more than 200 MBSE courses in France and abroad for more than 2000 engineers. Pascal Roques is the author of the first book ever published on Capella (ISTE – Elsevier) and regularly speaks at international conferences. More information about Pascal's book in *Arcadia and Capella Books* on page 373.

David Hetherington

David Hetherington is a leading Model-Based Systems Engineering (MBSE) consultant serving multiple defense and commercial industry sectors. He has extensive personal experience in designing and leading design teams for both software and hardware covering an unusually broad range of system types. These complex systems have varied from real-time control, to software internationalization, to offshore oil drill ships, to enterprise software applications, to automotive radar chipsets, to electronic publishing, and more. In addition to MBSE, he has a strong concentration of domain knowledge in safety, reliability, maintainability, and diagnostics. He uses this broad domain knowledge in combination with his MBSE skills to assist clients at two levels:

Figure C-2 – David Hetherington

- **Individual Skills –** He teaches engineers across a wide variety of disciplines to sharpen their systems thinking and produce higher quality work products by using MBSE tools and techniques to clarify the requirements and objectives of their specialized designs.

- **Organizational Transformation** – He supports executives faced with the challenge of bringing teams with legacy, siloed, "document-centric" development habits into the integrated digital engineering future.

David has a BA in Mathematics from the University of California San Diego and an MBA from the McCombs School of Business at the University of Texas.

David is an active member of:

- INCOSE – International Council on Systems Engineering
- IEEE – Institute of Electrical and Electronics Engineers
- SAE – Society of Automotive Engineers
- United States Naval Institute

David speaks Japanese and German fluently as well as some Chinese and Spanish. He lives in Austin, Texas.

Appendix D – References

- **[AgileM]** *Agile Manifesto* http://www.agilealliance.org/the-alliance/the-agile-manifesto/

- **[Beatty]** Beatty, Joy and Anthony Chen – *Visual Models for Software Requirements.* Microsoft Press, 2012
  See the entry for this book in *Model-Based Engineering Books* on page 376 for more information.

- **[Booch]** Booch, Grady and James Rumbaugh, Ivar Jacobson – *The Unified Modeling Language User Guide, Second Edition.* Addison-Wesley, 2005
  See the entry for this book in *UML Books* on page 374 for more information.

- **[CMS17th]** The University of Chicago Press Editorial Staff – *The Chicago Manual of Style, 17th Edition.* University of Chicago Press, 2017
  The Chicago Manual of Style is the primary source of style and formatting conventions used in this book.

- **[Cockburn]** Cockburn, Alistair – *Writing Effective Use Cases.* Addison-Wesley, 2001
  See the entry for this book in *Systems Engineering Books* on page 375 for more information.

- **[Delligatti]** Delligatti, Lenny – *SysML Distilled: A Brief Guide to the Systems Modeling Language.* Addison-Wesley Professional, 2013
  See the entry for this book in *SysML Books* on page 373 for more information.

- **[Douglass]** Douglass, Bruce Powel, Ph.D. – *Agile Systems Engineering, 1st Edition.* Waltham, MA, USA:Morgan Kaufmann, 2015
  See the entry for this book in *SysML Books* on page 373 for more information.

- **[Friedenthal]** Friedenthal, Sanford and Alan Moore, Rick Steiner – *A Practical Guide to SysML, Third Edition: The Systems Modeling Language.* Morgan Kaufmann, 2014
  See the entry for this book in *SysML Books* on page 374 for more information.

- **[IEEE-1362]** *IEEE Standard 1362-1998, IEEE Guide for Information Technology—System Definition—Concept of Operations (CONOPS) (superceded by ISO/IEC/IEEE 29148 in 2011)* https://standards.ieee.org/standard/1362-1998.html

- **[INCOSE]** *The International Council on Systems Engineering* https://www.incose.org/

- **[InSilicoMedicine]** *Advancing In Silico Medicine at FDA* http://www.omgwiki.org/MBSE/lib/exe/fetch.php?media=mbse:patterns:incose2018--morrison.pdf

- **[IoT]** *Internet of Things* http://en.wikipedia.org/wiki/Internet_of_Things

- **[ISO-80000]** *ISO80000 Quantities and units* https://www.iso.org/standard/30669.html

- **[ISO-29148]** *ISO/IEC/IEEE 29148 Systems and software engineering — Life cycle processes — Requirements engineering* https://standards.ieee.org/standard/1362-1998.html

- **[Leffingwell]** Leffingwell, Dean – *Agile Software Requirements: Lean Requirements Practices for Teams, Programs, and the Enterprise.* Pearson Education, Inc., 2011
  See the entry for this book in *Requirements Engineering Books* on page 377 for more information.

- **[MetaEdit]** *MetaEdit+ Modeler* http://www.metacase.com/mep/

- **[Peak]** *Simulation-Based Design Using SysML, Part 1: A Parametrics Primer* http://ei-slab.gatech.edu/pubs/conferences/2007-incose-is-1-peak-primer/

- **[Pohl]** Pohl, Klaus – *Requirements Engineering: Fundamentals, Principles, and Techniques.* Springer, 2010
See the entry for this book in *Requirements Engineering Books* on page 378 for more information.

- **[Rechtin]** Rechtin, Eberhardt and Mark W. Maier – *The Art of Systems Architecting, Second Edition.* CRC Press, 2000

- **[Roques]** Pascal Roques – *Systems Architecture Modeling with the Arcadia Method: A Practical Guide to Capella (Implementation of Model Based System Engineering), 1st Edition.* ISTE Press - Elsevier, 2017
See the entry for this book in *Arcadia and Capella Books* on page 373 for more information.

- **[Roques(UML)]** Pascal Roques – *UML in Practice.* Wiley, 2004
See the entry for this book in *UML Books* on page 374 for more information.

- **[SysML]** *Systems Modeling Language* http://sysml.org/

- **[SysML1.5]** *OMG System Modeling Language Specification™ Version 1.5* https://www.omg.org/spec/SysML/1.5/PDF

- **[SysML1.6]** *OMG Systems Modeling Language (OMG SysML™) Version 1.6* https://www.omg.org/spec/SysML/1.6/PDF

- **[UML]** *Unified Modeling Language* http://www.uml.org/

- **[UML2.5.1]** *Unified Modeling Language Version 2.5.1* https://www.omg.org/spec/UML/2.5.1/PDF

- **[Unicode]** *a standard that aspires to assign a unique number to every character of every human written language* http://unicode.org/

- **[UTF-8]** *Unicode Transformation Format 8. See section 2.5 on page 36 of the Unicode standard.* http://www.unicode.org/versions/Unicode7.0.0/ch02.pdf

- **[Visio-SysML]** *Visio Stencil and Template for SysML 1.0* http://softwarestencils.com/sysml
Note that at the time of the publication of this book, the currently approved version of the SysML standard is 1.5. As such, some parts of the stencil/templates might not reflect current SysML usage.

- **[Voirin]** Jean-Luc Voirin – *Model-based System and Architecture Engineering with the Arcadia Method, 1st Edition.* ISTE Press - Elsevier, 2017
See the entry for this book in *Arcadia and Capella Books* on page 373 for more information.

Appendix E – Glossary

- **Agile** – an umbrella term for a set of frameworks and practices for managing product development based on short cycles, frequent delivery, and continuous replanning. The concept was originally articulated for software development – see [AgileM]. Currently, the term is used in a wide variety of engineering and management disciplines.

- **codepage** – a table that maps the characters in one or more languages to a set of numerical values. [UTF-8] can be considered to be a codepage of [Unicode].

- **containment tree** – the hierarchical tree of packages within packages in a SysML model.

- **INCOSE** – The International Council on Systems Engineering See [INCOSE].

- **MBSE** – See "Model-Based Systems Engineering".

- **Model-Based Systems Engineering** – an approach to systems engineering that uses a model database to keep track of the relationships between elements in the system. Model-Based Systems Engineering (MBSE) usually combines the database of relationships with diagramming tools that allow an arbitrary number of diagrams to be created showing different combinations of model elements and their relationships. These diagrams are used to allow different stakeholders to understand and comment on the aspects of the system that are within their area of expertise without becoming confused by the detail of the other aspects of the system.

- **namespace** – In SysML (and UML) a namespace is defined by the node path, starting from the root node and including all the elements down to the current element. For example, using full namespace notation a *door handle* might be shown in a diagram as a *vehicle::body::door::door handle*.

- **requirements traceability** – The process of proving that a system fulfills its requirements by establishing links all the way from the initial marketing requirements down to the actual components that are involved in fulfilling each requirement.

- **SoC** – System on Chip. A highly integrated semiconductor device that combines one or more processing cores with multiple specialized accelerators and peripheral interfaces.

- **SysML** – Systems Modeling Language. See [SysML].

- **system** – INCOSE refers to [Rechtin] in offering this definition of a system: "A system is a construct or collection of different elements that together produce results not obtainable by the elements alone. The elements, or parts, can include people, hardware, software, facilities, policies, and documents; that is, all things required to produce systems-level results. The results include system level qualities, properties, characteristics, functions, behavior and performance. The value added by the system as a whole, beyond that contributed independently by the parts, is primarily created by the relationship among the parts; that is, how they are interconnected".

- **toolchain** – a set of software applications, often provided by different suppliers, that are used in sequence to produce a product.

- **UML** – Unified Modeling Language. See [UML].

- **Unicode** – a standard that aspires to assign a unique number to every character of every human written language. See [Unicode].

- **UTF-8** – an encoding scheme for [Unicode] that uses 1 to 4 bytes of information to represent every character. See [UTF-8].

- **Waterfall** – traditional development project flow. Phases start with customer requirements, moving to architecture, moving to high-level design, then to low-level design, and so on. A chart of the project phases resembles a left-to-right waterfall.

End Notes

Chapter 1 – Introduction

- [1] – For more information on these books, see the *Arcadia and Capella Books* section on page 373.
- [2] – The *Kindle Cloud Reader* is available here: https://read.amazon.com/

Chapter 2 – Quick Start

- [3] – If you have accidentally closed the activity explorer, see *What if the Activity Explorer isn't there?* on page 14.
- [4] – Here we are actually working around a small inconsistency in *Capella*. In other areas of the tool, the letter for the architectural level would be added to the prefix automatically.

Chapter 3 – Take a Deep Breath

- [5] – It is actually worse than this. The SoC alone may require as many as 20 different kinds of specialized design engineer. Also, we have not considered the verification and validation teams or the teams that would handle all of the automotive qualification testing. However, as we will soon see, even this modest underestimate of the number of stakeholders is difficult enough.
- [6] – The procedure to create a "[SAB]" is similar to the procedure to create a "[LAB]" shown in *Adding an Architecture Blank Diagram* on page 16 and will be covered in detail in Chapter 6 on page 123 . For the moment, don't worry too much about the details of the diagram and how to construct it. Instead, visualize the face-to-face discussion with each stakeholder and see how the contents of the diagram (not the intricate construction details) convey the concerns of the stakeholder.
- [7] – This goal – making diagrams that everyone can understand – sounds rather obvious. However, it is easy to get carried away and produce diagrams that your stakeholders don't actually understand.
- [8] – Chocolate Mousse recipe provided by Pascal Roques; David Hetherington has no idea how to make a Chocolate Mousse.
- [9] – Reproduced with permission from an Obeo presentation at **SESE tour 2021:** *Capella (3/3), Focus on Deployments & Community*
- [10] – https://www.eclipse.org/capella/adopters.html

Chapter 4 – Mission Capability Blank Diagrams

- [11] – Like most other MBSE tools, *Capella* allows for either "select and click" or "drag and drop" operation when creating new elements.
- [12] – This is one of the few odd cases in which *Capella* does not fully support the *Arcadia* architecture. Operational missions are described in [Voirin] – Chapter 17, section 15.

Chapter 5 – Data Flow Blank Diagrams

- [13] – It is also possible to create Data Flow Blank diagrams directly from the Project Explorer, by right-clicking on a capability, and then selecting "New Diagram - Data Flow Blank Diagram". The name of the diagram would then be automatically set with the name of the capability, and the diagram would be saved under the capability in the "Capabilities" package.

- [14] – Some modelers might not bother with this sort of trivial decomposition; we are just using this simple example to demonstrate the *Arcadia* subfunction decomposition concept.

- [15] – We definitely would **NOT** want to use the "Delete From Model" tool!

- [16] – As of version 6.1, *Capella* allows you to skip this menu step. Simply double-click on the blue box. *Capella* will create a "Functional Chain Description" and bring you to the same name panel.

- [17] – David Hetherington: My apologies for the odd usage of the word "involvement" here. This term really should have been "relationship". Unfortunately, the authors of *Capella* have hard-coded this odd English usage into the user interface for the tool. As such, we will continue with the word: "involvement".

- [18] – Note that the "Text Color" menu (the one next to the pitcher pouring icon) and the "Fill Color" menu (the one next to the letter "A") are right next to each other in the toolbar. We want the "Fill Color" menu.

Chapter 6 – System Architecture Blank Diagrams

- [19] – We defined the functions in Chapter 5.

Chapter 7 – Logical Architecture Blank Diagrams

- [20] – Be careful to click "Apply" rather than "OK". If you click on "OK", nothing will happen. (It is not clear what the intended function of the "OK" button is; it might be there to separate novice modelers from expert modelers).

- [21] – Notice that it is acceptable (best practice in fact) to have more than one functional exchange with the same name.

- [22] – Sharp-eyed readers will notice that in this diagram, the paths through the functions are not shown. This seems to be a bug in *Capella*. If we examine Figure 6-16 on page 133, we will see that architecture blanks sometimes show the paths through the functions. We have not yet been able to determine when the tool displays functional chains which way.

Chapter 8 – Physical Architecture Blank Diagrams

- [23] – This is slightly odd English. "Deploy" is a verb used for things that already exist. "Create" would have been better here.

- [24] – The French spelling of "Behaviour" seems to have slipped into the group name, but not into the entries which use "Behavior".

- [25] – These physical architecture level functional exchanges were created automatically when we transitioned the model from the logical architecture level.

- [26] – See Figure 6-39 on page 151 and Figure 6-42 on page 152.

- [27] – Something every modeler should do frequently!!

- [28] – "COTS" = "Commercial-Off-The-Shelf"

- [29] – Notice in Figure 8-30 that we have provided the image file in the examples package.

- [30] – You may wish to briefly return to the section *Component Exchanges* on page 140 to review the procedure for allocating function exchanges to component exchanges.

Chapter 9 – Breakdown Diagrams

- [31] – See: *Cloning Diagrams to Provide Different Views* on page 149.

Chapter 10 – Scenario Diagrams

- [32] – Strangely, *Capella* automatically adds the first letter only at operational level.

- [33] – Or don't bother selecting the actors and instead click on the double arrow (>>) to "add all available actors" to the diagram.

- [34] – You can see the name "Execution" in the semantic browser.

- [35] – Readers with any sort of programming experience will recognize the familiar "IF-THEN-ELSE" style of operation.

- [36] – and also fail model validation

Chapter 11 – Mode or State Machine Diagrams

- [37] – This fine distinction between "modes" and "states" is unique to *Arcadia*. The rest of the worldwide systems engineering community uses "states" and "state machines" or "state charts". If you are talking to a systems engineer who is not a *Capella* user, you can expect a blank stare if you mention a "mode machine".

- [38] – Notice that we have not yet selected "mode" or "state". That selection will happen implicitly the first time we add one or the other to the diagram.

- [39] – Many new modelers get very excited by the possibilities of describing modes and states and model far too much detail leaving their target audience utterly bewildered and unable to understand what the modeler was trying to express.

- [40] – Display of a name for the initial pseudostate is also not standard notation for UML/SysML.

- [41] – And when any model becomes "sufficient" or even "just barely good enough" you should stop modeling immediately!! The worst thing you can do for yourself is to continue modeling, and modeling, and modeling – adding extra detail that simply makes your model difficult for the users to understand and more difficult for you to maintain.

Chapter 12 – Class Diagram Blank Diagrams (CDB)

- [42] – Notice that the *Capella* palette for the CDB is a bit tricky. It may have listed all the basic types under "BooleanType" previously, but if you select "NumericType" the tool will decide that you are more interested in that choice and make that the top (container) tool.

- [43] – The library and the clock radio are separate projects. You will need to save each of them separately.

Chapter 13 – Challenges Ahead

- [44] – In the 1980s when David Hetherington was developing software for IBM, these sorts of behaviors were often referred to as "Feechurs", as in: "That's not a bug, it's a *Feechur*".

- [45] – One of the early leaders in the UML arena is said to have taken the position that models should be produced on whiteboards and that the whiteboards should be erased at the end of each discussion.

- [46] – See Scott Ambler's excellent discussion of "Just Barely Good Enough" models: http://agile-modeling.com/essays/barelyGoodEnough.html

Appendix A – Planning, Ordering, Installing, Getting Help

- [47] – The speed of the hosting server does not seem to be a concern. David Hetherington has fiber-to-the-home service in Austin, Texas and was able to download the file in about 30 seconds.

- [48] – CP 1252 was last updated by Microsoft for Windows 98 to include the Euro symbol. This codepage only covers English, French, Spanish, and German. This codepage has not been in common use by Windows itself or web pages on the internet for 20 years or so. You really probably don't want to be using CP 1252 for English, French, Spanish, or German anymore either.

- [49] – UTF-8 is a method of encoding that preserves 7-bit ASCII compatibility for plain English and punctuation at the cost of using two bytes per special European character, three bytes per East Asian character, and four bytes for some special supplementary characters. In this manner, UTF-8 can encode the full range of human text languages in the Unicode standard while maintaining (at least English) backward compatibility with the decades of existing English language text data and software. UTF-8 is now the dominant codepage for the worldwide web.

- [50] – The emoji "fingers crossed" is Unicode Ü. See: https://emojipedia.org/hand-with-index-and-middle-fingers-crossed/

Appendix B – Further Reading and Information

- [51] – Many of these disillusioned modeling experts were key early founders of the Agile movement and now resist documentation in any form, especially any sort of modeling.

- [52] – Requirements elaboration is a very basic topic that gets surprisingly little treatment in some of the other larger books on requirements.

Index

This chapter will be about...